CITY OF LAKE AND PRAIRIE

HISTORY OF THE URBAN ENVIRONMENT

Martin V. Melosi and Joel A. Tarr, Editors

CITY OF LAKE
AND PRAIRIE

Chicago's Environmental History

Edited by Kathleen A. Brosnan, Ann
Durkin Keating, and William C. Barnett

UNIVERSITY OF PITTSBURGH PRESS

Published by the University of Pittsburgh Press, Pittsburgh, Pa., 15260
Copyright © 2020, University of Pittsburgh Press
All rights reserved
Manufactured in the United States of America
Printed on acid-free paper
10 9 8 7 6 5 4 3 2 1

Library of Congress Cataloging-in-Publication Data

Names: Barnett, William C., editor. | Brosnan, Kathleen A., 1960- editor. |
  Keating, Ann Durkin, editor.
Title: City of lake and prairie : Chicago's environmental history / William
  C. Barnett, Kathleen A. Brosnan, and Ann Durkin Keating, editors.
Description: Pittsburgh, Pa. : University of Pittsburgh Press, [2020] |
  Series: History of the urban environment | Includes bibliographical
  references and index.
Identifiers: LCCN 2020027901 | ISBN 9780822946311 (cloth) | ISBN
  9780822987727 (ebook)
Subjects: LCSH: Chicago (Ill.)--Environmental conditions--History. | Urban
  ecology (Sociology)--Illinois--Chicago.
Classification: LCC GE155.I3 C58 2020 | DDC 307.760977311--dc23
LC record available at https://lccn.loc.gov/2020027901

ISBN 13: 978-0-8229-4631-1
ISBN 10: 0-8229-4631-9

Cover photograph by Mason Hopfensperger on Unsplash
Cover design by Melissa Dias-Mandoly

Contents

Acknowledgments

The editors are grateful for the support that they have received in completing this project. We thank Martin Melosi and Joel Tarr, both for their editorship of this series, the History of the Urban Environment, and for their leadership in founding and strengthening the field of urban environmental history. We also benefited from the guidance of Sandy Crooms at the University of Pittsburgh Press. We appreciate the efforts of the editorial and design staff at the press as well as the insights offered by the anonymous peer reviewers.

The American Society for Environmental History and its former executive director, Lisa Mighetto, generously offered us meeting space during the 2017 conference in Chicago, allowing us to workshop the individual papers and engage in meaningful collaboration with our contributors; we are grateful. During this same conference, the Newberry Library hosted an open session exploring environmentalism and environmental justice in Chicago; we thank the Library and the noncontributor participants, Jerry Adelmann and Kim Wasserman. We also appreciate our collaborators for their innovative, deeply researched essays and their efforts to meet various deadlines.

We are grateful to the editor and publishers of the *Journal of Historical Geography* for allowing us to include a revised version of Craig E. Colten's article, "Chicago's Waste Lands: Refuse Disposal and Urban Growth, 1840–1990," from its 1994 issue. We also share our appreciation of the editor and publisher of *Environmental Justice* for permission to add a revised version of Sylvia Hood Washington's article, "Mrs. Block Beautiful: African American Women and the Birth of the Urban Conservation Movement, Chicago, Illinois, 1917–1954," from its 2008 issue. Our collection of essays is all the stronger for the participation of these noted scholars.

We also want to thank our colleagues at North Central College and the University of Oklahoma. We are indebted to the Faculty Development and Research Committee at North Central College and the provost, vice president of Research, and History Department at the University of Oklahoma for their financial assistance in publishing this volume. We thank Curt Foxley for the index.

Finally, we want to celebrate our families and friends for their unwavering support and enthusiasm for our efforts.

CITY OF LAKE AND PRAIRIE

INTRODUCTION

WILLIAM C. BARNETT

Humans have been altering the natural world of prairies and wetlands along Lake Michigan ever since they first built communities here. From the opening chapter on native peoples expanding the tallgrass prairie to the closing chapter on twenty-first century efforts to restore the damaged Calumet ecosystem, this volume provides rich stories of people transforming this region. Many of these groups believed they were improving their community and its use of nature, even as they reversed their predecessors' efforts. First, native peoples expanded the tallgrass prairies, then early farmers plowed up those prairie lands, and later industrialists and suburban developers built where farms once stood. Assertions that these transformations represented progress, however, have been called into question in recent decades because of increased understanding of ecology and, later, of growing awareness of environmental justice, sustainability, and climate change. This trio of emerging concepts points the way to new types of human transformations that can improve built environments and restore natural landscapes by balancing economic prosperity with the health of ecosystems and human communities. People will continue to remake Chicago, but this constant process of urban transformation can yield healthier environmental relationships. To achieve this goal, it is critical to examine the history of the intricate and constantly evolving system of man-made and natural components that people have woven together to build the city of Chicago on this dramatically altered landscape of lake and prairie.

Chicago residents and visitors to the city each have their own unique understanding and personal map of the complex web of interconnected neighborhoods along the southwestern shore of Lake Michigan. And all of these maps are in motion, as Chicago is in an ongoing process of transformation and reinvention, one that is surpassed by few American cities. Longtime residents have multiple layers to their understanding of Chicago's urban geography, with memories of the built environments of their early years lying beneath a contemporary map that accounts for urban redevelopment. Historians and geographers possess an ability to dig deeper into the city's past, researching and writing about the ways that previous generations reshaped the urban environment, and revealing broad patterns such as industrial booms, transportation shifts, or one immigrant group replacing another. Chicago has been reconfigured by American migrants moving west, by ongoing waves of European immigrants, by African Americans migrating from the rural South to the urban North, and by more recent newcomers from Latin America, Asia, and Africa. Each group fought to make a place for itself, providing labor and ideas to fuel the city's economy, creating new communities and businesses, and reshaping neighborhoods. Today's Chicago is the result of the struggles and contributions of countless men and women, and the city's environmental history is a complex story of a landscape transformed by its people.[1]

Visitors and newcomers to Chicago soon become familiar with a few neighborhoods, but most have difficulty creating their own maps that reflect the complexity of this sprawling city. Outsiders typically start not with a micro-level understanding that emerges from childhood, but instead with a macro-level perspective that is less detailed but can show how Chicago differs from other cities. For the first-time visitor flying into O'Hare or Midway airports, the view from an airplane window is arresting—the topography is as flat as a table, with a cluster of skyscrapers rising from the lakeshore, and development radiating across the landscape. Especially at night, with the lake in darkness and streets lit up, the traveler looks down on a built environment that appears to be a perfect Cartesian grid, with straight lines extending north, south, and west, as regular as graph paper. Immigrant communities cannot be identified, and lines drawn onto the landscape by ethnicity, religion, class, and race are not visible. This flat plane can be viewed as a blank slate, a vast canvas on which to plan, build, and then rebuild a city. In the nineteenth century boosters such as land speculator Gurdon Saltonstall Hubbard and first Chicago mayor William Butler Ogden inscribed canals, railroads, and industrial facilities onto the prairies and wetlands, and twentieth-century city planners and political leaders such as the architect Daniel Burnham and Mayors Richard J. and Richard M. Daley sketched out new visions for the booming metropolitan region. For better and for worse, a myriad other Chicagoans have also left their marks on the constantly evolving urban landscape.

This collection of essays is an examination of the environmental history of Chicago by a diverse group of scholars trained in history, geography, and other disciplines. The book's contributors include Chicago residents who were born here, native daughters and sons who have moved out of the region, and recent arrivals who are analyzing their adopted home. Our shared goal is to research and narrate stories about the physical transformations that have occurred in one of the nation's most influential cities across its history, with an emphasis on reciprocal relationships as people have altered nature and the natural world has shaped the city. Some of us teach courses on Chicago history and were trained in urban history, a field with deep roots in local universities, while others identify first with the newer field of environmental history. We believe more conversation between urban and environmental historians, geographers, and other scholars will benefit our shared understanding of Chicago's past, present, and future.

This book is a new contribution to the fast-growing subfield of urban-environmental history, and it is the latest release in the History of the Urban Environment series published by the University of Pittsburgh Press. Starting with Craig Colten's 2000 book *Transforming New Orleans and Its Environs*, this series has published books edited by leading environmental historians and geographers such as Colten, Joel Tarr, Martin Melosi, and Char Miller that offer path-breaking examinations of the complex interconnections between people and nature in cities.[2] Those volumes gave readers notable insights into New Orleans, Pittsburgh, Houston, and San Antonio, and more recent additions to the series have offered valuable analysis of Los Angeles, Phoenix, Boston, Philadelphia, and Sacramento.[3] While the West, South, and Northeast are well represented, there is no book on a midwestern city, even though Chicago, Detroit, and other Great Lakes cities have had a profound impact on urban America. Many of these books identify a single feature as central to the city's environmental history, such as the energy industry in Houston, the Mexican border in San Antonio, sunshine in Los Angeles, and the limited water in Phoenix. But Chicago's story is more varied and fluid than Pittsburgh's long reliance on steel or Houston's continuing links to oil. Chicago's constant process of evolution and reinvention closely parallels national economic patterns. The city's ability to remake itself, as in its shift from Rust Belt decline to postindustrial prosperity, makes Chicago difficult to label but also suggests it can function as a microcosm of urban-environmental changes and offer insights relevant to a large number of American cities.

This series has long lacked a volume on Chicago, due in part to the city's size and complexity, but also because the seminal book combining environmental and urban history was a study of Chicago—William Cronon's *Nature's Metropolis: Chicago and the Great West*. Cronon's fascinating and authoritative book, published in 1991, provided an exciting blueprint for scholars studying a single city's environmental relationships. Six years later Andrew Hurley's *Common Fields: An Environmental History of St. Louis*, published before the University of Pittsburgh Press series began, provided a model for a collection of essays on a single city that examined the creation of place over time from multiple perspectives.[4]

Nature's Metropolis forced readers to see cities as built on and within nature, not separate from it, and made scholars rethink their understandings of linkages between cities and their hinterlands. Cronon explains in his introduction that as a child going through Chicago on car trips to rural Wisconsin, he viewed the city as an outsider and was deeply uncomfortable with its industry and pollution. As an adult, however, he realized "that the urban and rural landscapes I have been describing are not two places but one. They created each other."[5] Cronon's remarkable book excels in using an outsider's macro-level view, and the city he depicts can be viewed as an organism that thrives on its ability to pull in resources from its hinterlands. This approach explains how Chicago grew into a great industrial city by processing the grain, meat, and lumber brought in by railroads from the Midwest and West. Cronon's analysis is powerful and persuasive, and shapes the way all of us teach and write about urban and environmental history.

We do not see this edited collection as a corrective to *Nature's Metropolis*, which recently celebrated its twenty-fifth anniversary, but instead as an opportunity to restart the conversation about Chicago while adding new perspectives and voices. Our book draws more from the model laid out by Andrew Hurley in his edited collection on the creation of place over time in St. Louis. We seek to tell a variety of more narrowly focused stories, examining specific locations and analyzing the roles of different groups and social movements in transforming the city. *Nature's Metropolis* explains the city's rapid nineteenth-century growth, culminating in the 1893 Columbian Exposition, while this volume begins with Native Americans before European settlers arrived, and then analyzes nineteenth-, twentieth-, and twenty-first-century Chicago, including varied efforts to grapple with the pollution from earlier industrial booms. Cronon structured his study around flows of commodities and capital, which has great explanatory power, but he is the first to admit that it results in a book with few individual people beyond boosters and capitalists. The chapters that follow offer closer analysis of transformations created by everyday Chicagoans, such as African American migrants from the rural South and Mexican American steelworkers.[6] They also tell stories about conservation and environmental restoration efforts in specific places far from the business district, such as Skokie's lagoons, Cook and DuPage Counties' suburbs, and the Calumet region's brownfields.[7]

The physical landscape underlying the built environment might be viewed as a blank slate on which to erect the city of Chicago, but this mixture of land and water is not as uniform as it appears, and knowledge of its natural systems is required to build effectively. While the metropolitan area has only one major body of water and its numerous rivers are quite small, the abundance of freshwater in the region plays a complex and changing role in its environmental history. Lake Michigan has contributed to the region's development in quite different ways as the lake port became an industrial center and then a postindustrial city.[8] The region's low elevation alongside Lake Michigan and its high water table resulted in a diverse natural landscape of tallgrass prairie, islands of oak savanna, wetlands, and dunes. This subtle but richly varied patchwork of ecosystems has been altered and remade by humans over and over, dating back to precolonial Indians whose interventions encouraged the expansion of the tallgrass prairie.[9] Multiple authors in this volume examine efforts to remake wetlands, joining conversations in environmental histories such as *Discovering the Unknown Landscape* by Ann Vileisis, *An Unnatural Metropolis* by Craig Colten, and *The Fall and Rise of the Wetlands of California's Great Central Valley* by Philip Garone.[10]

This location at the meeting of lake and prairie offered commercial opportunities because of its geographic advantages—advantages that the region's Native Americans long recognized. The site stood at the juncture of two of North America's principal watersheds. The Chicago River and other smaller streams drained into the Great Lakes hydrological system. Holding nearly twenty percent of Earth's freshwater, the Great Lakes provided access to the Atlantic Ocean via portages and the St. Lawrence River.[11] Other Chicago-area rivers, including the Des Plaines, flows into the Illinois River, which is part of the Mississippi River drainage. The Mississippi stretches 2,300 miles from its Minnesota headwaters to the Gulf of Mexico, and, with its tributaries, drains more than forty percent of the contiguous United States.[12] Native Americans portaged the short distance across the low-lying divides that separated the watersheds in dry seasons, and the traverse was even easier when floodwaters covered the higher areas. The linkage between these two watersheds generated a fertile trading ground connecting peoples across broad stretches of land and water. Indians lived on the Chicago River's banks for centuries, and the arrival of Europeans in the late seventeenth century expanded the regional fur trade and led to a métis settlement that existed through the early nineteenth century.[13]

Efforts to control the region's water, whether for canal transportation, industrial use, waste removal, recreational purposes, or flood prevention, are a major theme in Chicago's environmental history. The tallgrass prairie landscape was unattractive to nineteenth-century settlers until they developed the ability to plow the land. Then farmers poured in for the rich soils and abundant water, and the region's agricultural economy boomed alongside the city's commerce and industry. Connecting the Great Lakes to the Mississippi River by canal was a critical step in using engineering to improve upon the region's water systems. As industries such as meatpacking used and polluted enormous quantities of water, the reversal of the flow of the Chicago River to protect residents' health became one of the most notable environmental interventions in American urban history.[14] By the twentieth century little tallgrass prairie was left, and scientists and conservationists, including Henry Chandler Cowles, Jens Jensen, and May Watts, wrote about the beauty of vanishing midwestern ecosystems and called to protect them from urban development and suburban sprawl.[15] These reformers believed access to green space such as prairies would improve the well-being of Chicagoans. In examining links between health and environment, our contributors are in dialogue with recent environmental works ranging from *The Health of the Country* by Conevery Bolton Valencius and *Inescapable Ecologies* by Linda Nash to *Toxic Bodies* by Nancy Langston and *On Immunity* by Eula Biss.[16]

This book's title, *City of Lake and Prairie*, highlights the fact that Chicago sits on a strategic location in the middle of the continent astride two crucial ecosystems in the Midwest. It developed at the western edge of the Great Lakes, on the prairies that begin near the Illinois-Indiana border and stretch across the Mississippi River to the Rocky Mountains. The people of Chicago have altered their environmental relationships many times, but over and over their city has functioned as a point of connection linking distant communities and environmental systems. In the early days of European settlement, it was a fur-trading entrepôt, and water routes traveled by canoes and ships linked native peoples hunting in North American landscapes to European markets. In the canal era Chicago's Illinois and Michigan Canal joined the Great Lakes and the

Mississippi River, and canal boats and ships connected Illinois farmers with New York markets. Railroads extended Chicago's reach to the Far West, making it the crucial hub between the nation's Atlantic and Pacific seaports. Chicago's agricultural processing businesses and its vast industrial facilities pulled in crops, livestock, lumber, metals, fossil fuels, and other commodities from national and international locations. Today Chicago's new postindustrial businesses and service economy continue to make advantageous use of this mid-continent location.

The city of lake and prairie has maintained its position as the nation's second or third largest urban area since 1890, adapting to broad economic patterns and new technologies so continuously that it can be seen as a microcosm of urban America's key shifts from the late nineteenth century to today. Chicago is now the most diversified economy in the nation, according to World Business Chicago, and its economic mix most closely represents the U.S. economy as a whole. Only New York City is the headquarters of more Fortune 500 corporations, and the Chicago region's economy is large enough that it compares favorably with leading global economic hubs.[17] The Chicago metropolitan area continues to generate great wealth, but it also remains the site of alarming economic inequality, as prosperous green suburbs and revitalized downtown neighborhoods lie a short distance from districts of persistent poverty that lack access to adequate jobs, housing, safe streets, stores with healthy food, and green space. The political troubles in the city and the state, particularly the problems of underfunded pensions and government corruption, impede progress on these troubling economic and environmental disparities. Thus Chicago continues to be an international example of America's business successes and its social justice failures. Solid environmental histories of urban centers outside the United States have been published recently, such as *London: Water and the Making of the Modern City* by John Broich and *Metropolitan Natures: Environmental Histories of Montreal* edited by Stéphane Castonguay and Michèle Dagenais, and our book seeks to engage with these international urban-environmental case studies.[18]

While the city's flat physical landscape and its businesses are surprisingly varied, the fundamental economic philosophy of Chicago's farmers, merchants, industrialists, and commodities traders has been more unified than might be expected. Capitalism is the central principle, leading to a drive to control and impose order on the natural world. Strenuous efforts to increase agricultural and industrial production and efficiency have defined the city's growth, and these business practices have etched inequality into the urban landscape. The broad patterns that emerged in the nineteenth century and carried through the twentieth century include constant economic innovation, a series of important shifts in transportation technology, significant environmental damage, and dramatic cycles of economic growth and decline. Carl Sandburg's 1914 poem "Chicago" depicts a brutal but vibrant and proud city based on blue-collar pursuits such as railroads, meat, and wheat. Similarly Nelson Algren's 1951 *Chicago: City on the Make* captures the mix of vitality and greed that defined the city in the Cold War years and continued to yield great wealth and abject poverty.[19] Chicago began as a shipping center and benefited from river and canal transportation, but truly took off as a railroad and telegraph hub, and now is defined by interstate highways, air travel, and data networks. The city has seen a series of great economic booms, with an early focus on meatpacking and agricultural processing, followed by manufacturing of agricultural machinery, steel, and electrical equipment by corporations from McCormick to Western Electric.[20] Real estate, retail empires such as Sears Roebuck and Montgomery Ward, and printing companies like R. R. Donnelley and Rand McNally have stretched across multiple eras, while in recent decades service economies such as financial services, information technology, and tourism have surpassed manufacturing.[21] Our authors are in dialogue with other recent environmental histories that critically examine modern American capitalism, including Andrew Needham's *Power Lines*, Christopher Jones's *Routes of Power*, and Ted Steinberg's *Down to Earth*.[22]

These cycles of industrial expansion regularly remade Chicago's urban geography, with the booms creating successful commercial and residential districts and huge industrial facilities, and the ensuing downturns yielding bleak areas of poverty. The downtown business district has been the site of remarkable architectural advances, as Louis Sullivan, Frank Lloyd Wright, Ludwig Mies van der Rohe, and others influenced urban design around the world. Huge facilities such as the Union Stock Yard, Pullman, U.S. Steel, and Western Electric were sources of blue-collar jobs, civic pride, and wealth, but Chicago's manufacturing decline eventually shuttered all these industrial sites, even as Houston and Detroit continued to process petroleum and make cars. Chicago's neighborhoods have always included distinct landscapes of wealth and poverty, with fashionable elite enclaves such as the Gold Coast as well as neighborhoods full of poor immigrants and African Americans such as Back of the Yards, Bronzeville, Lawndale, and Little Village. Many important industrial sites have been nearly erased from the landscape, and so have infamous areas of poverty such us the demolished Cabrini-Green and Robert Taylor Homes.[23] Newcomers to Chicago have never seen those high-rise public housing projects, and students already view them as bleak but hazy stories from a past almost as distant as Jane Addams's Hull House, Upton Sinclair's Back of the Yards, and the rigidly segregated Bronzeville. This volume's close examinations of Chicago's complex, layered landscapes seek to connect with other environmental histories chronicling the battles that shape urban environments, including Michael Rawson's *Eden on the Charles*, Catherine McNeur's *Taming Manhattan*, Matthew Klingle's *Emerald City*, and Christopher Wells's *Car Country*.[24]

How Chicago's working classes adapted to the city's challenging urban environment is another key theme in this volume. Waves of immigrants from every continent, as well as internal migrants from America's rural South, provided much of the labor for the industrial metropolis. These varied immigrants did not draw up the blueprints, but they performed the work that reshaped nature into city. Blue-collar workers embodied Chicago, which Carl Sandburg called the "City of the Broad Shoulders," and they built its urban infrastructure by digging canals, paving streets, forging steel, and erecting skyscrapers. The urban poor had a more intimate and intense relationship with nature than the wealthy, being more likely to endure heat and cold, and to live and work close to animals in the slaughterhouses and in poor neighborhoods.[25] The city's working-class residents were also more likely to die of cholera and other diseases linked to a lack of clean water, and to be killed in workplace accidents.[26] The living conditions in densely packed neighborhoods, the varied foods that immigrants ate, and the ways they structured their free time, including seeking recreation in urban green spaces, are all important pieces of Chicago's environmental past. In examining the environmental experiences and struggles of working-class urban residents, particularly in Part III of this volume, our authors seek to engage with other environmental histories that focus on labor and class, including *Killing for Coal* by Thomas Andrews and *Making a Living* by Chad Montrie.[27]

Chicago's growth into an industrial giant with high levels of inequality led to a wide variety of reform efforts from the Gilded Age and Progressive Era forward. Early pressure for change came from grassroots activism by the working classes, including labor unions and anarchist groups.[28] Middle-class and wealthy residents also organized to address economic, social, and environmental problems in the early twentieth century, and Hull House and the University of Chicago were key hubs for these reform efforts. Since Daniel Burnham's 1909 *Plan of Chicago*, urban planners have laid out large-scale responses to the challenges facing the Midwest's greatest industrial city.[29] The Great Depression provided new opportunities for reform and conservation, with New Deal agencies such as the Civilian Conservation Corps planting trees and draining wetlands to create Cook County's Forest Preserves.[30]

Meanwhile, Chicago's heavy industries continued to create widespread and intense air and water pollution, and the booming city generated huge quantities of solid waste.[31] The flat and watery landscape of the expanding metropolitan area complicated the city's sanitation efforts from its beginnings and continued to be used as a sink for a range of waste products, from the Calumet region's steel industry to suburban landfills.[32] In the decades after World War II, cities around the Great Lakes grappled with deindustrialization as the region became known as the Rust Belt. The industrial giant that had emerged in the midnineteenth century faced great challenges in the mid-twentieth, and the city's population declined due to the loss of industrial jobs and "white flight" to the suburbs. During the 1960s and 1970s, in the era of Rachel Carson and Earth Day, the intensity of the response to environmental problems grew in the region and across the nation. Grassroots activists included women and men, from students to retirees, in both urban and suburban neighborhoods; and notable nonprofit groups such as Businessmen for the Public Interest and OpenLands emerged in Chicago.[33] This volume seeks to engage with works on twentieth-century environmental reform, including Neil Maher's *Nature's New Deal*, Dorceta Taylor's *The Rise of the American Conservation Movement*, and Robert Gottlieb's *Forcing the Spring*.[34]

Chicago has made a more successful transition from manufacturing center to postindustrial city than many of its Rust Belt neighbors, and its complex story of reinvention continues to unfold. In the past several decades environmental cleanup and restoration projects have made progress in a range of neighborhoods, but a huge array of problem areas continues to dot the metropolitan landscape. Three new ways of understanding the fundamental relationships between Chicagoans and nature in the twenty-first century have emerged, with the first being a rising awareness of environmental inequality. The core idea in environmental justice, as scholars such as Robert Bullard and Sylvia Hood Washington have explained, is that the poor and people of color are exposed to environmental risks such as polluted air and water, unsafe working conditions, substandard housing, and lack of waste removal at far higher rates than middle-class or white Chicagoans.[35] This legacy has sparked grassroots efforts among African Americans and Mexican Americans to improve environmental conditions and seek social justice in Chicago.[36]

A second new concept is the idea of sustainability, and in recent years political and business leaders, including Mayors Richard M. Daley and Rahm Emanuel, have worked to make Chicago a "greener" city while promoting it as a destination for visitors. Millennium Park, which is built over rail yards and is one of the world's largest rooftop gardens, is an excellent example of a new public landscape highlighting urban renewal and sustainability for a city based more on tourism than industry.[37] Recent efforts to improve energy efficiency, water use, and recycling and to promote economic development and green space have come from political and business leaders in addition to grassroots activists. On one hand, Millennium Park and the planned Barack Obama Presidential Center in Jackson Park on the city's South Side build upon a legacy dating back to the 1893 Columbian Exposition in which public spaces such as the Museum of Science and Industry and the Field Museum are sited near the lakefront. But these two projects, together costing about $1 billion, with funding from a new group of philanthropists such as the Pritzker and Crown families, also represent a new era of urban design based on sustainability principles that projects a new image of Chicago to visitors. Such ongoing efforts to engineer a "greener" Chicago seek to realign the city with its 1830s motto, *Urbs in Horto* or "City in a Garden."[38]

A third recent idea, with links to environmental justice and sustainability, is a growing acknowledgment that climate change has the potential to profoundly alter relations with nature in the metropolitan area in the twenty-first century. The vast store of freshwater in Lake Michigan is an asset with increasing value, and the Chicago River is dramatically cleaner today, but the region's abundant water also has liabilities. The likelihood of increasingly intense rains due to climate change creates major challenges for the highly developed and flood-prone region.[39] So far, Chicago has escaped the devastation from flooding seen in New Orleans and Houston, although heavy rains caused widespread flooding in April 2013 when the Chicago River was diverted into Lake Michigan for the first time in over a century. Chicagoland's flood risk calls for new interventions, as the flat landscape of prairies and wetlands that once absorbed excess rain is now a patchwork of urban areas with increasingly impermeable roads, parking lots, commercial developments, and suburban sprawl. The Tunnel and Reservoir Plan, also called Deep Tunnel, is a massive water management system of underground tunnels and reservoirs to be completed in 2029, but this multi-billion-dollar project will not be able to handle rains such as Houston experienced in 2017 during Hurricane Harvey.[40]

For nearly two centuries Chicagoans have erected, demolished, and rebuilt layers of built environment on top of this landscape of prairies, wetlands, and dunes while also constructing unseen technological systems belowground and underwater.[41] The infamous 1871 Chicago Fire caused a substantial reshaping of the city's landscape, with extensive land-making along the lakeshore, but the capitalist imperative won out over more far-reaching environmental reforms.[42] The recovery from the fire followed on the heels of the massive grade-raising project of the 1850s and 1860s that used jackscrews to raise buildings, sidewalks, and streets out of the lakefront's mud. The city of Chicago that flourished in the century after the fire reshaped built environments across the nation by exporting new architectural styles and innovative urban planning ideas, including the Chicago School of architecture and the City Beautiful Movement. Less visible interventions into nature include the Stickney Water Reclamation Plant and the Jardine Water Purification Plant, which are among the world's largest wastewater facilities and water filtration facilities. Together these plants serve several million people in the city of Chicago and in scores of suburban communities, uniting a huge metropolitan region with ignored but crucial environmental infrastructure.

Twenty-first century efforts to address environmental justice, sustainability, and climate change goals have great potential but they require a deep understanding of the complex layers of Chicago's environmental history. Generations of Chicagoans have profoundly altered the landscape of prairies and wetlands along the lakefront, and these changes will continue. The city's stunning industrial growth in the late nineteenth century transformed the region, creating vast environmental problems that reformers and urban planners in the Progressive Era were able to lessen but not solve. Subsequent generations have continued this pattern, focusing on the primary goal of growing the city's economy and expanding its built environment and transportation systems, with a distinctly secondary goal of ameliorating damage to the natural world and to human health. Fifty years ago an emerging understanding of ecology led more Chicagoans to question that status quo. Skepticism about the long-accepted definition of progress has increased in the new millennium due to a growing awareness of environmental justice, sustainability, and climate change. These important concepts offer a road map to new types of urban transformation and landscape restoration that prioritize improving the health of ecosystems and human communities. This book tells an array of deeply researched stories about how people have reshaped the landscape of Chicago to help us understand the road traveled to this point. The next chapter of the environmental history of the city of lake and prairie has yet to be written.

PART I

WHERE PRAIRIE
MEETS LAKE

Geography is not destiny, but Chicago's location, with the Great Lakes to the east and the prairie to the west, shaped the city's history. Chicago sits at the intersection of the eastern woodlands and the western prairie. When New Yorker Charles Butler visited in 1833, he described the prairie as "one great unoccupied expanse of beautiful land, covered with the most luxuriant vegetation—a vast flower garden—beautiful to look at in its virgin state and ready for the plough of the Farmer." Butler envisioned the bounty of American farmers sent to eastern markets via the Great Lakes.[1] However, in his essay on Native Americans Robert Morrissey contends that this region of tallgrass prairie was not virgin soil when the city of Chicago was founded. Rather, he observes, it was an "anthropogenic landscape" shaped by Indian activity for millennia.

Chicago sits on a subcontinental divide separating the Great Lakes from the Mississippi River watershed. Over centuries Native Americans had portaged between these water systems. With the arrival of more Europeans in the eighteenth century, the Potawatomi and other regional Indians increasingly engaged in a far-reaching fur trade.[2] By the early nineteenth century they faced a United States engaged in a settler colonial project that sought land. White Americans worked to dispossess Indian communities and replace them with racially restrictive societies. Claiming authority, the new arrivals transformed supposedly "virgin" or "underused" indigenous lands into private property.[3]

Chicago's founding is grounded in such a dispossession. Regional waterways connected Chicago to these political, economic, and social transitions and, as Ann Keating and Kathleen Brosnan also observe, to global environmental changes, including nineteenth-century cholera pandemics. Over time, as Chicagoans built their city at the junction of lake and prairie, they responded to a series of cholera outbreaks by gradually shifting from privatized, temporary services to greater municipal control of the sanitation infrastructure and public health.

The first plat at the mouth of the Chicago River by James Thompson, for the Canal Commissioners, began the process of making the land into real estate. As part of the larger western capitalist enterprise, Chicago helped to mold distant hinterlands by controlling trade in the region's rich natural resources, commodifying the land, lumber, crops, and livestock. Domesticated animals, Katherine Macica argues, were a part of nature that was interwoven into the city's built environment from the start, providing energy as well as food, until reformers worked to remove livestock from the twentieth-century city.

CHAPTER 1

Native Peoples in the Tallgrass Prairies of Illinois

ROBERT MORRISSEY

It would be hard to overstate the achievement and far-reaching impact of William Cronon's landmark 1992 book, *Nature's Metropolis*. It was a monumental history of Chicago's rise. It was a methodological tour-de-force. It was a reflection of Cronon's tremendous gifts as a writer, demonstrating his genius ability to turn an improbable subject such as nineteenth-century commodity flows into a page-turning history classic.[1]

Perhaps the book's most enduring contribution, however, was the way it used the case study of Chicago to challenge deeply held and problematic ideas about the relationship between city and countryside in U.S. environmental history. As Cronon argued brilliantly, much of American thinking about nonhuman nature has always imagined a fundamental disconnect between the "natural" rural landscape on the one hand and the city on the other. Reorienting our view, Cronon used the story of Chicago's rise to show how deeply connected city and hinterland were and are. The city was not "artificial," nor the countryside "natural." Rather both were part of a profoundly humanized landscape, a hybrid "second nature" containing the imprint of both human and nonhuman agency. This point, which Cronon later extended in his seminal essay on the American idea of wilderness, usefully critiqued bedrock assumptions about "nature" in American culture and shaped the environmental humanities down to the present.[2]

As valuable as Cronon's lesson about second nature was, however, it contained an assumption, more implied than explicit, about the environmental history of the Midwest *before* Chicago. Narrating the complex rise of second nature in the mid-continent, *Nature's Metropolis* suggested a predecessor, a "first nature" that previously characterized the region. Although not at all a focus of Cronon's book, the concept of "first nature" was nevertheless an important presence in his argument, the imagined alternative against which his concept of "second nature"— the linked processes of the city *and* countryside—could be contrasted.[3] Speeding along to tell the story of the rise of the city, Cronon emphasized what constituted this "first nature": the nonhuman forces like glaciers that created the landscape of the Midwest; the "natural legacies" that created soil types and the "vegetational geography."[4] Before Chicago, it was not people and their markets but nonhuman forces—"the land," "energy flows," "natural patterns," and local "direct ecological adaptation"—which were responsible for shaping the landscape.[5] Representing a profound break with the past, Chicago "redefined and reordered" the landscape in the nineteenth century, giving local ecologies "an ever larger human component," and leaving them increasingly "improved towards human ends."[6]

Cronon's implied idea of "first nature" echoed an assumption that Americans commonly make about the landscape of the West, namely that it was largely unaltered wilderness until nineteenth-century white settlement, farm making, species shifting, and industrial development transformed it. Missing from this, however, are all the profound ways in which the landscape of the Midwest and West was *already* deeply humanized well before the rise of center cities such as Chicago. To be sure, Cronon did not ignore Indian people and their presence on the land, and it would be foolish to suggest that the author of *Changes in the Land* (a book in which one central purpose is exploring ways that Indians shaped the landscape of North America before contact) was ignorant of or incurious about midwestern Indians' impact on their environment. Still, in *Nature's Metropolis* Cronon's narrative focus was surely on discontinuity as he contrasted Indian land use with the anthropogenic "second nature" of the nineteenth century.[7] Largely underplaying all the critical ways that Indians shaped the region, Cronon followed a common simplification of midwestern Indians by casting them as "users" rather than as "shapers" of the land, and in any case not as consequential in their impacts as the market makers that followed them.[8] Emphasizing discontinuity between the pre-Chicago past and its modern rise, he reified a chronological boundary which is arguably as problematic as the distinction between city and countryside he so effectively challenged.

The premise of this chapter is that the region in which Chicago grew up was a thoroughly anthropogenic landscape—a profoundly altered example of "second nature"—well before Chicago began.[9] Indeed, much of Chicago's rise as the central "city in the garden" was owed to the ways in which the long history of human occupation in the region had shaped and conditioned the soil, cropped the landscape with its distinctive mosaic-like ecozones, and encouraged its particular fauna. The environmental history of Chicago needs to begin with a recognition of the great continuity of anthropogenic change in the Midwest. The city had a profound effect on nonhuman nature through its marketplace and its capital infrastructure, of course, but it only continued—and in many ways was shaped by—a previous trajectory of human exploitation, modification, and alteration. Environmental historians have done useful work in breaking down the binary between city and hinterland, teaching us not to think of the one as "natural" and the other not. But let us go further and challenge a still-powerful *chronological* binary that separates the modern from the premodern: the notion of a profoundly altered "second nature" that Chicago created and the largely nonanthropogenic "first nature" that preceded it. At a time when historians are debating the concept of the Anthropocene, it may be instructive to contemplate ways in which the Midwest in the Holocene is second nature all the way down.[10]

Chicago grew in a distinctive borderland, at the edge of eastern oak-hickory forests, western grasslands, and within reach of northern pine forests accessible by means of Lake Michigan. The most distinctive feature of its regional ecology was surely the place where these bioregions came together, the edgy ecotone of the tallgrass prairies.[11] The prairies were a landscape in transition, receiving enough moisture to support tree growth and tending toward forest over time. Given high summer precipitation, the prairies were also super-productive, creating more organic material every year than the ecosystem could consume. For this reason annual nutrient cycles resulted in the gradual accumulation of excess biomass; waste materials from frequent fires and annual decomposition, together with windswept loess, deposited layer upon layer over glacial till. Over time that accumulation constituted several feet in most places, comprising some of the richest soil on the planet. Given the presence and viability of species drawn from both grassland and woodland taxa, and climate conditions alternately favorable to either vegetation zone, the tallgrass was a distinctively biodiverse mosaic.[12] I have recently argued that the edgy, transitional prairie landscape shaped human history in important ways in the period before and after contact.[13] But the direction of influence was not one way.

Up until recently, nature writers, natural historians, and observers have often considered the prairie as an utterly nonhuman landscape, the product of purely "natural" forces such as climate, glaciers, wildfire, and wind. Humans, of course, destroyed much of the prairie in the nineteenth century, but the original creation of what many considered a "primeval" midwestern landscape was believed by many writers to have nothing to do with them, and indeed to date much further back than human history in the region. In her classic *Shaping of America's Heartland*, for example, the geographer Betty Flanders Thomson celebrated the relic prairies that survive in rural cemeteries in Illinois, Iowa, and Minnesota. Such relics give nature-lovers the consolation of knowing what the land *should* look like in its "natural form," since they are "purely natural and not the product of human manipulation." For Thomson and many other writers, the tallgrass prairies were an ancient and "wild" landscape, not anthropogenic.[14]

But scientists have begun to tell a different and much more complex story about the origin of the prairie and its ecology. Exploiting natural archives from pollen deposits to tree rings, scientists can reconstruct past vegetation history with increasing specificity. In the case of the Midwest, pollen records from bogs and ponds help to pinpoint the evolution of the grassland and surrounding vegetation communities in the distant past, while rare ancient oak trees provide low-resolution temperature and rainfall series for the region over a more recent period. Interested environmental historians can gain powerful insights about regional (and sometimes local) scale vegetation change from these datasets. In addition to specific landscape histories, the paleoecological literature provides a general lesson about the ubiquity and complexity of landscape change and the perils of viewing past landscapes as stable or timeless.[15]

Indeed, in contrast to old stories about the prairie as a timeless, "primeval" landscape, scientists now understand that the eastern prairies were formed relatively recently, clearly during the Holocene. Far from ageless, the prairie landscape in Illinois was probably formed around 8,000–6,000 years before the present (ybp), according to recent palynological study.[16] Summarizing a new consensus, Roger Anderson argues that the landscape of Illinois was covered by hardwood forests prior to and then again after the ice ages, and only came to be occupied by grasses in the mid-Holocene, probably coincidentally with the mid-Holocene warm period, or hypsithermal.[17] Although popular wisdom used to hold that the main factor in prairie formation was the fifty-five-million-year-old Rocky Mountain rain shadow and resultant aridity across the mid-continent, scientists have known since the 1930s that the prairies received plenty of summer rainfall to support trees. An emerging consensus now holds that other factors were responsible for the formation of the prairies and their eastern extent and persistence.[18]

The most important factors were probably periodic drought and fire. Recent charcoal studies confirm that a rise in fire frequency accompanied the rise of the grassland ecotone and its eastern push. A marked spike in non-arboreal pollen (grasses and forbs) is coincidental with a dramatic increase in charcoal concentrations in sediment samples from Chatsworth Bog, in Illinois, around 6,000 ybp.[19] Fire, and not just climate, seems to have been determinative in creating the prairie.[20]

Many scholars agree that this increased fire frequency was almost certainly anthropogenic, the mark of Archaic period peoples on the landscape.[21] Indeed, as Anderson writes, while the droughts of the hypsithermal may have helped drive the eastern movement of the prairie, they cannot explain its persistence during a much wetter period beginning five thousand years ago. The emerging consensus is strong: "Most believe . . . that for the last 5,000 years, prairie vegetation in the eastern United States would have mostly disappeared if it had not been for the nearly annual burning of these grasslands by the North American Indians."[22] In other words, humans may not have created the grasslands, but they seem to have been the primary maintainers of the prairie peninsula for well over 80 percent of its life.[23]

It is impossible to know about the specific human ecology of the prairie eight thousand years ago, but contact-era records give us a sense of how Native peoples managed the land in the region well before Chicago.[24] They burned huge swaths of the uplands each year, usually in the fall. In the seventeenth century French observers noted that the fire regime resulted in intricately diverse patches, including the upland prairies and the lowland woods. As Jesuit Priest Louis Vivier put it, the prairie fire regime created huge steppes and groves, as well as considerable dense woodlands. Meanwhile, the table-flat landscape left by the glaciers in the eastern reaches of the prairies also ensured that water could not easily drain from the land, resulting in great wetlands around the slow-moving rivers. Describing the region around the Mississippi River in southern Illinois, Vivier wrote:

> Both banks of the [river] are bordered throughout the whole of its course by two strips of dense forests, the depths of which varies [*sic*], more or less, from half a league to four leagues. Behind these forests the country is more elevated, and is intersected by plains and groves, wherein trees are almost as thinly scattered as in our public promenades. This is partly due to the fact that the savages set fire to the prairies toward the end of

autumn, when the grass is dry; the fire spreads everywhere and destroys
most of the young trees. This does not happen in places nearer the river,
because, the land being lower and consequently more watery, the grass
remains green longer and less susceptible to the attacks of fire.[25]

Contact-era observers understood that the prairie was a man-made
landscape; Jesuits pointed out that the ethnonym for the Mascouten
Algonquians living just west of Lake Michigan in the late seventeenth
century reflected their identity as prairie creators, since it meant both "a
treeless country" and "fire nation."[26] French observers also understood
the distinctive motivation for the Indians' fire regime. Indians did not
value grassland for its own sake, since relatively little of its vegetation
was useful as food. Nor did they use prairie as agricultural land. Evidence suggests that Native farmers did not begin cultivating corn in
the Midwest until 1000 CE, and not in the prairie uplands themselves,
since dense root structures of the tallgrass ecosystem made clearing the
field too laborious before John Deere's steel plow.[27] Instead, they used
the prairies as a huge game reserve.[28] The key purpose of the fire regime
practiced by midwestern Indians was to attract and hunt grazing mam-
mals, particularly elk and especially bison, which historic era records
suggest were a key part of the subsistence of Native peoples in the eastern
prairies. As Louis Hennepin observed among the Illinois in 1680, the
prairie Indians maintained the prairie as bison habitat by burning them
"every year," causing the prairie to flourish and the numbers of animals
to "multiply in such a manner, that notwithstanding the great numbers
they kill every year, they are as numerous as ever."[29]

Assuming this is all true, our picture of the tallgrass ecosystem as
an anthropogenic creation becomes even richer. Recent research from
Konza Prairie in the Flint Hills of Kansas has established much about the
relationship between grazers, grass, and fire in the tallgrass ecosystem.
Grazing had many important functions in maintaining the biodiversity and productivity of the prairie. Bison were a keystone species whose
grazing acted to select certain herbaceous species and joined fire in keeping forest succession at bay. Once enticed into the tallgrass prairie by fire,
larger herds of bison probably helped to maintain it—increasing growth,
affecting fire intensity, and preventing the overgrowth of dominant grass
species.[30]

But it is important to note that despite the key role they played in
maintaining the ecosystem, the bison were not at all a "wild" resource.
Indeed, absent a purposeful fire regime, it seems unlikely that bison
would have found the fast-growing and cellulose-rich tallgrasses and
forbs attractive. Tallgrass is definitely suboptimal habitat for grazers

such as bison, especially when compared with short-grass ecosystems farther west on the continent.[31] It was only in the presence of midwestern Natives' purposeful fire regime that the bison population expanded in tallgrass ecosystems, allowing the huge ungulates to play their essential roles in maintaining the grassland ecosystem. While here is not the place to explore the specific mechanisms of this relationship, the lesson is clear: We need to think of the ecology of the tallgrass not as a case of "first nature" shaped mostly by "direct ecological adaptation." Rather, the bison were, to use Aldo Leopold's language from his classic book *Game Management*, "artificialized"—their habitat was controlled to enhance (if never maximize) their productivity.[32] Here was a strongly hybrid ecology. In the tallgrass, people, grass, grazers, and fire fundamentally *made one another*. Indigenous lifeways in the tallgrass environment were not just intimately related to animals (as was the case for almost all pre- and early-modern peoples); they were a *partnership* in which humans and animals together made their niche in the eastern prairies.[33]

Given all this, we need to think about the prairie landscape in which Chicago grew up in a different way. Ecologists once thought that the soils of the Midwest made the prairie. In fact, the opposite is more nearly true.[34] The presence of such rich soils in the eastern prairies—the region that later became the corn belt—was the direct result not only of glaciation, but also of thousands of years of human occupation, of human selection, of ungulate grazing, and of intentional burning.[35] Although less sudden than the anthropogenic change represented by Chicago, the shaping of the Midwest by Native peoples over generations was really no less consequential.[36]

To be sure, it was different. Indians' shaping of the mid-continental landscape resulted in a complex mosaic of ecozones, not a series of simplified monocultures dominated by certain selected species. Their shaping of the land may have increased certain measures of biodiversity rather than suppressed it, as the industrialized farming driven by Chicago surely did. These effects amount not just to matters of scale but to *qualitative* differences in the way that Indians and Euro-Americans shaped their landscapes, and we need to recognize these serious differences and take their measure, lest we arrive at false equivalences across time. But recognizing differences should not obscure a basic continuity: the midwestern landscape has been shaped by people for a very long time.

And there is a reason why this matters to midwestern history in particular. To be sure, the basic concept of this article should not be surprising to environmental historians, even if they know few specifics about prairie ecology. After all, a main lesson of American environmental history over two generations has been that Indians shaped the land

profoundly.[37] Indeed, it was William Cronon himself who taught us the
extent to which Indians in New England had so profoundly shaped the
landscape before the colonists' arrival, maximizing the output of a diverse set of resources and moving about on the patchy landscape to create a stable ecology. The resulting mosaic of forests, cultivated clam beds,
and other hybrid ecosystems fostered by the Native peoples of New England were no more "natural" than the landscapes the colonists imposed
as they settled the land in the seventeenth century.[38]

But while this story of Indian ecology is familiar in the places where
colonial historians have traditionally focused, it is less well understood—
and sometimes ignored—in the case of the Midwest. The easiest explanation for this oversight might stem from the same reasons that early
and indigenous midwestern history is poorly known overall. Given that
the Midwest's role in "national" history begins only in the early nineteenth century, the presence and significance of indigenous peoples in
the Midwest are often obscured, offstage from the "main action" of early American history. By the time their region began its moment in the
national spotlight, Native peoples of the Great Lakes and Midwest were
weakened and quickly conquered. For all these reasons even historical
narratives that have made room for the Wampanoags and the Powhatans
sometimes fail to incorporate the Potawatomis or the Foxes or even the
Mississippians.

A transnational approach to early American history has corrected
some of these faults over time, and Richard White's enormously influential framework of "the middle ground" did much to center attention
on the indigenous history of the region. And yet the hold of nationalist
historiography still tempts Americans into thinking of the "beginnings"
of the Midwest as the nineteenth century.[39]

This must be part of the reason why it is still easy to overlook the way
in which Native peoples lived on the land and shaped it well before the
settler colonial story began in the nineteenth century. It is an unfortunate and ironic situation, since so much evidence suggests that Indians
shaped the Midwest perhaps even more profoundly than they did the
East, owing to the special dynamics of an ecotone landscape which was
particularly susceptible to burning. Moreover, as environmental human-
ists debate the concept of the Anthropocene, the history of Indian land
use and ecological creation in the Midwest is especially salient. For proponents of the Anthropocene idea, human history is in a fundamentally new era, a period when any kind of independent nonhuman nature
has ceased to exist. In the new era, all is hybrid, humans have become a
geophysical force, the entire world is an artifact, and there is no nature
outside of human agency.[40]

The deep history of the Midwest does not challenge the premise that we live in a hybrid world, or that cities such as Chicago have been making a profound and fundamental impact on the Earth's climate and even geology since the nineteenth century, or whenever we place the "golden spike." But a deep history of the Midwest *does* suggest that none of this is wholly new.[41] The politics of the Anthropocene imagine a fundamental discontinuity and binary between modernity and premodernity that is somewhat akin to the distinction between second nature and first nature. The history of the Midwest supports the voices of anthropologists and ecologists who, in emphasizing human impacts on ecosystems as a constant across global environmental history, critique the assumptions on which this imagined binary rests.[42] Native peoples of the Midwest did not simply use the nonhuman environment as they found it, and never in the Holocene was the landscape of the region primarily or simply shaped by "energy flows" or "ecological adaptation." Instead, the primary agents in shaping the midwestern landscape since the ice age were people, the architects of prairie, who created a world which was in many respects as hybrid a place as the world of the Anthropocene. The history of the Midwest encourages us to take Cronon's binary-shattering project a step further. Just as people in the city were never disconnected from and outside nature, as a naive environmentalism once held, premodern people were never merely inside it. Beginning nearly with the arrival of people in the region, the Midwest is second nature all the way down.

CHAPTER 2

Cholera
and the Evolution
of Early Chicago

ANN DURKIN KEATING AND KATHLEEN A. BROSNAN

Cholera appeared in Chicago a year before the town incorporated. On July 10, 1832, General Winfield Scott and his troops arrived on a steamship on their way to engage the Sauk warrior Black Hawk in western Illinois; they brought cholera ashore. The disease originated in the Indian subcontinent and spread across the globe via international trade routes in a series of pandemics during the nineteenth century.[1] Part of the second pandemic, the 1832 episode was the first to plague Chicagoans. Cholera had appeared at a time when local residents lacked the resources and institutions to prevent its ravages. When cholera next returned in 1849, Chicago was in the midst of a population boom and its residents appeared more interested in boosting their city than developing permanent means of disease control. By this time the link between cholera and contaminated water was clearer, but many commentators still blamed individual moral inadequacies for the spread of disease, particularly among the poorest and most vulnerable citizens. When cholera again erupted in 1866, Chicago had begun to construct a sanitation infrastructure, but many residents found the city's efforts inadequate and pushed for a permanent Board of Health. As Chicago evolved from an Indian Country outpost to a world metropolis, its responses to these cholera outbreaks reveal an often judgmental civic culture that sought private solutions to broad interrelated social and environmental challenges.[2]

CHICAGO'S FIRST EPIDEMIC

When General Scott arrived, Chicago was not yet an urban place. Some twelve log homes stood near Fort Dearborn's barracks and lighthouse. French traders, their Potawatomi wives and métis children, plus a small but growing number of American families made up the community, some living in tents.[3] Situating the fort on the swampy southern bank of the Chicago River, near its mouth, the U.S. government captured the expectations of many white Americans who envisioned a metropolis at Lake Michigan's southwestern tip. The location offered commercial opportunities because of its geographic advantages. It stands at the juncture of the Great Lakes system and the Mississippi River system. Native Americans had long used a low-lying passage across the continental divide separating the watersheds to generate a fertile trading ground. Indians lived on the banks of the Chicago River for centuries. The arrival of European settlers in the late seventeenth century and the expansion of the fur trade led to this métis settlement.[4]

Established in 1803 and rebuilt in 1816 following its destruction during the War of 1812, Fort Dearborn was one of a few U.S. outposts in the western Great Lakes.[5] Potawatomi, Ottawa, Sac, Ho-Chunk, Miami, Kickapoo, and Delaware controlled the land beyond Fort Dearborn. They claimed and contested the region's rich resources. During summers these primarily Algonquin-speaking people lived in villages at regular intervals along the rivers to the east (St. Joseph), south (Wabash), and west (Fox, Kankakee, and Illinois).[6] Women farmed, while men hunted and fished from local streams. The communities shifted to smaller encampments in winter. For more than a century they participated in a robust fur trade: first with the French, then the British, Spanish, and Americans. Indian men hunted beaver, otter, deer, and bear, using the meat while exchanging furs for products such as iron pots, cloth, flour, liquor, and guns.[7] By the 1820s environmental, economic, and political changes threatened this Indian Country. After decades of overhunting, fur-bearing animals grew scarce.[8] The discovery of western Illinois lead mines, the invention of the steel trap in 1823, and the American Fur Company's cutthroat tactics curtailed the fur trade.[9] The opening of the Erie Canal in 1825 and the introduction of steamships on the Great Lakes and the Mississippi linked the region more closely to the eastern United States; travel times from New York City to Chicago dropped from six to two weeks.[10] American settlers increasingly pressured the federal government to negotiate more land cessions from Indians in the western Great Lakes, while federal surveyors laid the gridded survey, transforming land into real estate.[11]

The changes antagonized many Native Americans. In April 1832 Black Hawk returned from Iowa with one thousand followers intent on farming lands allegedly ceded in a disputed 1804 treaty. Hostilities with American settlers escalated, and the U.S. Army moved to assist recently organized Illinois and Wisconsin militia.[12] Black Hawk sought support among the Potawatomi. While many sympathized with him, their leaders feared that an alliance would result in an "uncompensated removal." The Potawatomi understood that, under the 1830 Indian Removal Act, the U.S. government would not tolerate Indian lands east of the Mississippi and hoped to make the best possible deal.[13] Consequently Potawatomi leaders warned American settlers of the impending danger, although some young warriors joined an attack southwest of Chicago that killed fifteen white men, women, and children.[14] These events created panic among American settlers at and near Chicago. Many took refuge at Fort Dearborn which "was crowded to overflowing."[15] The first regular troops under Major Whistler arrived in June, forcing settlers from the fort. With inadequate rations, many refugees went hungry, but most remained nearby. One family, who claimed land several miles south of Chicago, did not return home "for fear of the Sacs." They sheltered in a barn, but soon "were ordered out . . . It was needed as a military storehouse." Hundreds of Potawatomi also camped around the fort, having "been ordered in by the government lest they should be killed as hostile to the government." All awaited the arrival of additional soldiers to "terminate the war speedily."[16]

In response to Illinois governor John Reynold's request for support, President Andrew Jackson ordered Winfield Scott to lead one thousand troops from Virginia. Their journey took them along the Erie Canal from New York City to Buffalo; cholera was present in both cities as they traveled through. Although not fully understood at the time, cholera is a waterborne disease caused by a comma-shaped bacillus and transmitted between humans by the fecal–oral route. Symptoms include severe cramps, diarrhea, and vomiting. In the nineteenth century some 50 percent of those infected died within a few days, their skin turning bluish-gray from rapid dehydration.[17] The bacillus flourishes in warmer weather.[18] Scott chartered four steamships at Buffalo. The *Sheldon Thompson* and the *Henry Clay* left on July 2, carrying officers and troops. The *William Penn* and the *Superior* departed two days later with provisions. Outbreaks of cholera on the *Henry Clay* and *Superior* forced them to abandon their journey. Captain Augustus Walker of the *Sheldon Thompson* left three ailing soldiers on Mackinac Island. Twelve more perished before the steamer reached Chicago.[19] Scott "had never felt his entire helplessness and need of Divine Providence as he did upon the lakes in the midst of the Asiatic Cholera."[20]

When Scott's steamer arrived at Chicago, sailing vessels already sitting offshore laid anchor upon learning of the cholera, "hoping to escape the pestilence."[21] Whistler's men left Fort Dearborn to Scott's sick soldiers and encamped eight miles away. Over the next few days more than two hundred of Scott's men received treatment; fifty-eight died. It was a chaotic place with "soldiers calling to each other, dogs barking, Indians hallooing, rendered it an awful scene never to be forgotten." On July 18 the *William Penn* brought more cholera-stricken soldiers.[22] The several hundred métis and Americans living at or near Chicago faced a perilous choice: flee military protection or remain in the midst of disease. His family forced from a shelter, Lemuel Bryant captured the fears of many: "I shall always remember the scene. The prairie on one side with enemies on it at probably no great distance, the other Lake Michigan on it the dark steamboat like a dark putrid mass."[23] There was no civil authority at Chicago to regulate the evacuation or care of civilians. No one counted the American settlers who fell ill or succumbed or the Potawatomi casualties, although cholera did spread beyond Scott's troops.[24] Once the army treated its sick, it turned west to defeat the Sauk warriors. By August 1832 Chicago's epidemic had run its course.[25]

Like the sailing vessels in the harbor, cholera revealed the region's growing environmental and economic connections with the nation and the world. The United States was a settler colonial state, and more than peace, white Americans sought land and the "elimination of the native."[26] A year after the cholera crisis, under the Treaty of Chicago, most of the Potawatomi and their allies exchanged 5 million acres of land in northeast Illinois and southeast Wisconsin for 5 million acres, first in Missouri and then in Iowa. White Americans purchased the fields once farmed by Potawatomi women.[27]

The shift from Indian Country to U.S. territory marked the entrance of American local government and civil authority. After Chicagoans organized their town in 1833, trepidation about cholera's return or the appearance of other contagious diseases led to some health regulations. Early ordinances made it unlawful to dispose of dead animals, dung, or "any other offensive substance" in the Chicago River or the streets. By 1835 the state of Illinois granted the town the power to organize a Board of Health to "prevent the introduction of the dread and fatal disease, the cholera, into said town." The town authorized the acquisition of a building to serve as a cholera hospital, designated cemeteries for each of the town's three divisions, and endorsed the private provision of clean water. The city periodically inspected ships, but focused on maintaining commerce. These measures relied on voluntary cooperation and proved temporary.[28]

Chicago's first charters of 1833 and 1835 made it difficult to implement citywide public works projects. They provided for a Board of Trustees elected at large which covered the costs of public works through general property taxes and adult male residents' labor, but offered the city few revenue sources. In its first four years the city surveyed and drained only four streets; it constructed no sewer or water systems. This limited activity raised public health concerns, especially given Chicago's topography. Situated only four feet above the level of Lake Michigan, the city sat on flat soil that absorbed little moisture. Profuse putrid puddles offered breeding grounds for mosquitoes and other disease vectors and produced "miasmas" that nineteenth-century medicine associated with illness.[29] The city's continued growth—its population reached four thousand by 1837—pressured the limited services. While there were no recorded cases of Asiatic cholera during these years, an epidemic "similar to Asiatic cholera in the summer of 1838 . . . forced stoppage of work on the [Illinois and Michigan] Canal." Overall, efforts remained interim and haphazard. As one historian observed, "Chicago appointed a new board of health whenever cholera was seen coming down the road (or down the lake or up the river)."[30]

A new charter in 1837 gave the city special assessment powers for street improvements. As the historian Robin Einhorn observes, "By the mid-1840s, the use of special assessment would transform Chicago's government, replacing the citywide booster system with a segmented decision-making process."[31] The urban real estate market, through such assessments and private subscriptions, dictated which streets were surveyed and drained. Thus in the city's first decade and a half, its physical infrastructure failed to meet the needs of all residents. Few streets had planking. Water supplies and waste removal were inconsistent and ineffective. The recurrence of cholera in 1849 would challenge this limited government.[32]

CHICAGO'S SECOND EPIDEMIC

During the 1840s harbor improvements and ongoing construction of the Illinois and Michigan Canal provided easier access to the Chicago River. In 1848 the canal opened and the first railroad arrived, enhancing the city's natural transportation advantages. With nearly thirty thousand people by decade's end, Chicago was the commercial center for the burgeoning agricultural and timber regions of the old Northwest. Lake steamers brought goods from the East, while canal boats connected the city to the Mississippi River. Immigrants from Ireland, Germany, and Norway streamed into the city, along with easterners in search of success. The metropolitan area oriented along the rivers in a region with considerable low-lying marshes.[33] Cholera followed these waterways when it returned in 1849 as part of another global pandemic.

Despite Chicago's growing stature, various problems persisted and worsened. Clark Carr recalled his first view in 1850: "The City of Chicago . . . was low and flat, the buildings were small, and beyond them there was nothing to relieve the eye but more low flat land. . . . The plank street coverings were covered with mud, and only seemed to keep the foot-passengers from sinking out of sight."[34] Vegetable and animal waste mixed with mud and the garbage people heaped in streets and alleys. The city hired private scavengers in 1845 to remove waste, but it was an inadequate effort.[35] Despite boosters' efforts to promote Chicago's healthfulness, the city earned the opposite reputation.[36] Other city services remained privatized, intermittent, and poorly distributed. For example, except for one public well, Chicagoans initially bought water from peddlers who drew barrels from Lake Michigan and sold buckets door to door. In 1842 the Chicago Hydraulic Company, a private corporation, opened its waterworks, quenching only its subscribers' thirst. "Its one pump connected with the water of the lake by means of an iron pipe, laid on a crib-work pier extending into the water about 150 feet." The service did not reach all neighborhoods nor did the pier stretch far enough into Lake Michigan to escape the waste that flowed from the river.[37]

In early 1849 Chicagoans read about cholera's return to North America. Their worst fears were realized when the disease arrived on April 29 on the canal boat *John Drew* with travelers whose journey originated in New Orleans.[38] In response to this new epidemic, civic leaders still sought informal, voluntary solutions, a phenomenon that the historian Sam Bass Warner Jr. described as "privatism." Rejecting public-oriented civic responsibility, communities were composed of "a union of such money-making accumulating families."[39] For instance, while Chicago appointed a city physician in 1849, he served "without salary."[40] The Board of Health recommended ordinances regulating public nuisances thought to exacerbate the contagion, including garbage, sewage, and dead animals in streets, alleys, empty lots, and waterways. Claiming that "every effort was put forth to cleanse the city," the Board appointed forty-five unpaid assistant health officers (one to each city block), who checked for violations and encouraged neighbors to clean up their properties and adjoining streets. Believing that covering manure and other "filth" with lime was an expedient way to prevent disease, the board purchased one hundred barrels for citizens to cover waste that could not be carted away.[41] These measures were insufficient. Eager to keep commerce flowing, the city did not implement a quarantine until 1854.[42]

YEAR	ESTIMATED CHOLERA DEATHS
1849	687
1850	405
1851	216
1852	630
1853	25
1854	1,424
1855	147

TABLE 2.1 Estimated Cholera Deaths in Chicago, 1849–1855.

The municipal government tread lightly even when collecting data. During the 1849 outbreak, the Board of Health asked private physicians to report cases on a weekly (and for stretches, daily) basis. The information requested is revealing: date of death and the patient's name, age, sex, nation (nationality), occupation, location, and habits. The dates and locations of deaths provided crucial intelligence about cholera and its course, and later contributed to plans for a new water system. Other requests betrayed a long-standing tendency to blame patients for their illnesses.[43] Speaking during the earlier cholera pandemic, the New York State Medical Society president observed, "We may therefore safely rank among the predisposing and exciting causes, intemperance, uncleanliness, and profligacy."[44] Chicago's board seemed determined to find that certain nationalities were more prone to disease. Some commentators explained cholera by the presence of large groups of Irish, German, and Norwegian immigrants.[45] Equally troubling was the query about habits. Given the quick onset of symptoms, it was difficult for physicians to assess habits. Some left the category incomplete, but others answered with phrases ranging from "regular habits" or "temperate, a very hardworking man" to "irregular and intemperate."[46] In several cases physicians also described caregivers: "The only care the patient had was from men and they were much intoxicated," or a one-year-old child who was "kept very filthily."[47] In the end, no clear pattern emerged with respect to background and behavior, but in asking the questions, the board hoped to place some responsibility with patients' "sinfulness," especially alcohol abuse, instead of the city's failure to resolve environmental problems.[48]

In 1849 the board supported a temporary hospital in July and August for "indigent persons attacked by cholera." The city opened similar facilities during virulent outbreaks in subsequent years.[49] However, family networks and charitable organizations cared for most cholera patients.[50]

Women were particularly important. Working through religious organizations, they provided nursing care and founded hospitals and orphanages. Mother Agatha O'Brien and a small cadre of Sisters of Mercy responded to cholera outbreaks beginning in 1849. Having arrived from Ireland three years earlier, these nuns, who already ran three schools and other programs, began nursing at the temporary hospital at the request of local physicians. In 1851 the sisters assumed management of the Illinois General Hospital; rechartered as Mercy Hospital, it became the city's first permanent hospital. The nuns also established the city's first Catholic orphanage. In 1854, the worst cholera year, the nuns "set aside all their other duties to nurse the sick and dying."[51] O'Brien, who once wrote that dying in the service of others was "the recompense we may hope to receive if we are faithful to our holy calling," succumbed to cholera in July. Within a few days three more nuns had died.[52]

Protestant women also responded. Because the city had no provision for children orphaned by cholera, except to place them in the almshouse, local churches intervened. By August 1849 the number of orphans overwhelmed ad hoc measures, and church representatives organized the Chicago Orphan Asylum. Separate boards of men and women drew members from Chicagoans who built the city's first generation of churches. While men provided overall administration, dedicated women began to create a home for orphans. A Swedish girl, whose parents died on arrival at the city's railway station, was the first child admitted. Over the next three years successively larger structures housed more than one hundred orphans.[53] In 1854 the Asylum's board petitioned Cook County for public funds, arguing that the institution accepted children who otherwise would be consigned to the poorhouse. The county government denied the petition because the asylum was a "voluntary benevolent organization."[54] Consequently the enterprise remained funded by the many fairs sponsored by the female directors and supported by the communities' "church ladies," plus intermittent donations from male directors. The operation rested on the director and other female volunteers who canvassed for supplies, supervised staff, and managed admissions and placements.[55] Chicagoans responded to cholera with institutions undergirded by private philanthropy.

Privatism persisted even with infrastructure. Urban inhabitants across the United States and Europe did not know about the cholera bacillus, but recognized that the spread of the disease was associated with poor water and waste systems.[56] By 1851 the Chicago Board of Health recommended a change "owing to the great fatality by cholera in 1849 and 1850 in localities where well water was used, particularly on the north side."[57] The Illinois legislature chartered the Chicago City

Hydraulic Company to provide lake water more systematically to residents. A quasi-public corporation with three elected commissioners, it received city funds and bond issues, but was not under city control. The company initially laid water pipes in a prescribed area that included Chicago's North Side, where many cholera cases occurred. It installed public hydrants so that residents could access the system even without direct connections into their property. Annual water rents for private taps allowed for expansion with limited public support. By 1856 the company installed nearly one hundred miles of pipes and reached 80 percent of Chicago. Five years later much of the city was connected. While bonds covered the central pumping station and lake intake tunnel, commissioners levied special assessments to extend the pipes.[58]

In 1855 the city council asked the legislature to approve the Board of Sewerage Commissioners, using the Chicago City Hydraulic Company model to some extent. Commissioners hired Ellis Chesbrough to design a new system that drained both rainfall from streets and wastes from buildings. The commissioners created special districts to raise funds, but could not charge continuing fees like water rents, revealing privatism's limitations and slowing system expansion. With circumscribed funds, commissioners chose the least expensive option Chesbrough offered for outlets: dumping the discharge into the Chicago River.[59] (Chesbrough believed the optimal and eventual solution would involve deepening the Illinois and Michigan Canal to reverse the river.) Given the topography, the city could not place sewer pipes deep underground. To ensure sewers drained properly, the best option was to elevate the street level and raise the city's buildings. George Pullman designed a method to lift buildings, allowing foundations to be built underneath. Workers placed storm sewers on top of the streets and filled them with dirt up to the front doors of raised buildings. In 1860 a crew of six hundred operating six thousand jackscrews elevated half a city block on Lake Street nearly five feet over five days. Businesses occupying the stone structures, some four or five stories, remained open throughout. Privatism continued as building owners bore the elevation costs.[60] The new sewers did not resolve all problems, however. Some owners delayed connections because they entailed adding plumbing fixtures. Finally, Chesbrough's design included two flushing conduits on the river's north and south branches to ensure dilution of sewage before it reached Lake Michigan. The commissioners eliminated the conduits to reduce costs. Despite dredging, ongoing population growth, increased sewage loads, and greater industrial activity overwhelmed the river.[61]

It took six consecutive years of cholera to force the city to begin its sewerage system. Thereafter infrastructure improvements continued, but Chicagoans, as they had in the past, lost interest in public health. Facing the economic depression that gripped the nation in 1857, "little was done by the Board . . . as there was no cholera here in the meantime and not much small-pox—the only diseases which, in the opinion of the authorities, seemed to require municipal supervision." The city held no elections for Board of Health members between 1858 and 1861. "The duties of the Board gradually became merely nominal, until its functions imperceptibly ceased."[62]

CHICAGO'S THIRD EPIDEMIC

During the Civil War Chicago continued to grow, with eleven railroad lines serving the city. Preoccupied with the war, and in the absence of epidemics, the city gave limited attention to public health, leaving issues to the police board. At the same time, the city worked to complete a sewer system that already seemed inadequate. Noting persistent challenges in cleansing the river in his June 1862 report to the Chicago Medical Society, Dr. Nathan Davis observed, "I know of no city, except Chicago, with a population of 110,000, that has neither a Health Officer, Board of Health, or any other official sanitary organization."[63] A charter member of the Chicago Medical Society and the American Medical Association, Davis had campaigned since 1850 for permanent rather than expedient sewerage solutions.[64]

Others also demanded a fixed Board of Health to monitor sanitary conditions and advocate for infrastructure improvements. However, in 1865, as the city watched a new pandemic advance across Europe and braced for the "expected visitation of cholera," the council did not heed these calls. Instead it authorized a health officer, supervised by the police board, to print thirty thousand "nuisance notices," one-third in German, about the anticipated epidemic. Such circulars often were the only way poorer Americans learned about home hygiene. Late in 1865 the council appointed a voluntary committee, chaired by Davis, to investigate sanitary conditions. The committee recommended forcing landowners to connect to the sewers and inspecting privies, slaughterhouses, and other nuisances. The city added seventeen temporary health officers in January 1866 as an ad hoc measure, but it proved too little, too late. By July Cincinnati and St. Louis faced full-blown epidemics, and the first cases appeared in Chicago. With a temporary cholera hospital constructed on the South Side, local physicians met trains to identify and quarantine infected passengers. In September the city appropriated an additional $10,000 to fight what was now an epidemic. Only in December did "meteorological conditions favorable to life" end the siege. In all, officials confirmed 990 cholera deaths in 1866.[65]

Chicagoans' response to the 1866 outbreak reflected new attitudes about sickness. In the 1840s and 1850s many attributed cholera to its victims' background or habits. By 1866 officials demonstrated that most deaths occurred among immigrants in poorer neighborhoods, but did not include data on personal behavior.[66] A year after the 1854 epidemic, British physician John Snow had published a study that undermined misdirected moralism. He confirmed a connection to contaminated well water by mapping clusters of cholera surrounding a particular London pump. Nonetheless, a debate raged in scientific circles between contagionists, who believed that germs caused cholera, and anti-contagionists, who stressed environmental factors. German microbiologist Robert Koch, who identified the bacillus *Vibrio cholera* in 1883, joined Snow and other scientists who advocated quarantines and disinfectants, while the German hygienist Max von Pettenkofer and others worked to ameliorate unsanitary conditions. In 1866 Chicago's officials straddled a middle ground between the differing approaches. They quarantined patients in the hospital and dispersed various disinfectants. At the same time officials collected data about hot and wet weather, groundwater conditions, and unsanitary streets.[67]

Calls for a permanent, independent Board of Health grew during the 1866 epidemic. By November, the *Chicago Tribune*, writing about New York's Metropolitan Sanitary District, opined "we need such a measure for Chicago."[68] By March 1867 the city council and state legislature had created a permanent board. The board immediately began sanitary inspections and an education campaign that perhaps contributed to a significantly lower loss of life in 1867 (ten deaths), although the disease was less virulent as well that year.[69] A number of factors contributed to the long-term commitment to public health. The success of the Northwestern Branch of the U.S. Sanitary Commission (USSC) in mobilizing aid for the Midwest's wounded soldiers during the war offered a model for moving away from informal, ad hoc responses. Initially families, congregations, and business networks responded to soldiers' needs. Some women who had been instrumental in caring for cholera patients, including the Sisters of Mercy and Jane Hoge (whose two sons served in the war), offered the USSC nursing expertise, fund-raising abilities, and managerial talents. Over the course of the war, however, the USSC developed into a more formal organization with separate, clearly defined roles for men and women.[70] Having observed its administrators' growing professionalism, residents in Chicago and other American metropolises demanded greater public responsibility for their health and safety. "The greatest crisis in American history, the Civil War, permitted unprecedented play for sanitarians," according to the historian Harold Hyman, "for it generalized among patriots the sanitarian's basic proposition that public power must protect public interests."[71]

The Civil War also hastened changes already under way in Chicago's economy. "A new social order . . . emerged with amazing rapidity during the 1860s. . . . rooted in the transformation of the city from a commercial center into a dynamo of industrial capitalism."[72] The environmental toll of tanneries, slaughterhouses, and breweries, plus the spewing steam engines in manufacturing and transportation, demanded a more active government to protect public health and business. These challenges arose not just during epidemics, floods, or fires, but every day as Chicago's population grew to almost 350,000 by 1870. The scale of industry and Chicago's expanding market sphere pushed city government to respond on a larger, wider and more permanent basis. Public water and sewerage systems replaced private, household-based services. Chicagoans increasingly viewed these as public goods to be provided through general taxes and bond issues, instead of private improvements funded via special assessments. As the public sphere expanded, however, the role of women became more circumscribed. As male-dominated institutions and government agencies took up public health and responded to medical crises, women's informal relief efforts within the private sphere declined in importance. Increasingly urban residents identified public health professionals, particularly physicians, as experts with the requisite skills to protect the city's well-being.[73]

As the historian Walter Nugent observes: "A society is often regarded as having become truly modern when fewer of its people die of contagion than from chronic causes." Cholera returned to Chicago one last time in 1873, killing 116, but the city did not pass Nugent's threshold until the 1920s. Smallpox, diphtheria, scarlet fever, whooping cough, tuberculosis, pneumonia, typhoid, and dysentery loomed even after cholera abated.[74] In sheer numbers, far more Chicagoans lost their lives to these other contagions, but cholera had scourged Chicago first, even before it was a city. Its death rates were higher because of smaller populations in 1832, 1849–1854, and 1866 than in later years. And the speed and ferocity with which it struck its victims proved terrifying.[75] In addressing the disease and the fear it engendered, the city's public health infrastructure, read in broad environmental terms, evolved. From an absence of civic responsibility in 1832, the City of Chicago offered limited, generally privatized responses to outbreaks between 1849 and 1855. Even the construction of water and sewerage systems utilized semiprivate financing mechanisms. By 1866 those systems increasingly fell under public control, and city residents had compelled their governments to establish a permanent, professional Board of Health to enforce sanitation as a public good.

CHAPTER 3

Animals at Work in Industrializing Chicago

KATHERINE MACICA

Legend holds that late on the night of October 8, 1871, a cow belonging to Catherine O'Leary kicked over a kerosene lantern, sparking what would become the Great Chicago Fire. High winds swept the flames across O'Leary's working-class neighborhood on the Near West Side to engulf the city. The fire burned for twenty-four hours, destroying more than eighteen thousand buildings and killing three hundred people.[1] In the aftermath of the fire, rumor spread that the blaze originated in O'Leary's barn. Although the official inquiry into the fire's causes uncovered no evidence of Mrs. O'Leary's guilt, Chicagoans nonetheless found a scapegoat in the working-class Irish family, and their cow became one the most infamous animals in the city's history.[2]

Despite the popular notion that Mrs. O'Leary's cow started the Great Chicago Fire, city officials did not take action to regulate the presence of animals in the city. Animals were a ubiquitous part of the landscape in American cities in the nineteenth and early twentieth centuries. Human residents of industrializing Chicago shared the city with a wide array of domestic animals.[3] Indeed, because of the stockyards, Chicago had the largest proportion of animals compared to any other city in the United States in 1900, with livestock equal to ten percent of the human population.[4] Chicago never established a commons within its borders, unlike older cities such as New York and Boston. As a newly incorporated city in 1837, Chicago sought to delineate its borders as urban space by prohibiting livestock from roaming within city limits, but has never banned the keeping of livestock outright.[5] Instead, laws changed over time to reflect the cultural and economic value of working animals alongside the perceived drawbacks of sharing urban space with animals.

As Chicago transformed over the course of the nineteenth and early twentieth centuries, so too did relationships between humans, animals, and their environment. Chicagoans adapted the ancient practice of using working animals to a new environment: the industrial shock city.[6] Working animals were integral to Chicago's industrializing economy and environment. The labor of horses and cattle provided transportation, setting the pace for the flow of people, goods, and services throughout the city. The bodies of horses and cattle became calories and saleable commodities to the residents and businesses of Chicago and beyond. Working animals shaped and were shaped by the urban environment. Horses had to adapt to the sights, sounds, and terrain of city streets. At the same time humans had to contend with the vast quantities of waste generated by working animals. Human and animal residents adapted to the dynamic urban environment in different ways, based on their places within the economic and physical landscape of the city.[7] This essay examines the ways in which Chicagoans used and regulated working animals, and the changing relationships between animals and the built environment from the nineteenth century through the 1930s, when working animals were no longer commonplace elements of the city.

COWS AND CATTLE

Catherine O'Leary's notorious cow numbered among the thousands of dairy cows and beef cattle that lived and died in nineteenth- and early twentieth-century Chicago. More than twenty-two thousand head of cattle could be found in Chicago in 1900, mostly within the city's stockyard complex but also scattered throughout the neighborhoods.[8] These bovines provided food and a means of livelihood for Chicago's working class, serving important roles in the industrializing city. Chicagoans regulated and used dairy and beef cattle in various ways, reflecting cattle's value and place in the urban environment.

In the first half of the nineteenth century the city placed few restrictions on cattle ownership. But as the city grew, laws changed to reflect the difficulties of sharing urban space with large livestock. In 1856 an ordinance prohibited cattle from running loose within the present-day Loop and limited the number of cattle that could be kept outside of a slaughterhouse to ten. The city further reduced that number to three in 1881 and required cattle owners to obtain a permit from the Health

Department.[9] Although municipal ordinances in the late nineteenth century restricted cattle ownership, nonetheless many Chicagoans maintained cattle in the city well into the twentieth century.

Many working-class Chicagoans raised dairy cows for milk. Most of the milk was kept for home use, but some surplus was sold as a source of income. Dairy cows primarily lived in the rural areas along the periphery of the city, but many also inhabited the small residential lots of densely populated neighborhoods. In 1900 more than five thousand dairy cows lived in Chicago, and nearly two decades later a survey reported around two thousand cows within the city limits.[10] Although waste from these animals often proved a public health nuisance, nonetheless family-owned cows provided an important source of nourishment at a time when access to fresh, clean milk was a luxury few working-class Chicagoans could afford.

The supply chain for milk became increasingly complex as Chicago's population boomed in the second half of the nineteenth century. As more people moved to Chicago to work in the city's growing industries, residential and industrial land uses pushed agricultural production farther away from the city. By the turn of the century most of Chicago's milk supply was shipped via railroads from dairies as far as 123 miles away. The expanding distance between cow and consumer created a greater potential for contaminated or spoiled milk. Milk could be up to four days old between milking and consumption, often without refrigeration.[11] Researchers identified impure milk as a major contributor to childhood mortality rates as early as the 1860s and connected outbreaks of cholera, scarlet fever, and tuberculosis in Chicago to contaminated milk.[12] Public health officials concerned with this problem proposed three possible solutions: pasteurization, strict regulation of sanitary conditions throughout the supply chain, or encouraging milk production closer to sites of consumption in the city.[13]

The idea of expanding milk production within Chicago neighborhoods gained traction among some policy makers, including the city's health commissioner, Dr. William Evans. In 1908 the *Chicago Daily Tribune* reported on Dr. Evans's plan to bring dairy cows into the city under the supervision of the Health Department. Evans estimated that Chicago would need around eight thousand cows to supply enough milk to feed the city's young children. Cows would be distributed across neighborhoods in small numbers, placing the source of milk as close to the consumer as possible to reduce the likelihood of contamination.[14] Evans's plan built upon an existing tradition of dairying in Chicago and represented "that happy combination of city and rural life," but the scheme proved at odds with the increasingly crowded urban landscape. Instead, by 1916 the city embraced pasteurization to ensure a clean supply of milk for its citizens.[15]

Rather than being scattered throughout the neighborhoods, most of Chicago's livestock was concentrated into one area within the city: the Chicago Union Stock Yard. The stockyard emerged as a "combination of city and rural life," although it was not a happy place for the thousands of animals that found themselves moving through the complex each year. Established in 1865, the Chicago Union Stock Yard served to consolidate the city's livestock marketing and slaughtering into one location, and to eliminate the necessity of driving cattle through the growing city. The stockyard became the conduit between the urban and rural, receiving more than one billion animals from the rural West over the course of a century, transforming the grasses of the plains into meat through cattle, hogs, and sheep.[16] The Union Stock Yard and adjacent slaughterhouse complex was a city unto itself. When it opened in 1865, the Union Stock Yard covered 60 acres but grew to 475 acres by 1900, or 560 acres including the slaughterhouses of adjacent Packingtown.[17] The 250 miles of railroad tracks and 25 miles of streets within the stockyards enabled livestock to move from railcar to pen to slaughterhouse. The stockyard had a capacity of 461,000 animals by 1900, including space for 75,000 cattle, 300,000 hogs, 80,000 sheep, and 6,000 horses.[18]

Upon arrival to the stockyard via railroad, cattle, hogs, and sheep were herded to one of thirteen thousand pens where buyers for the city's meatpacking companies would inspect and purchase animals. Once sold, livestock were driven to the scale house, then on to the slaughterhouses. From there the animals became part of a disassembly line, first killed by a worker known as a "sticker," then their bodies dismembered into saleable commodities. Swift and Company, one of the largest meatpacking firms, could slaughter a cow in as little as thirty-five minutes, and had a daily capacity of processing 2,500 cattle and 8,000 hogs.[19] In its first thirty-four years of existence, Chicago Union Stock Yard received more than 59 million head of cattle, the majority of which ended up in the slaughterhouses.[20]

HORSES

Horses played an essential role in nineteenth- and early twentieth-century Chicago, providing the majority of transportation for urban residents and businesses. Horses set the pace of life in Chicago, both figuratively and literally, through speed limits codified in terms of gait. Chicagoans understood horses as "living machines" whose labor powered the city.[21] Their impact was made clear in 1872, when an outbreak of equine influenza brought the city to a virtual standstill. During that "week without horses," the *Chicago Daily Tribune* lamented that "the streets were deserted of horses, and a depressing silence fell upon the city."[22] The economic value of horses and their importance to the functioning of urban life ensured a place for them in the city until machine power replaced horse power in the first half of the twentieth century.

Many horses began their lives in Chicago in the Union Stock Yard. Horses had an entirely different experience than most livestock in the stockyards. They were valued for their potential labor, not the products of their bodies, and thus enjoyed a privileged position relative to cattle, hogs, and sheep. The status of horses was evidenced in the built environment of the stockyard. Whereas livestock destined for the slaughterhouse lived in open pens, the Union Stock Yard and Transit Company erected brick barns to house horses and a large amphitheater to show and sell them.[23] Dealers sold all variety of horses at the stockyard. Draft horses comprised around 40 percent of horses sold, and were generally used by business and industry for heavy work. Lighter work horses, for pulling smaller wagons or coaches, constituted about 35 percent of the sales, with the remainder including saddle horses, smaller working horses, and range horses.[24] At the turn of the century the Union Stock Yard sold as many as 180,000 horses each year, at the rate of 60 horses per hour. Horses that passed through the Union Stock Yard became working horses in Chicago and other cities in the United States and around the world.[25]

The lives of Chicago's horses varied greatly depending on their job and owner. Organizations and businesses of all types utilized horses in Chicago. Transit companies employed thousands of horses to pull omnibuses and streetcars throughout the city. Chicago's first omnibus debuted in 1852, and by 1890 the city's streetcar lines used 10,000 horses to pull the cars.[26] Horse cars continued to operate in Chicago until 1906, by which time electric streetcars had become the dominant form of mass transportation.[27] Police officers patrolled on horseback, and horses pulled police and fire vehicles.[28] Breweries, department stores, lumber companies, ice companies, and dairies used horse-drawn wagons to deliver their wares throughout the city. Marshall Field and Company built stables near their downtown store to house a fleet of horses and delivery wagons that grew from 100 horses in 1875 to 344 around the turn of the century.[29] The Mandel Brothers department store used horse-drawn wagons until 1916, and the *Chicago Daily Tribune* employed delivery horses until 1928. As late as 1927 nearly 5,000 horses were still at work delivering milk to homes throughout Chicago.[30] Horses connected the city center to the hinterlands and beyond, bridging the gap between railroads, docks, homes, and businesses, and facilitating the flow of commodities through Chicago.

Small businesses, households, and individuals relied on the labor of horses as well. Men worked as teamsters and drivers, employing all manner of horse-drawn vehicles, from buggies to large wagons, to move people and products around the city. Although most larger businesses owned the horses and vehicles that the drivers operated, numerous working-class Chicagoans owned their own horses and vehicles. Workers with enough money to purchase and maintain a horse could earn money as a peddler, selling produce, rags, and junk from carts along city streets. A 1900 survey of three working-class neighborhoods surrounding the Loop revealed that, out of 1,443 horses located there, about 1,000 were owned by individuals. The remainder belonged to factories or transfer companies.[31] Farmers in outlying areas of the city used horses for farm labor and to bring produce to city markets. From there, peddlers, storekeepers, and small dealers used their horse-drawn vehicles to further distribute the fruits and vegetables within the city.[32]

Chicago adopted numerous ordinances to regulate the use, movement, and treatment of horses. Unlike other working animals, horses were not confined to particular areas of the city or to enclosures. Rather, they shared the streets and labored together with people throughout the city. Accidents and traffic jams were common in nineteenth-century Chicago, as tired horses struggled under heavy loads or panicked horses bolted. Streets posed a hazard to horses, as uneven cobblestones and the accumulation of mud and ice on roadways could lead to falls, broken legs, and death.[33] In its 1837 municipal code, the city proclaimed its first traffic law, prohibiting the riding or driving of horses on sidewalks.[34] Later ordinances imposed a speed limit, at first measured by gait and later by miles per hour; required horses to be secured if left standing; and required the use of bells to warn pedestrians of a horse's approach.[35] In addition, an 1856 law imposed a fine on anyone "us[ing] any sport or exercise likely to scare horses, injure passengers, or embarrass the passage of vehicles." [36] These nineteenth-century traffic laws sought to impose order on the shared space of city streets and to make them as safe as possible for the city's equine and human residents.

Chicagoans also attempted to police the treatment of horses to ensure the humane handling of the economically valuable animals. Generally, horses owned by individuals were treated well. People whose livelihood depended on their horse tended to provide good care and training and to work to earn the animal's trust, leading to a closer relationship and greater compliance on the part of the horse.[37] Horses owned by large companies, on the other hand, received less care and training. The handlers themselves were often not experienced in horse care and were generally more concerned with hauling loads and meeting timelines than

tending to their equine charges.[38] The understanding of horses as commodities, especially when owned by corporations rather than individuals, encouraged a disconnect between humans and horses, leading many
handlers to mistreat their animals. Abuses, including beating, over-
working, or abandoning injured horses, were fairly common incidents in
nineteenth-century cities.[39]

By the middle of the century residents in Chicago and other major
cities with large equine populations became concerned with the inhumane treatment they witnessed on city streets. In 1866, the same year
that the American Society for the Prevention of Cruelty to Animals was
founded in New York, Chicago codified a new ordinance prohibiting the
inhumane or abusive treatment of animals. This law and the work of
local humane organizations were primarily aimed at drivers and companies who saw horses as "thoughtless and disposable" and believed that
it was easier to replace a horse than to treat it properly since "'horses
are cheaper than oats.'"[40] On the contrary, a well-trained horse was a
valuable commodity in industrializing Chicago. The work of these animals provided transportation and a means of livelihood for many people
and enabled the city to function. But as Chicago grew, horses and other
working animals came into more conflict with their human neighbors.

SANITATION AND HOUSING

Concerns about public health became acute as the city's population
boomed. One of the fastest growing cities in the nation, Chicago's population swelled from over five hundred thousand people in 1880 to more
than 2 million in 1910.[41] Just as the number of people in Chicago grew,
so too did the number of animals, particularly horses. In 1890 one hundred thousand horses lived in the city and produced between fifteen and
thirty-five pounds of manure per horse per day.[42] The waste products
from horses and other livestock became a critical public health concern
for city officials and Progressive Era reformers.

For much of the nineteenth century medical science relied on the
miasma theory of disease. Thus, if something appeared dirty or smelled
bad, it could spread disease.[43] Indeed, the city's 1893 Health Department
report explained that "filth . . . is itself the active cause of disease, and
that little else is essential to the production of certain infectious diseases
than to deposit a certain amount of filth, or to allow filth to accumulate
within the premises occupied by a given population in order to generate
a pestilence."[44] Reformers and those concerned with public health took
aim at the filthy conditions that they believed were responsible for poor
health and epidemics in Chicago. Manure was particularly troubling,
considering its perceived role as a disease producer and the number of

animals living in the city. Even as germ theory eclipsed miasma theory, manure was still a target, since the flies and other insects that carried disease thrived in the accumulated manure found in Chicago's streets and alleys.[45]

Reformers believed the manure and filth problem was connected to the larger issue of unsafe housing conditions in the city's densely populated, working-class communities. To combat these issues, concerned citizens, including Jane Addams, formed the City Homes Association in 1900. That year the organization undertook a survey of three areas of the city that were representative of the general condition of Chicago's working-class neighborhoods. The report generated from the survey, *Tenement Conditions in Chicago*, provided concrete evidence of the unsanitary conditions found throughout the city. Reformers used this data to push for stronger sanitation laws and building codes, and actual enforcement of those rules.[46]

The majority of ordinances regulating livestock were intended to promote a clean and healthy environment and prevent the spread of disease. Since 1881 Chicago municipal code had prohibited residents from keeping horses, cows, swine, sheep, and goats in tenement or lodging houses.[47] Despite this ban, however, City Homes Association researchers encountered numerous instances of livestock sharing living space with their human owners in the tenement districts. Environmental, economic, and cultural factors explain the close quarters shared by many working-class Chicagoans and their livestock. The city's poorest and most densely populated neighborhoods had little open space. Property owners regularly built multiple structures on one lot, moving an older cottage to the back of the lot and constructing a larger tenement building at the front. Such an arrangement often resulted in buildings covering 70 to 90 percent of a lot.[48] Despite the lack of space, many Chicagoans raised animals to feed their families and earn money outside the wage economy. Cows, goats, and chickens kept in tenement areas provided a continual supply of food and a means of generating income by selling their products to neighbors. Without the funds to move out of these neighborhoods, humans and livestock alike adapted to the environment. Researchers observed horses and cows stabled in the first floor or basement of tenements. Smaller livestock, including pigs, sheep, goats, chickens, geese, and rabbits, were found in lots, basements, attics, and even kitchens and bedrooms.[49] Likewise, people adapted to the animals' environment by constructing living quarters above barns. This practice was particularly common among single men who owned their own horse and worked as peddlers.[50] Most European cultures had a long history of housing animals and people in the same structure. Many recent immigrants from Europe who had previously lived in rural areas brought some of these traditional animal husbandry practices to urban neighborhoods.[51] Though not ideal when practiced in densely populated industrial Chicago, many Chicagoans nonetheless maintained livestock in their homes or lots.

Livestock kept in urban neighborhoods offered both benefits and drawbacks. Although animals contributed to the household income for working-class Chicagoans, maintaining livestock in close quarters proved potentially hazardous to the health of both the animals and their owners. Damp basements and poorly drained lots were unhealthy environments for livestock. Cows stabled in such conditions and denied proper pasturage did not produce quality milk, curtailing the benefit of locally produced milk. This environment was certainly not what Dr. Evans had in mind when he proposed his urban cow scheme. Indeed, the 1881 ordinance relating to livestock required that stables must be clean and that no cattle be housed where the "water, ventilation and food are not sufficient and wholesome for the preservation of their health, safe condition and wholesomeness of food."[52] While this ordinance sought to encourage a healthier environment for livestock and humans, most people in crowded working-class neighborhoods could not afford to create or maintain such an environment.

Chicago also had numerous ordinances relating to the collection and disposal of animal waste. As early as 1849 the city forbade people from throwing manure or "anything likely to become offensive" into the streets or alleys, or allowing any such matter to accumulate on a lot "in such a manner as to be offensive to the neighborhood."[53] The conditions that the City Homes Association researchers encountered demonstrated the lack of enforcement of these laws and regulations, especially regarding refuse removal. Researchers found alleys filled with refuse and manure, a stable floor covered in two feet of manure, and a manure pile seven feet high. More than half of the stables and manure boxes found in the surveyed districts were in bad condition.[54] The report asserted that "the conditions here show how backward, in some respects, the City of Chicago is. The reports on tenement-houses in other cities do not include studies of these conditions, for the simple reason that most other large cities would not permit to exist to such an extent these horrible and filthy insanitary conditions."[55] Chicago had statutes relating to sanitation, but the city failed to enforce them with any consistency.

The City Homes Association used the evidence they gathered in their study to promote tenement and sanitary reforms. A sweeping new tenement ordinance passed in December 1902, providing specific regulations for ventilation, open space on lots, plumbing standards for new tenement buildings, and reiterated the ban on keeping livestock in tenements.[56]

Jane Addams continued to pursue sanitary reforms after the success of the tenement ordinance, targeting the political corruption that enabled unhealthy conditions to persist. The corruption that Addams and her Hull House associates revealed prompted an investigation of several municipal agencies, including the Health Department, and spurred outrage among Chicago residents. Angry voters, internal investigations, and clear evidence forced city officials to address the failure of enforcement. A 1913 ordinance made the removal of animal waste and bedding the charge of the commissioner of health. An additional ordinance required manure bins to be constructed of impervious materials, with a fly-proof cover, and be placed in stables rather than alleys. The Health Department claimed that, by the 1920s, the "rigid enforcement" of this ordinance virtually eliminated the manure nuisance in the city.[57]

Chicago's success in reducing the manure nuisance was not entirely due to the enforcement of effective laws. The introduction of the automobile played a key role in making Chicago's streets cleaner by eliminating the need for horses. Edith Abbott's 1935 follow-up survey to the City Homes Association's report credited the replacement of horses with motor vehicles as having the greatest impact on improving the sanitary conditions of Chicago's streets, alleys, and lots.[58] In 1910 Chicago's horse population numbered around 77,000, with 58,114 registered horse-drawn vehicles. By 1920 the horse population had dropped to just over 30,000, and by 1932 the number of horse-drawn vehicles fell to about 6,000. At the same time, automobile and motor truck registration in Chicago climbed from 13,725 in 1910 to over 450,000 by 1932.[59] This shift from horse power to mechanical power extended across economic classes, as many working-class residents replaced their horse-drawn wagons with motor trucks.[60] Certainly Chicagoans did not trade their horses for motor vehicles in order to solve the city's sanitation issues, but the shift proved an inadvertently effective solution.

Increasing problems with overcrowding and unsanitary conditions in working-class neighborhoods, along with Chicago's territorial expansion in the 1890s, led legislators to consider different ways to manage the location of animals in the city by the early twentieth century. Rather than banning livestock outright, or restricting them to certain areas, as was the case throughout the nineteenth century, new ordinances relating to livestock in 1905 required owners to provide a certain amount of space for their animals. The new municipal code required one thousand square feet of outdoor space for each head of cattle and one hundred square feet of outdoor space per swine, unless the animals were kept in the stockyards bound for slaughter. The ordinance also forbade livestock from roaming free or being herded within the city limits.[61] Other cities, including Baltimore and Buffalo, adopted similar laws at the turn of the century to discourage residents from keeping livestock in the city.[62] Such laws requiring minimum space for certain animals ensured that only residents who could afford the space could legally maintain livestock. While this created the potential for better sanitary conditions for animals, it prohibited the city's poorest residents living in tenement areas from keeping the animals that many relied on for food and extra income. The minimum space ordinance helped to push livestock from the densely populated areas around the city center out to the periphery to rural neighborhoods and towns.[63]

Chicagoans and their animals adapted to the ever-changing urban environment in different ways based on their place within the economic and physical landscape of the city. Working animals, such as horses and cattle, served important functions in the city, providing transportation, energy, and a means of livelihood for Chicagoans. As the city's population grew, animals and humans came into more conflict and the need to control the use and movement of animals became urgent. The problems of animals in the city came to a head at the turn of the twentieth century as reformers publicized the unsanitary conditions in which people and animals lived in the city's poorest neighborhoods. Despite the new laws regulating animals and sanitation that came as a result of the reformers' efforts, Chicago never officially evicted livestock from its city limits. However, the rising cost of land, changing economic and cultural values of animals, and development of new technologies ultimately exiled working animals from Chicago, but not until several decades into the twentieth century.

PART II

A FRESHWATER CITY

Encountering the Chicago area for the first time, many found the flat expanse of lake and prairie boring. In 1843 Margaret Fuller, having traveled across the "vast monotony of the lakes," confronted prairie, a "monotony of land with all around a limitless horizon."[1] But the lake offered connections to distant places. In a 1925 novel Willa Cather described Lake Michigan as "the open door that nobody could shut. The land and all its dreariness could never close in on you. You only had to look at the lake, and you knew you would be free."[2] Nineteenth-century residents saw Lake Michigan as an essential transportation route. And as Chicago grew, residents sought refuge in the lake. Capturing shifting urban visions of Lake Michigan, Theodore Karamanski's essay explores the transformation of this inland sea from commercial asset to an aesthetic and recreational resource.[3]

Lake Michigan offers more than commerce or escape. Chicagoans draw their drinking water from it. Lakefront cities like Chicago have an abundance of freshwater, unlike seaports, but they are challenged to keep from fouling this resource with their wastes. Local residents' decision to use the Chicago River, which flows into the lake, as an open sewer proved problematic even as the city began its sewerage system in the 1850s. By 1900 the city's engineers famously reversed the Chicago River. Matthew Corpolongo explores how this engineering feat involved restructuring and simplifying the region's riverine ecology while simultaneously constructing the complex built environment of the Sanitary and Ship Canal.

Chicago's responses to protecting its drinking water and disposing wastes would continue to rely heavily on engineering, such as extending water intakes out into the lake through cribs and tunnels. Continued growth of the built environment in the twentieth century required new solutions to the problem of excess water, and Harold Platt's chapter examines efforts to grapple with the frequent flooding prompted by suburban sprawl and climate change. By 1965, facing a grassroots movement of homeowners on one hand and conservationists concerned with forest preserves on the other, local governments turned again to another large, radical public works project—the Tunnel and Reservoir Plan (TARP)—in an attempt to alleviate all of these issues.

New infrastructures addressed the needs of metropolitan Chicago and placed the region in larger global contexts. Daniel Macfarlane and Lynne Heasley interrogate the designed obfuscation of three intersecting Great Lakes infrastructures—the Chicago Diversion, the Enbridge Lakehead pipeline system, and the electric fish barrier system. By hiding or disguising these "built ecologies and technological megaprojects," their creators obscured environmental risks to aquatic ecosystems and adjacent communities.

Intriguingly these essays speak to what was once perceived as a divide in the field of environmental history but has now emerged as a strength. In the lead article of a 1990 round table in the *Journal of American History*, Donald Worster "offered a straightforwardly materialist agenda for environmental historians."[4] Recognizing the cultural turn in the larger field of history, respondents Richard White and William Cronon emphasized the role of ideas, values, and beliefs. As the historian Paul Sutter observes, however, subsequent environmental historians rejected this bifurcation, seeing environments as hybrids—as mixtures of nature and culture.[5] Thus within this section Karamanski's emphasis on ideas of the lake contrasts with but also supplements the materiality expressed in the other essays.

CHAPTER 4

An Inland Sea?

Coming to Terms with Lake Michigan in Nineteenth-Century Chicago

THEODORE J. KARAMANSKI

In 1837 the newly incorporated municipality of Chicago organized a committee of businessmen, led by the city's first mayor, William B. Ogden, to design a symbol that would lend dignity to a town one contemporary described as "a miserable collection of huts." These businessmen came to the fledgling city on Great Lakes sailing ships. Their design reflected the important relationship they saw between Chicago and Lake Michigan. Prominent was the image of a ship under full sail. Atop the municipal seal was an infant in a seashell, a classical allusion to the city born of the surf and a prediction of Chicago as the future pearl or "gem of the lakes."[1]

Chicago's seal reflects a truth nineteenth-century Chicagoans took for granted but their twenty-first-century inheritors ignored—Chicago was the child of the great inland seas of North America, and Lake Michigan in particular. Among the Great Lakes only Lake Michigan is entirely within the United States, oriented on a north–south axis some 307 miles in length. It intersects distinct environmental zones, from southern prairie to northern boreal forest. Lake Michigan shipping became a way to send grain belt harvests east and forest resources south and west; Chicago was the key transshipment point. The lake is a structural element in Chicago history, its *longues durees* stretching from the time of the Pleistocene creation of the broad flat plain upon which the city rests to contemporary land-filled lakefront parks where citizens play.

Nineteenth-century Chicagoans perceived and utilized the lake as an essential commercial asset. In the twentieth century the lake became an amenity and aesthetic backdrop. Between these two divergent perspectives is much of the environmental history of Chicago.

The maritime dimensions of Chicago history have largely been limited to the study of the building of the Illinois and Michigan Canal (1836–1848) and the Sanitary and Ship Canal with its reversal of the Chicago River (1888–1900). Environmental historians of the Great Lakes have delved into some issues of water pollution, environmental degradation, and overfishing. Maritime historians have focused primarily on shipwreck narratives. This essay reorients Chicago's environmental history toward the lake. It explores how Chicagoans managed and manipulated their inland sea and how that broad restless basin influenced economic production and social constructions. Lake Michigan was a white-capped wilderness, and an important part of the city's early history was the "settlement" of that watery frontier. Central to this story are the so-called antebellum market revolution, the actions of the federal government, the transfer of maritime technology from saltwater to fresh, and the development of a distinctive regional maritime culture.[2]

Indigenously designed and powered canoes traversed Lake Michigan and its sister Great Lakes for several thousand years. Rigged sailing ships, beginning with Le Griffon in 1679, dared its water for more than 150 years. Yet, given the limited information on the shape of its shores, its depths, shoals, currents, tempests, and tides, the lake was a wilderness for white American settlers, a dangerous place for terrestrials to tarry. Like any great environmental feature, it proved difficult to bend to human will.[3]

The first and perhaps the most enduring challenge to the European American settlement of the Midwest's maritime frontier was intellectual, what might be called the social construction of the Great Lakes. The first Europeans to see the lakes dubbed them "La Mer Douce," the freshwater sea, but later actors challenged the notion that these great inland bodies of water—ninety thousand square miles—should be equated with the size and power of the seas. The pilot who navigated Le Griffon, an experienced saltwater sailor, disparaged Michigan waters as "a nasty freshwater lake." Experienced sailors' tendency to regard the Great Lakes as less than the oceans echoed through the centuries. In James Fennimore Cooper's novel The Pathfinder (1840), an ocean mariner remarks upon his first sight of the Great Lakes, "Just as I expected. A pond in dimensions and a scuttlebutt in taste." Just two years after Cooper's novel, however, a French naturalist caught in a Lake Michigan storm was fully persuaded that the lake had the power of a true sea. "I have seen the storms of the

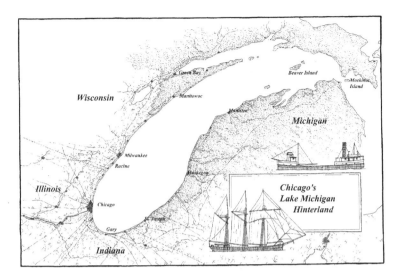

Fig 4.1 Lake Michigan. *Source*: Theodore J. Karamanski, *Schooner Passage: Sailing Ships and the Lake Michigan Frontier* (Detroit: Wayne State University Press, 2001), 18–19.

Channel, those of the Ocean, the squalls off the banks of Newfoundland, those on the coasts of America, and the hurricanes on the Gulf of Mexico. Nowhere have I witnessed the fury of the elements comparable to that found on this freshwater sea." For people who experienced Lake Michigan firsthand, the beauty or utility of the lake was mixed with a dread of its overwhelming natural power. Herman Melville reflected this disquiet in *Moby Dick* (1850), when Ishmael states the lakes "possess an ocean-like expansiveness, with many of the ocean's noblest traits" and "they are swept by Borean and dismasting blasts as direful as any that lash the salted wave; they know what shipwrecks are, for out of sight of land, however inland, they have drowned full many a midnight ship with all its shrieking crew." Crossing the lakes in a steamer a generation later, Rudyard Kipling still became unnerved, observing they were "like a fully accredited ocean, a hideous thing to find in the heart of a continent." These countervailing currents posed a question for federal policy makers: Were the Great Lakes inland seas deserving the treatment afforded the nation's coastline?[4]

If Lake Michigan and her sisters were indeed inland seas then the utilization of them required the technology developed to sail the world's oceans, as Rene Robert Cavalier, Sieur de La Salle, concluded in 1679 when he recruited European shipwrights to build *Le Griffon*. In the following 150 years, nearly all ships built on the Great Lakes modeled the proven designs of Atlantic shipping. The transfer of technology included the infrastructure necessary to sail ships including harbors, lighthouses, and other navigational aids. Unfortunately, even as Chicago merchants imported the ship technology, the federal government could not decide if the Great Lakes constituted a maritime region meriting such infrastructure. The notion that the lakes were something less than a sea had particularly dire consequences for harbor development at Chicago and other Lake Michigan towns.

There are relatively few natural harbors on the Great Lakes and none in the southern half of Lake Michigan. With no islands, few points or peninsulas and a low, flat topography, the only places to shelter a ship were the numerous rivers that flowed into the lake. The slow, meandering streams of the watershed, however, could not punch a passage through the sandbars deposited by the wave action of the lake. Therefore, each town developing along the shores of Lake Michigan required major civil engineering projects to create commercially viable harbors. Visiting Chicago in 1823, the geologist William Keating predicted that "the dangers attending the navigation of the lake and the scarcity of harbors along the shore must ever prove serious obstacles to the commercial importance of Chicago." In 1828 the Fort Dearborn garrison tried to build Chicago's first harbor, digging a fifteen-foot-deep channel through the sandbar that blocked the mouth of the Chicago River, but new sand deposits quickly blocked this improvised passage from the lake to the river's protected waters. Five years later Congress finally appropriated $25,000 to clear a channel into the Chicago River. Workers built twin piers into the lake, set two hundred feet apart, to block the flow of sand. By July 1834 these wood and rock structures allowed the schooner *Illinois* to sail from the lake into the river. As she docked next to Fort Dearborn, a cheering crowd gathered, and the city fathers popped forty-eight bottles of champagne to celebrate the birth of Chicago's harbor. The *Illinois* was one of 180 ships to enter the harbor that first year, but the celebrations were premature. The sand soon returned, necessitating more dredging and pier work. In 1839 with expenditures exceeding $100,000, the U.S. Army Corps of Engineers declared victory over the sandbar.[5]

While the Chicago harbor moved forward, communities from Lake Erie to Lake Michigan lobbied for federal engineering support. A coalition of strict constitutionalists in the Democratic Party and southern planters suspicious of an activist national government blocked most Great Lake harbor projects. From 1840 until after the Civil War, the Democratic Party's national platform contained the following clause: "Resolved that the constitution does not confer upon the general government the power to commence and carry on, a general system of internal improvements." Consequently, little lake navigation funding passed the U.S. Senate where southern states had the power to block legislation or override a presidential veto. Midwesterners lamented that in these southerners' hands, the constitution became a "salt-water instrument" allowing only seacoast improvements. In May 1840 two large steamboats sank on Lake Michigan and several schooners incurred severe damage because they could not find refuge from a sudden spring gale. This disaster prompted one businessman to comment, "There has been enough property lost within the last ten days on Lake Michigan, to have built three good harbors." He bitterly added, "What a pity" that the lost ships were not "loaded with Senators and members of Congress." The leaders in Washington simply did not appreciate the scale of commerce on the inland seas. Equally frustrating, as one mariner observed, was that "the idea that the lakes were little more than a 'goose pond' prevailed in Congress."[6]

Storms were only one of the natural challenges posed by unimproved navigation of the lakes. Isaac Stephenson, a shipmaster and later U.S. senator, argued, "Sailing a ship was not unlike blazing a way through the forest. With conditions wretched as they were the navigator was practically without charts and the master figured his course as nearly as he could, estimating the leeway and varying influence of the winds." To thrive under such conditions, midwestern mariners adapted saltwater ship designs to lake conditions. For example, William Wallace Bates, a shipbuilder initially based in Manitowoc, Wisconsin, and later in Chicago, redesigned the classic Baltimore clipper, giving it a shallow draft and an almost flat bottom to manage the unimproved lake waters. To ensure greater stability and control under sail, he fitted a retractable centerboard to the keel. His vessels became the prototype for the Great Lakes schooners that in their hundreds carried people and products to the heartland. When the schooner *Illinois* arrived in Chicago in 1834, she carried 104 immigrants, their baggage, supplies, and even a wagon with its wheels tied to the rigging.[7]

In addition to the want of harbors, navigational aids were sorely lacking in the 1830s and 1840s. The same Congress that refused to acknowledge the need for harbors authorized the first lighthouse on Lake Michigan in 1831 on the grounds of Fort Dearborn, one of thirty-four built on the lakes in the 1830s. These early lighthouses tended to be poorly built and maintained. The day contractor Samuel Jackson presented the completed fifty-foot Chicago tower for inspection, it collapsed. During the thirty years in which Stephen Pleasonton, an auditor in the Department of the Treasury, administered American lighthouses, poor construction was a common problem. His penny-pinching and his possibly corrupt relationship with subcontractors ensured that even when Great Lakes lighthouses did not collapse, they were too short and equipped with inadequate lights to be effective navigational aids. Eventually the maritime community forced Congress to remove Pleasonton, and in 1851 the Lighthouse Board of army engineers and naval officers began reforming the system.

The lack of charts proved equally vexing. As a passenger in 1836, the British social reformer Harriet Martineau commented: "The navigation of these lakes is, at present, a mystery. They have not yet been properly surveyed. Our captain had gone to and fro on Lake Huron, but had never before been on Lake Michigan; and this was rather an anxious voyage to him." In unknown waters, the captain traveled only eighty miles before he ran his ship onto a sandbar that took the better part of a day to get off. Five years later Congress finally authorized the Army Corps of Engineers to begin the decades-long process of surveying the six thousand miles of Great Lakes shoreline. Due to budget cuts and a shortage of skilled personnel, they did not release the first partial charts until the eve of the Civil War.[8]

Despite the federal government's indifferent policies and the hundreds of drownings due to the lack of adequate harbors and navigational aids, commerce continued to proliferate. By 1846 nearly one hundred steamboats and more than four hundred sailing vessels operated on the lakes. Chicago's congressman, "Long John" Wentworth, marshaled an impressive array of statistics to sway his legislative colleagues to help tame the lake frontier. Some seven thousand sailors worked on the Great Lakes, and their vessels annually carried 250,000 passengers, mostly immigrants heading west. The value of their cargoes topped $125 million at a time when, he claimed, America's international trade (mostly southern cotton exports) was only $114 million. Wentworth's statistics combined with the swelling population of the Old Northwest states to give the region new political clout. In July 1846 Congress approved a $500,000 River and Harbor Bill. Celebrations were short-lived, however. One month later President James K. Polk vetoed the bill, claiming the projected improvements were "local in character." He dismissed lake harbors as "comparatively unimportant objects." Unlike Atlantic harbors that received substantial federal support, Polk incorrectly complained that the Great Lakes were not "connected with foreign commerce, nor are they places of refuge or of shelter for our navy or commercial marine on the ocean or lake shores." All but one southern Senator voted to uphold Polk's action, narrowly defeating an attempt to override the veto.[9]

Polk's veto sparked a political firestorm that burned particularly bright in Chicago. Although still a city of only sixteen thousand people, its leaders rallied nationwide support to renew efforts for a harbor fund. In July 1847 they held one of the most important political meetings in antebellum America. The largest American political gathering up to that time, the River and Harbor Convention attracted between four thousand

and ten thousand visitors to Chicago. Southern statesmen shunned invitations, deriding the gathering as "humbuggery." This response was particularly galling because only two years before southern planters interested in shipping cotton on western rivers held their own convention and resolved that "the Mississippi River is entitled to be considered an inland sea and in regard to appropriations . . . to be placed on the same footing as the Atlantic coast." One frustrated midwesterner complained, if the Mississippi was an inland sea, "what may our great chain of lakes be called." Chicago's convention drew almost every person prominent in the future development of the Republican Party, including Abraham Lincoln. Save for the issue of slavery, which was not mentioned, the convention's resolutions became cornerstones of that party. Great Lakes towns were already economically tied via the Erie Canal to northeastern capitalists. The South's uncompromising opposition to federal support for the midwestern maritime infrastructure further pushed the western states into a political alliance that eventually doomed Dixie and its "peculiar institution." Stephenson commented, "Every sailor on the lakes became a Whig and afterwards a Republican."[10]

The men and women traversing the Great Lakes faced the consequences of an ever-expanding number of vessels utilizing their waters with no navigational charts and few harbors of refuge. What Polk never understood was that the lakes were in many ways more difficult to navigate than the ocean. The lakes were big enough to experience sudden, sustained gales, but confined enough that mariners on freshwater had little opportunity to simply run before the storm as the ocean sailor could. All too often a lake vessel driven by a storm came upon the lee shore, and if her anchor could not hold to the sandy bottom, faced destruction in the surf. The death toll on the lakes rose in proportion to the scale of regional trade. In the seven years that followed the initial opening of the Chicago harbor, 118 people died on Lake Michigan. Between 1848 and 1855 financial losses to Great Lakes shipping increased from $404,830 to $2,797,839. In 1851 alone 431 people died on the lakes; a fact reported to the nation by the Whig president Millard Fillmore in his State of the Union message. Hailing from the Great Lakes city of Buffalo, Fillmore was the only president from 1840 to 1861 who enacted meaningful harbor and navigation improvements. Finally, the outbreak of the Civil War and the ascendency of the Republican Party made Great Lakes navigation improvements a national priority.[11]

With the city's economic life dependent on lake navigation and that commerce fraught with danger, antebellum Chicagoans nursed a healthy respect for the inland seas and dreaded its storms. Most vessels wrecked on Lake Michigan were driven ashore and often could be refloated. But when a storm raged, the shore offered no escape. This was the case with the wreck of the steamer *Lady Elgin*. As the tumultuous 1860 presidential race neared its climax, Chicago hosted an immense rally for the Democratic nominee and hometown hero, Stephen A. Douglas. Party faithful came by boat and rail from across the region. When the speeches and fireworks ended, several hundred Irish from Milwaukee joined the passengers and crew of the *Lady Elgin* for the trip home. After several hours a lumber schooner appeared out of the blackness, colliding amidships with the *Lady Elgin*. The schooner careened on, both captains thinking it had taken the worst of the incident. Yet the steamer quickly took on water and broke apart in less than an hour. Three hundred and eighty-five passengers and crew went into the lake; only ninety-eight staggered ashore. Most died in the raging surf off Evanston and Glencoe, Illinois, just north of Chicago. For weeks, bodies washed up along the Chicago shore and newspapers told of the widows and orphans left in the wake of the wreck. Even with civil war looming the tragedy captured the imagination of every Lake Michigan community dependent on the inland sea.[12]

During the antebellum years Chicago outstripped its Lake Michigan rivals, Milwaukee and Green Bay, in business and population, in part because of its improved harbor and the Illinois and Michigan Canal connection to the Illinois prairie. Milwaukee obtained an improved harbor in 1851 only during the brief window of the Fillmore administration. Even then that city's business leaders put up the lion's share of the improvement costs. Racine and Kenosha, Wisconsin, did the same. In 1854 sand again began to clog the Chicago River mouth, sinking four ships, taking seven lives, and prompting the city to act on its own to keep its harbor viable. In lieu of federal funding, the Chicago Board of Trade petitioned Secretary of War Jefferson Davis to "borrow" the army's steam-powered dredge. When Davis refused, disgusted Chicagoans seized the equipment in an act of rebellion and cleared the river mouth.[13]

In the wake of the Civil War the federal government was finally free to apply system and science to "settle" the Midwest's blue-water frontier. It finally was free of southern opposition to recognize the inland seas as a vital national coastline. President Ulysses S. Grant ordered the Army Signal Corps to collect the data necessary to issue marine weather forecasts. Another Civil War veteran, Orlando Poe, supervised the construction of a series of brilliant lighthouses which guided vessels to major ports and warned them of dangerous shoals. Perhaps most importantly, the Army Corps of Engineers embarked on a dramatic series of harbor and channel improvements that finally began to offer lake shipping a safe haven from storms. With maritime commerce better supported, traffic surged into Chicago's harbor. In 1871 more ships entered and left

Chicago than other great ports such as New York, San Francisco, Philadelphia, or Baltimore.[14]

The narrow, sinuous confines of the Chicago River berthed all the
white-winged ships flocking to the burgeoning metropolis. Flowing
through the heart of the city, the river was a waterway in three parts.
The main stem flowed into Lake Michigan, but less than a mile from its
mouth, the river split into two channels, the North Branch which was
navigable as far north as Goose Island and the South Branch that skirted
the central business district and reached Bridgeport and the start of the
Illinois and Michigan Canal. Slips excavated from the river into neighborhoods further expanded the reach of lake vessels. Towed by one of
the ubiquitous tugboats, a sailing ship or steamer could wind its way
through most parts of the city. This was a boon to businesses relying on
ship-borne cargoes of lumber, coal, grain, stone, or fish. It was a nuisance, however, for terrestrial traffic. In the 1870s the city needed thirty-
seven bridges for access to the central business district. These spans
swung open more than a hundred times per day to allow vessels to pass.
Delayed pedestrians and teamsters fumed in frustration.

As Chicago became the world's greatest primary grain port and lumber market, Lake Michigan shipping accounted for its burgeoning "big
shoulders." For many nineteenth-century Chicagoans the lake and what
happened on the lake defined their relation to their environment. To the
Board of Trade that in 1864 began to trade commodities "futures," the
lake was first and foremost an economic highway. To the leather apron–
clad Bohemian "lumbershovers" who unloaded pine boards down from
Wisconsin or Michigan, or to the grain trimmers in the towering elevators along the Chicago River, or to the Irish bridge tenders who swung
open their spans for vessels entering or exiting the harbor, Lake Michigan was primarily an economic asset. To the ordinary citizen working in
the central business district, lake shipping often was a nuisance. "Sorry
boss, I was bridged" was a common refrain for workers late to the job.
And the lake was fundamental to the neighborhoods where the families
of schooner masters and lake sailors awaited the return of loved ones
from the broad blue horizon. Yet as the city grew in population and
wealth, a new relationship with the lake began to be cultivated. While it
remained important as Chicago's doorway to the world and as a source
of drinking water, city dwellers increasingly articulated an aesthetic appreciation of Lake Michigan.[15]

The architect Louis Sullivan captured this alternate vision. In his
autobiography he recalled the tremendous impression made by his first
sight of Lake Michigan in 1873 from the train carrying him from Boston:
"Soon Louis caught glimpses of a great lake, spreading like a floor to the

far horizon, superbly beautiful in color, under a lucent sky. Here. . . . was power, naked power, naked as the prairies, greater than the mountains." Sullivan was impressed with the willful boasts of Chicagoans about their great trade in grain and lumber, their busy harbor, and their "dream of commercial empire." According to his memoir, he drew inspiration from Chicago's natural setting, "for his eye was ever on the boundless prairie and the mighty lake." Sullivan appreciated the "impelling purpose" that commerce gave to Chicago, marveling at the forest of masts that crowded the Chicago River harbor and the daily fleet of schooners." Yet at its heart Sullivan's vision of the "very wonderful lake" was aesthetic not mercantile. "Above the rim of its horizon rose sun and moon in their times, the one spreading o'er its surface a glory of rubies; its companion, at the full, an entrancing sheen of mottled silver. At other times far to the west in the after-glow of sunset the delicate bright crescent poised in farewell slowly dimmed and passed from sight." Sullivan sometimes baffled his fellow ar- chitects with his poetic prose, but his designs reflected the belief that "all functions in nature" conveyed power. The design motifs on his facades reflected this insight. Thus the mantra of the emerging Chicago School of Architecture, "form follows function," was born by and of the Lake Mich- igan shore.[16]

Sullivan shared with Chicago's antebellum businessmen the concept of the city as a threshold open to both the Great Lakes and the prairie. Building the harbor, the Illinois and Michigan Canal, and later rail- roads provided the means to speedily pass grain, lumber, and manufac- tured goods through that door. The grand arches that graced Sullivan's most iconic constructions reflect his embrace of this image. Yet as the nineteenth-century Garden City gave way to a densely inhabited indus- trial metropolis, the city's relationship with the prairie became more ten- uous. By the twentieth century the word *prairie* meant a vacant urban lot to most Chicagoans. Lake Michigan, however, remained a source of nat- ural inspiration. By the 1880s and 1890s, as a new harbor and heavy in- dustry center was developed south of the city at Lake Calumet, maritime traffic seemed less important, and the lakeshore itself became instead a threshold separating a hectic city from the serene horizon of the lake.[17]

For poets and novelists, Lake Michigan offered an antidote to the city's congested human and animal throng. Carl Sandburg shared this sentiment in his thirty or so poems about the Great Lakes or the ocean. In "The Harbor," Lake Michigan is an uplifting alternative to the city. Sandberg takes the reader through the "ugly walls" of the city, past door- ways where women looked "from their hunger-deep eyes, / Haunted with shadows of hunger-hands." But because of the lake, the atmosphere changes dramatically, "I came sudden, at the city's edge, / On a blue burst of lake." Here waves break in the bright daylight and the heart is lifted by "a fluttering storm of gulls. . . . wheeling free in the open." The harbor, once the engine of city commerce, became in Sandburg's poem a "blue burst of lake" offering hope and freedom.[18]

Lake Michigan's liberating relief spawned a new commercial use. The heyday of the lake excursion steamers occurred from 1890 to 1925. By the 1920s as many as eight hundred thousand Chicagoans annually used the lake boats. The Chicago elite escaped by ship for weeks-long vacations to northern Michigan at Mackinac Island's Grand Hotel or Little Traverse Bay, where hay fever sufferers formed lakeside resort "achoo clubs." The working class took shorter day trips to escape the city's fumes and street filth. Embarking on a Sunday morning at a Chicago River dock and later from Navy Pier, excursionists enjoyed a five-hour cruise across the broad lake, arriving for lunch in St. Joseph, Michigan. The Silver Beach Amusement Park, restaurants, and a dance hall kept Chicagoans busy until the night boat returned them home. In his poem "Picnic Boat," Sandburg described a brightly lit steamer emerging from a lake as dark as a "stack of black cats" with "the rhythmic oompa of the brasses playing a Polish folk-song for the home-comers." Michigan City, Indiana, was another popular destination. A Saturday afternoon sailing accommodated those who worked a half-day. Closer to Chicago, Michigan City lacked Silver Beach's excitement but offered a zoo at the landscaped Washington Park. Round-trip tickets cost as little as $1.00 because the steamers also carried mail, freight, and especially fresh fruit from Michigan orchards to Chicago's vast Water Street produce market.[19]

As the new century began, the older commercial vision of Lake Michigan clashed with these growing aesthetic and recreational uses. A series of events contributed to new perceptions of Chicago's shore. In 1892 the U.S. Supreme Court opened new possibilities for the Michigan Avenue shoreline when it blocked the Illinois Central Railroad's plans to develop areas adjacent to its lakefront tracks.[20] The 1893 Columbian Exposition, located on the lakefront at Jackson Park, sparked the environmental imagination of the city and nation. In 1894 the architect Normand Patton proposed a new landscaped civic center complex on the lakefront. Supporting this notion, influential aldermen advocated a lakefront dedicated to public use rather than the "ugliness, the bustle, the turmoil, and the uncleanliness" of a harbor. The more famous Daniel Burnham came late to the lakefront debate, but quickly incorporated ideas like Patton's when the City Club of Chicago hired him in 1906 to develop a plan to bring order and beauty to the city's chaotic growth. A lakefront dominated by parkland became central to Burnham's vision of the future Chicago.

At the same time, Mayor Fred Busse formed the Chicago Harbor Commission to find a way to preserve the city's domination of the Great Lakes trade, an urgent issue for its economic future. In 1900 Chicago was the fourth busiest port in the world, trailing only London, New York, and Hamburg. However, the growing size and sophistication of lake vessels made the Chicago River increasingly untenable as a harbor. Giant steel freighters, five hundred to six hundred feet in length, rapidly replaced old wooden schooners. Maritime Chicago required a major infrastructure facelift to avoid a rapid decline.[21] The harbor commission noted that most of the world's great port cities were undertaking major construction projects to update their harbors. Chicago must do the same: "Our future is as great as we are wise enough to make it." Unfortunately, they noted, most Chicagoans regarded lake commerce with "indifference." Detailed studies eventually identified two possible courses of action. One was to construct a wider, deeper channel allowing modern vessels to enter the heart of the city, either by expanding the Chicago River or by excavating a new channel from the lakefront to its South Branch. This plan had the virtue of bringing ship traffic to consumers and existing rail yards. The other alternative abandoned the Chicago River as a port and proposed using breakwaters to create an outer harbor that would be serviced by several large commercial piers projecting from the lakefront. Burnham supported the latter proposal, with some modifications, and the city adopted it. Significant among the outer harbor plan's flaws was the fact that it isolated the shipping terminal from Chicago's railroad network. It also stranded passenger ships far from public transit lines.[22]

Burnham presented his vision of a park-filled lakefront in dramatic color paintings by the artist Jules Guerin, capturing the imagination of the public and civic leaders. His plan became the guide, if not the blueprint, for lakefront planning for the next century. At the same time the appetite for a major reengineering of Chicago's harbor faded. The cost of building new channels would have been nearly as expensive as the Sanitary Canal which had protected the city's water supply. And the issue of limited access ultimately sank the outer harbor alternative. Inviting rail networks and warehouses to the water's edge would create, in one editorialist's words, "a cesspool of filth." Burnham paid lip service to the city's need for a new harbor by projecting two long piers from his formal gardens, but the city built only one, Navy Pier, which from the start was unsuited to commercial transportation.

The Burnham Plan completed the city's transition from a commercial to an aesthetic or amenity-oriented appreciation of Lake Michigan. The lake no longer shaped Chicagoans' cognitive landscape in the way it had when shipmasters influenced the city's economy. The lake no longer dominated the business of the Board of Trade, commission house agents, or lumber yard managers. Burnham's landscaping contributed to Chicago developing a shallow, liminal relationship with Lake Michigan. Parks, lakeshore paths, and public beaches opened miles of shoreline for recreation. Lake Michigan became a blue backdrop to gaze upon rather than a resource to be directly engaged By the late 1920s Lake Shore Drive and a museum campus expanded accessibility and popularity, while Chicago's share of lake commerce continued to decline. Even excursion ships lost popularity in the face of improved roads and growing automobile ownership. The lakefront parks included moorings for several thousand private sailboats and motorboats; docked for much of the week, these recreational vehicles rarely ventured more than a mile from shore.

While perceptions shifted, nineteenth-century Chicagoans and their descendants shared in common a desire to bend the inland sea and its shoreline to their will. The city's maritime landscape is testament to what William Cronon called "second nature," the reshaping of land and water by human engineering. Chicago's outer harbor, breakwaters, and majestic lighthouse were conscious attempts to transform the lake into an economically productive space. Similarly, the landfilled parks and carefully managed beaches constituted a later generation's attempt to make the lake serve the city's recreational needs. Yet while the waterfront where Chicagoans played became "second nature," Lake Michigan still heaves and recedes unbridled beyond the breakwater. It may seem tamed—after all there has not been a major shipwreck since 1960—yet the lake still shows its primal face hundreds of times each year when it snatches to its cold embrace the careless or the hapless who venture off the beach.[23]

In the twentieth and twenty-first centuries, most Chicagoans engage Lake Michigan vicariously, from the perch of a downtown skyscraper, a beach blanket, a car stuck in Lake Shore Drive traffic, or perhaps most unthinkingly with a drink of tap water. Chicagoans appreciate that at its doorstep is a "fully accredited ocean" only when the storm surge of a November gale throws waves across roadways, high water levels erode the backyards of north shore mansions, or rip currents grab summer swimmers. For a city built on its ability to transform nature, Lake Michigan's ever-changing face is a reminder of the mercurial natural world upon which all human life depends. Lives, buildings, whole neighborhoods rise and fall to the earth yet the lake abides. In the writer Ben Hecht's view, the lake reduces "to puny proportions the routine by which people live and. . . . delude themselves into admiring." Lake Michigan reveals the transient human hustle but also offers to Chicago a limitless horizon upon which to dream.[24]

CHAPTER 5

Cleansing Chicago

Environmental Control and the Reversal of the Chicago River

MATTHEW CORPOLONGO

By 1900 the city of Chicago famously reversed the flow of its namesake river to improve drinking water and ameliorate wretched sanitation conditions. Contaminated water threatened public health and the city's commercial dominance. Yet only six years later the South Branch of the Chicago River remained an open sewer. Repulsed by its condition, social-ist activist and author Upton Sinclair documented the city's horrid envi-rons in his novel *The Jungle*. Chicago's Union Stock Yard was a "square mile of abominations" which "stank and steamed contagion" from the tens of thousands of cattle crowded into wooden pens.[1] Nearby slaugh-terhouses dumped their waste into the river. As sanitation systems failed, decaying animal flesh, fecal matter, acids used by meatpacking compa-nies to dissolve animal carcasses, and festering garbage transformed the Chicago River into a bubbling stream of hot blood that "smelt like the craters of hell" and "defied a breath of fresh air to penetrate."[2] Despite concerted efforts to improve the city's sanitation, the ambitious policies of sanitarians and elected leaders underserved those who endured the worst of industrial pollution.

Prior to its reversal, the Chicago River both facilitated and impeded the city's meteoric rise. It offered a historic portage that linked the Great Lakes to the Mississippi River system. Yet throughout much of Chica-go's history the river presented the city with its most enduring and bitter sanitation struggle. Since the city's incorporation in 1833, poor water

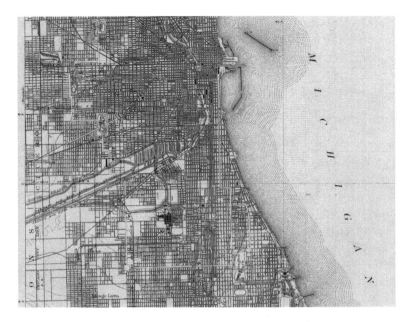

FIG 5.1. South Branch of the Chicago River, and the Chicago Sanitary and Ship Canal (labeled), flowing to the south and west of the city. The South Fork of the Chicago River's South Branch, known as Bubbly Creek, extends due south from the opening to the Sanitary Canal in 1900. *Source*: U.S. Geological Survey, Portion of Topographic Sheet, Chicago Quadrangle, U.S. Geological Survey, March 1901.

quality endangered Chicago's economic status. Cleaning the "cesspool of filth" that was the Chicago River secured the city's success and continued existence over the following decades.[3] The reversal also reflected a larger ideological motive beyond the diversion of water from Lake Michigan toward the Illinois River. For the engineers charged with ensuring clean drinking water, the reversal represented the chance to better residents' lives. Success for the city's elected leaders meant not just financial growth but improved public health through efficiency and technological ingenuity. It was at the nexus of engineering, public health, and politics that the construction of the Chicago Sanitary and Ship Canal (Sanitary Canal) proceeded in 1890. The river reversal had improved drinking water quality, but it also created an illusion of comprehensive success that masked and in turn reinforced the continued pollution of the river's South Branch, negatively affecting working-class immigrant communities nearby.

NATURE'S WATERWAYS

The intersection of the Chicago Portage with the Illinois and Des Plaines Rivers, provided an early transportation network that made canal construction logical and cost-effective.[4] These natural waterways served many peoples across generations, linking the Great Lakes with the Eastern Seaboard and the Great Plains. Indigenous societies, including the Cahokia and Illinois, first used this ecological highway. European explorers, regional traders, and later U.S. officials also relied on this riparian system to establish regional commerce.

With the arrival of Europeans in the eighteenth century, the importance of this connection intensified.[5] The Chicago River and Lake Michigan attracted investors and supported the city's construction. These portages linked the Illinois and Des Plaines Rivers to the Chicago River and then to Lake Michigan, inviting commercial transactions. Heavy traffic from fur traders and merchants using the Chicago River and portages carved deep gutters into the landscape, creating a channel between Mud Lake and the Des Plaines River west of the city. Water from the Chicago River eventually flowed through these gutters, directly connecting the river and portages, while creating conditions for a reversal. The gutters made year-round travel possible without the reliance on heavy summer rains to cross the portages. More importantly, they reinforced the need for a canal.

The Illinois and Michigan Canal (I&M), completed in 1848, was an early infrastructural upgrade for Chicago and the region. Although built for transportation, the canal effectively reversed the Chicago River, without shifting its current. Nonetheless, eastward and westward travel became possible. The I&M eased transit to Chicago and invited a population explosion that stunned urban planners. As the city's populace swelled from eighteen thousand in 1840 to sixty thousand by 1850, the streets and sidewalks amassed both human and animal waste. Without effective sewerage, people simply discarded garbage and bodily fluids in public spaces.[6] The I&M, serving as the city's default sewer, accepted the coagulated waste that washed into the Chicago River during heavy rains. Deeply contaminated water flowed freely into the canal toward the Illinois River. Eventually feces, urine, garbage, and industrial refuse filled the channel to capacity. This material, having no other place to go, flowed into Lake Michigan via the river. Although the I&M improved Chicago's transportation infrastructure, engineers never planned to use it as a drain.

The I&M represented Chicago's first artificial waterway, an avenue toward greater economic prosperity. It also served as the city's primary sewer. As industrial refuse spread, water-borne illnesses infiltrated the city via its new canal. Privy waste in cramped, working-class neighborhoods had seeped into the dirt and clay of Chicago's topsoil and into private wells, unleashing the catastrophic cholera outbreaks of 1849 to

1854 that afflicted thousands and killed some 3,500 people.[7] While early sewerage advancements marginally improved the city's water over the course of twenty years, residents demanded a more reliable alternative for its distribution.[8]

In 1851 the first rail lines arrived in Chicago, attracting even more capital, people, and waste to the city. As industry exploded, offal and acids colluded with biological wastes to contaminate the city's only sources of drinking water. An unbearable stench in the cramped, South Side neighborhoods remained an oppressive reminder of the city's deteriorating habitability. Chicago's relatively low elevation contributed to flooding and allowed refuse to drift, challenging both travel and sanitation.[9] Standing water also lured disease-carrying insects and rats to the river. The Chicago River, which had brought the city prosperity, was threatening its survival.

Civic and economic leaders attempted several strategies to clean streets and divert sewage. Wooden planks offered a quick solution, but they handled foot traffic poorly and merely absorbed steaming wastes. Thereafter a variation of gravel and sand paved most of the city's streets until 1890.[10] Improved bridges and drains attracted larger industrial businesses to the city including tanneries, breweries, brick mills, and meatpacking plants. Infrastructure advancements caused Chicagoans to place greater trust in elected officials and a technological elite to provide sanitation services.[11] As pollution spread, civic leaders soon realized the need for a subterranean sewer system to divert contaminated waters from the river.[12]

In 1861 Chicago's Board of Sewerage Commissioners hired Ellis Chesbrough, impressed with his work on Boston's sewer system.[13] Born in Baltimore, Maryland, in 1813, Chesbrough had humble origins that differed from the educational pedigree enjoyed by most "experts" and reformers. Contemporaneous accounts, including the pamphlet *Biographical Sketches of Chicago's Leading Men*, illustrate the engineer's life and his contributions to engineering reform in Chicago and throughout the country. Born to a working-class but well-connected family tied to the burgeoning railroads of the Eastern Seaboard, Chesbrough sought a career in engineering, despite more limited educational opportunities.

Chesbrough earned his first engineering credentials in 1828, working under his father on the Baltimore and Ohio Railroad.[14] While working with railroads in Maryland, Ohio, Pennsylvania, New York, and New Jersey, Chesbrough garnered the attention of military engineers, including U.S. Army colonel John H. S. Long, who recommended him for public works projects in Boston.[15] According to *Biographical Sketches*, Chesbrough's commitment to public service remained unparalleled, and he "cheerfully" went to work on Boston's new sanitation and water distribution aqueducts.[16] Chesbrough not only designed Boston's first waste-removal system, but was instrumental in revolutionizing the city's engineering bureaucracy.

In Chicago Chesbrough set about surpassing his earlier accomplishments. In the words of Daniel Burnham, famed Chicago architect and urban planner, Chesbrough made "no small plans."[17] Viewing sanitary structures as interconnected organisms, Chesbrough devised a massive underground disposal system that required raising the entire city ten feet.[18] Chicago's sanitation infrastructure proved inadequate to expel the wastes converted from raw materials consumed by the city's industries and populace. As Chicago's new chief sanitarian, Chesbrough seized upon the opportunity to revolutionize sanitation in the nation's fastest growing city. His first solution, a large tunnel with a filtration crib, brought water from Lake Michigan to residents. Unfortunately for Chesbrough, it failed to keep pace with the city's growing population. The sewer system, upon its completion, promised greater efficiency but despite its early success, many South Side residents still complained about foul drinking water. Although Chicago possessed an improved waste-removal system, designed by one of the nation's most promising engineers, access to clean drinking water remained inconsistent.

Noxious odors, wafting northward from the city's South Side industrial sector further alarmed affluent Chicagoans. Irritated North Siders formed the Citizens' Association of Chicago in 1876 composed of all well-connected residents concerned about packinghouse pollution.[19] This organization held public demonstrations and meetings to pressure authorities, thus hastening the response to water and air quality in industrial neighborhoods. Citizens' Association president Murry Nelson, an attorney and entrepreneur, petitioned the city to improve cleanliness and eradicate "offensive smells." Nelson's leadership of the Citizens' Association afforded him great political and financial clout in the city. Eventually he attained an advisory role within Chicago's sanitation reform cadre and even obtained a seat on the Board of Sewerage Commissioners. Building the rapport of his association, Nelson noted the "self-sacrificing zeal" of the association's activists, who "turned out night after night during the winter months to trace the location and origin of the nuisance."[20]

Finding the "origin of the nuisance" proved easy and the Citizens' Association pushed the municipal government to take direct action on pollution in the Chicago River's South Branch. The previously ineffective efforts of Chicago's sanitarians, which Nelson described in town hall addresses as "great embarrassments," provided the association president with an arsenal of political ammunition. Nelson continued his public barrages stating that previous sanitation solutions constituted a "recklessness, born of impunity." According to Nelson, earlier failed strategies allowed the "terrible scourge" that affected the city's southern and western sections to persist, which "rendered residence . . . almost intolerable."[21] The ecological damage wrought by meatpacking plants and glue factories represented an environmental and social quandary. Working-class Chicagoans confronted an immense sewage problem produced by a meatpacking industry that employed thousands of workers. The same riverine ecology that supported Chicago's mighty industrial capacity threatened its citizens' health and economic survival. Using his seat on the Board of Sewerage Commissioners, Nelson, with no formal engineering experience, drew upon Chesbrough's work and suggested a radical, ridiculous new idea: a reversal of the Chicago River, known contemporaneously as "diversion."[22] Board members, however, met Nelson's idea with skepticism and outrage despite agreement that Chicago needed a sweeping and permanent sanitary solution.[23] Diversion, although individually proposed by Nelson and his associates, went largely ignored by the city's engineering community.

Political tension quickened the city's sanitation response but an important environmental event in 1885 bolstered support for diversion. In August a massive rainstorm that stretched from northern Minnesota through southern Illinois, inundated Chicago and most of the surrounding area.[24] Although many residents suffered flooded basements, the artificial waterways that extended the Chicago River to the Illinois River protected the city from wastewater and sewage that would otherwise have swamped its most populated sections.[25] The rushing floodwaters forced Chicago's filth away from its populace and vulnerable water distribution system. Diversion, as a sanitation strategy, garnered increased favor from the city's politicians, business leaders, and some engineers. Those same leaders, however, wanted a controlled diversion. A natural disaster convinced Chicago's political and financial leadership to favor a radical, technological solution to defend the city from its tremendous waste.

Convinced that a drainage canal was necessary, city and state officials enacted swift changes. The Illinois General Assembly simplified Chicago's sanitary bureaucracy, dissolving the Board of Water Commissioners and the Board of Sewerage Commissioners in 1889, combining the responsibilities of both agencies in a new organization: the Sanitary District of Chicago (SDC). As a state agency, the SDC had jurisdiction in both Chicago and all surrounding areas affected by Chicago River pollution. In practice the SDC superseded Chicago's municipal government. After reviewing Chesbrough's subterranean drainage tunnels, the SDC's

Board of Trustees found that they could not adequately move the quantity of wastes needed to improve the city's sanitation. By 1890 the SDC determined that only a large canal could move the necessary amount of sewage.

The SDC's founding revealed that although Chesbrough's ambitious plans failed to adequately satisfy residents, he succeeded in raising expectations. The city's new subterranean sewer showed Chicago's political, economic, and engineering leadership that technological solutions held the potential to address environmental problems. In the view of sanitation experts, natural waterways, which had assisted in the city's founding, required improvement. Engineers believed that they could bend nature to their will and make the regional ecology both profitable and controllable.

PEOPLE'S WATERWAYS

Despite this support, municipal leaders still doubted diversion as a solution.[26] The process seemed impossible as no prior precedent existed for such a project. A reversal also represented a tremendous financial expense, which further concerned those in the city's government. They believed that simply dumping sewage into a waterway remained easier and less expensive.[27] Although the SDC received ample state funding, politicians feared selling the project to taxpayers. Opposition from rural residents along the Illinois River also made a potential reversal unpopular in both the town and country. This dilemma created two factions within Chicago's municipal leadership: one that favored the continued disposal of pollutants into Lake Michigan, and another that supported diversion.[28] The SDC had to prove the need for a new drainage canal.

For that proof, the SDC's governing body, the Board of Trustees, turned to Dr. John H. Long of the Illinois State Board of Health. Long assessed water from one South Side hydrant every Saturday during 1885 and 1886 and found that pollution peaked during the summers.[29] That public hydrant drew water from Lake Michigan, contaminated by the Chicago River. Long discovered nearly one hundred thousand bacterial organisms in one glass of water. In all, the SDC authorized 152 chemical analyses and 880 water level measurements.[30] The stated purpose of this study was to illustrate the extent of pollution in the Chicago River and Lake Michigan, while prescribing appropriate solutions. Ultimately, SDC engineers and scientists concluded that the sanitary situation demanded a reversal of the Chicago River. Armed with a year of Long's biological data, the SDC substantiated its support for diversion.

Chicago's problem, however, was twofold: city leaders had to provide clean drinking water, while diverting sewage. Removing contaminated water was one issue, but disposing of pollutants required greater engineering expertise and additional funding.[31] Hoping to find a cheaper alternative, many Chicago sanitarians, researchers, and civic leaders sought a solution that did not involve a new canal.[32] SDC engineers theorized that aeration of sewage and wastes in the I&M would adequately degrade sewage. Diversion, therefore, would be unnecessary. Long, however, warned that without purification and mechanical pumping, the I&M, or adjoining rivers, would prove ineffective in diluting waste— existing waterways simply did not have the required current. Should the City Council approve a drainage canal, the SDC also would need to guarantee its navigability. Therefore, when waters froze in the winter, pumps would need to facilitate water flow for diversion and navigation.[33]

In 1889, after proposing the canal to the Chicago City Council and the Illinois General Assembly, the SDC worked to secure funding. Initial funding for the project came from the city of Chicago and adjoining townships in exchange for promises of toll revenue from increased shipping traffic on the Illinois River. Government bonds and donations from private entities, specifically the Northern Trust Company of Chicago, whose leaders saw a long-term commercial benefit to building a healthier city, also supported diversion.[34] No matter the amount of money issued to the SDC, however, the entire project relied on the manipulation and alteration of the landscape. Lyman E. Cooley, who replaced the retired Ellis Chesbrough as chief engineer, believed in his abilities as an engineer to mold that landscape to his will. He sought to continue his predecessor's success.

Cooley's arrival represented a dramatic shift in Chicago's approach to its sanitary crises. Although Chesbrough initiated that turn, Cooley was simply in the right place at the right time. When the Board of Sewerage Commissioners hired Chesbrough, the city had an overly complicated sanitary management bureaucracy. Chesbrough also lacked the dilution and construction technology needed to realize a successful diversion. Cooley had a state agency devoted to Chicago's sanitation and worked among individuals who believed in technological solutions to the problem of industrial pollution. Unlike Chesbrough, Cooley possessed a formal engineering education and was armed with the confidence of expertise, a faith that Chesbrough had founded. Cooley's presence on the SDC leadership reflected that confidence in human infrastructures as a solution to environmental problems. People could improve and control nature.

The SDC gave Cooley little time to relish his situation. Hoping to begin construction quickly, Cooley set to work formulating a plan for the reversal.[35] Dredging and excavation would literally tip the Chicago

River's water in the opposite direction. The SDC also integrated regional rivers to increase the current of a new channel paralleling the I&M.[36] Once completed, the $33 million Sanitary Canal would extend forty miles westward between the South Side Bridgeport neighborhood to the town of Joliet.

Despite earning the General Assembly's authorization and ample funding, political strife threatened the project early on.[37] According to the SDC's daily proceedings, the Board of Trustees squandered more than one million dollars of public money to begin work.[38] Richard Prendergast, a Chicago lawyer and financier, won the presidency of the SDC's Board of Trustees in 1889. Controversial events soon followed Prendergast's election. Cooley first ordered additional surveys of the construction area that wasted time and money. Under Cooley's supervision, officials duplicated many findings from previous surveys and proposed unnecessary modifications to the canal design. As an engineer, these mistakes were not only inexcusable, but inexplicable. The SDC's Board of Trustees, at Prendergast's request, fired Cooley as chief engineer after only two years of work and hired his chief lieutenant, William E. Worthen, to avoid an investigation by the Illinois General Assembly.[39] Media criticism amplified these problems.

Covering the SDC's meetings and Cooley's supposed blunders in 1891, the *Chicago Daily Tribune* challenged the agency's competence. According to the paper, the SDC "promised decisions and action . . . but has yet to take any," citing "trivial debates amongst board members."[40] The loss of valuable time, rooted in Cooley's wasted ventures, claimed more casualties in the agency. Prendergast himself lost his post as SDC president amid the board's controversies and Illinois politics.[41] In 1892 Cooley seemingly exacted his revenge for being fired. Despite being a trained engineer, Cooley was a far better politician. In that year's midterm elections, Republicans regained control of the General Assembly. Having chartered the SDC, the General Assembly could nominate and promote individuals on the SDC's Board of Trustees. Cooley, a Republican, gained favor and secured a place on the board, despite his removal as chief engineer. This pressure forced Prendergast, a Democrat, to resign amid his party's loss of power.[42] Prendergast lost the board presidency and, although failing as chief engineer, Cooley had remade himself as a politician. These controversies occurred at the expense of improved sanitation.

Remarkably, construction continued in 1892 amid this political wrangling and growing public skepticism.[43] Regardless of the clash of egos that plagued the SDC Board of Trustees, journalists, politicians, business leaders, and residents doubted that a complete river reversal was possible. It seemed far too unwieldy. The SDC sought to build public enthusiasm and held an inaugural ceremony on September 3 near the Cook County line. Speakers, including new Board of Trustees president Frank Wenter, a political moderate, supported the reformist ideals of politicians and their "unwavering commitment to the public interest" through attention paid to citizens' concerns.[44] The *New York Times* sent a contingent of reporters to cover the project's inauguration, highlighting its importance to both Chicago and the prevailing national sanitation movement. According to the *Times*, the excavation of the proposed waterway represented an "enterprise that will rank, when completed, with the most important modern marvels of engineering."[45] The media were also quick to document the controversies surrounding the project. Nonetheless the *New York Times* brought the story national attention but emphasized the intrigue and instability in a city competing with the East for economic preeminence.[46]

Media coverage intensified the SDC's pressure to succeed. Engineers quickly began with excavation along a ridge that straddled the city's western perimeter, sometimes initiating work without adequate equipment. This excavation created a stronger current in the canal between the Chicago and Des Plaines Rivers.[47] Auxiliary channels also allowed for water-level adjustment based on sewage flow and weather. Workers detonated several tons of TNT on a small hill west of the city to lower the surrounding elevation to create space for the canal bed.[48] Construction occurred in phases, beginning first with excavation and then digging. The SDC also approved several miles of new canal space to connect the Chicago and Des Plaines Rivers westward beyond the city. In September 1894 the SDC moved toward primary construction of the main drainage channel. The project represented a regional event that coalesced human and natural resources from the entire Illinois River Valley. Contracted laborers from companies in and around Chicago dug one-mile sections of canal floor, beginning at Willow Springs, just southwest of the city.[49] Workers moved more than 26 million cubic yards of soil per year at the height of work, plus another 12 million cubic yards of solid rock using steam shovels and hoists.[50] Additional equipment including wagons, light-gauge trains, and bridges helped move up to a hundred yards of earth an hour.[51] Engineers employed the new cantilever conveyor to dislodge boulders and assist with equipment transfer. This machine had wheels that allowed for 360-degree rotations and had two cantilever arms that carried dirt onto a conveyor belt which paralleled the ditch. At full capacity the cantilever conveyor moved more than five hundred cubic yards of earth in an hour.[52] Although large, the bridge required only one operator, making it both efficient and inexpensive.

SDC reports celebrated the "most popular construction methods of the period," and touted the "expertise and logistical prowess of the project's architects."[53] The SDC saw technological innovation as another competitive advantage that Chicago possessed over its eastern rivals. National media coverage reflected this competition. Eastern Seaboard cities saw a benefit in criticizing the reversal, whereas Chicago's business and elected leadership held a vested interest in diversion's success. That competition informed the SDC's calculations. Should Chicago successfully reverse its namesake river, it would prove the city most effective in technologically controlling nature, while bestowing needed reforms on its residents. Both were highly marketable qualities. The new chief engineer, Isham Randolph, under direction of Board of Trustees president Wenter, selected equipment that expedited construction. The board also debated how technology could mitigate unnecessary expenditures, reflecting their concerns about bad press and the public skepticism over reversal.[54]

Hydraulic dredging was one innovation that resulted from these debates. This process involved the use of small boats with attached shovels that carved out sections of the riverbed underwater. Although ultimately more expensive and laborious than the SDC anticipated, this process moved 2,500 cubic yards of earth in ten hours, making it the most effective method of underwater excavation.[55] When dislodging heavy rock, workers used explosives and a new steam drill to dig, while the cantilever conveyor moved the rock and dirt out of the trough onto railcars for transport.[56] Between 1894 and 1899 SDC engineers and workers installed mechanical pumps and auxiliary channels to improve management of water flow. These additions ensured that the canal could handle the necessary volume of sewage while diluting waste as it traveled downstream. Ditches, levees, dams, and spillways also helped manage increased inflows and provided an infrastructure prepared for significant refuse removal. Technology made even the slightest variations in the area's environment predictable and simplified a complicated riparian ecology.

The Sanitary Canal opened on January 17, 1900, amid subdued praise.[57] Unlike its inauguration, completion of the $33 million canal was unceremonious and welcomed only about a dozen spectators. Controversies, delays, and disputes that had plagued the SDC from the project's beginning stifled celebration. The *Washington Post* reported that "probably never before has the completion of a public work of this magnitude, been marked with such absolute lack of ceremony."[58] While acknowledging Chicago's engineering accomplishment, the paper emphasized the city's worsening living conditions, particularly in South

Side neighborhoods near the Union Stock Yard, where large eastern European immigrant communities resided.[59] Once the canal was opened, thirty thousand cubic feet of water per second (cfs) poured from Lake Michigan through its locks. Within a week, the canal reached its flow capacity of three hundred thousand cfs.[60] The mechanized current flushed water from the lake, westward through downtown, and then southwest toward the canal adjoining the river's South Branch. Regardless of media scrutiny or lingering doubts, the Sanitary Canal marked a significant achievement in civil service and environmental engineering. The presence of a riverine sewer, however, reflected the desperation of local and state governments confronting economic and political peril.

Despite a completed reversal of the Chicago River, the larger canal system remained incomplete. The Sanitary Canal required improved locks and pumping facilities to accommodate inflows of polluted water.[61] As industries continued dumping waste in the Chicago River, many South Side neighborhoods endured the worst of the city's industrial prosperity, as they had before. Residents confronted coagulated sewage, noxious odors, and blackened water at home and on the job through the 1920s.[62] Public wells remained useless as refuse had congealed in the piping systems. South Siders, therefore, had to walk to cleaner wells on the North Side and carry water in barrels back to their families to drink.[63] In response, the SDC installed new water purification plants near the Union Stock Yard to clean the infamous "Bubbly Creek," the South Fork of the Chicago River's South Branch that merged with the Sanitary Canal.[64]

The social reformer and labor organizer Mary McDowell resided in the heart of the Packingtown neighborhood, which bordered both Bubbly Creek and the Union Stock Yard.[65] She worked alongside Jane Addams, the famed settlement activist and Hull House founder. Both women adopted an acute interest in the area's persistent sanitation problems and attempted to organize cleaner living conditions for the largely eastern European residents. In her book *Twenty Years at Hull House*, Addams recalled a conversation with McDowell, who possessed a steadfast determination to organize South Side workers around the issue of sanitation. Speaking from the perspective of area residents, McDowell remarked that "if you lived near Bubbly Creek, into which the five largest slaughterhouses in the world discharge their refuse, you would be so interested in garbage you would talk about it at luncheon or any other time."[66] Addams assured McDowell that she too was "interested in garbage," and the negative impact it had on the community. Noting the widespread concern of residents, McDowell replied that if you lived in Back of the Yards, "conversation about waste was not disagreeable, regardless of the occasion."[67]

Both stalwart social justice advocates recognized that engineered solutions remained ineffective and underserved the most vulnerable people. Although the reversal represented an achievement unparalleled in ingenuity, it proved inadequate in addressing the needs of citizens closest to the points of pollution: those employed by the same industries responsible for Chicago's sanitary crisis.[68] Addams's settlement work and McDowell's efforts with packinghouse unions show that the reversal provided merely the illusion of reform and instead neglected immigrant, working-class Chicagoans. Both activists sought a radical reform where workers' environmental experiences, felt in the home, merited direct assistance.

Through improved sanitation and assurances of a clean urban environment, city officials claimed a resounding victory. Industrialists continued to establish factories in Chicago, which increased the demand for laborers. Ultimately the canal succeeded in combating pollution and improving Chicago's drinking water for most residents. Nonetheless, immigrant working-class residents living in neighborhoods near the Union Stock Yard lacked effective sanitation and waste-removal services despite the technological marvel built just blocks from their homes. The success claimed by the city's political, technological, and commercial leadership, rather, was an illusion. The reversal marked one stage of an incomplete process in addressing the city's horrid environs. Political and economic pressure placed on municipal officials from business owners and others worried about the image of a contaminated Chicago River ignored the human toll caused by industrial refuse. The reversal was, instead, a marketable solution that presented a cleaner image for a city competing to maintain regional dominance.

The Sanitary Canal reversed the Chicago River and immensely improved water quality for most of the city. Prior to its completion, skeptics, many of them engineers, deemed the Chicago River's diversion impossible. Instead, the reversal did prove possible. In no other instance in the United States did a municipality, state, or federal entity reverse a river's flow to forge an entirely new sanitation system. The Sanitary Canal also expanded the application of artificial waterways beyond travel and transportation to include sanitation at a time when railroads made canals seem irrelevant. While many American engineering projects, including the rerouting of western rivers or the dredging of the Hudson River, were enormous civic undertakings, they did not share the reversal's scale. The Sanitary Canal was the largest earth-moving project in North America, exceeded entire state budgets in cost, and mobilized people across the Great Lakes region, all while realizing the faith in technological expertise. Diversion finally mounted an effective assault on a sordid reputation, reassuring outside investors of Chicago's cleanliness and commercial viability.[69] The Sanitary Canal also marked an unexpected political shift in the city, where entire municipal agencies devoted budgets and resources to sanitation management. The Sanitary Canal established an infrastructure that shaped the physical aspects of life in the Windy City. Although the project achieved its stated goals, the reversal of the Chicago River did not end the city's sanitation problems. Industrial slaughterhouses, mills, and foundries instead dumped waste with revitalized confidence, assuming it would all simply wash away. But the project did solidify the grip of technological and political elites on Chicago and its growing, diverse population.

Today the Chicago River symbolizes the city's safety, security, efficiency, and cleanliness—many of the defining characteristics boosters sought to market as early as the 1840s. However, Upton Sinclair's evocative illustrations of an industrial Chicago enveloped by the smoke, steam, and stench of the slaughterhouses endure. The solidified, black sludge that coagulated in the Chicago River, oozing, bubbling, and sizzling in the summer heat, is an image not easily cleansed. Nonetheless, recent efforts to make the Chicago River safe for recreation reflect continued confrontations with the river's ecological presence and the environmental history that both threatened and secured the city's legacy.

CHAPTER 6

Too Much Water

Coping with Climate Change and Suburban Sprawl in a Flood-Prone Environment

HAROLD L. PLATT

BABY BOOMS AND CLOUDBURSTS

After 1945 Chicago suffered increasingly damaging floods, which were caused by a combination of suburban sprawl and climate change. When the war ended, the baby boom began. Searching for housing, Americans found that buying a dwelling outside the city limits was cheaper than renting one inside them. Paving over prairie wetlands kept raising the speed and volume of storm runoff in an expanding metropolitan area. At the same time six decades of below-average rainfall turned into a still on-going, extra-wet period. Flood-damaging events have continued to rise in frequency and severity. Analyzing recent weather patterns, scientists predict a "future [of] intensification of hourly precipitation extremes."[1]

Chicago is flood-prone because it was built on flat land with poor drainage and an impervious layer of clay just below the surface. Rainwater quickly saturates the soil, inundating the land and turning it into a "sea of mud."[2] While floods are natural, flood damages are human made. Heavy rains caused Chicago's combined sewer and drainage system to overflow into its six river watersheds and Lake Michigan, the ultimate sink of the metropolitan region. Contaminated with raw sewage, the storm runoff surges also back-flowed into people's basements and the city's streets, blocking underpasses and sinking low-lying neighborhoods. In response, sanitary engineers proposed to widen, deepen, and

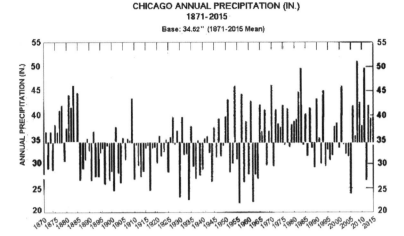

FIG 6.1. Annual Precipitation in Chicago (in inches), 1871–2015. *Source*: Climate Station, Jan. 2016, https://www.climatestations.com/wp-content/uploads/2016/01/CHIPRCP.gif.

channelize the rivers as the best way to cope with suburban development and its ever-expanding infrastructure of wastewater management.

In the early twentieth century, however, these rivers had gained a special status in both the public law and imagination in Chicago as cherished, natural landscapes to be saved at any cost from destruction. Since the successful crusades of Progressive Era conservationists, forest preserve districts have protected the areas along the riverbanks. Forming a thirty-five-thousand-acre (14,164-ha) greenbelt around the city, their picnic grounds, nature trails, and open spaces became immensely popular, especially during the hard times of the Great Depression and the crammed housing of the war years. After 1945 family life, expressways, and a youth-oriented culture of consumption would add millions more visitors. Collectively they formed a powerful, grassroots constituency of proto-environmentalists. They supported forest preserve officials who opposed the engineers' plans to turn the suburban rivers into open sewers, like the Chicago River.

The following case study of environmental politics examines this policy debate. The contestation among policy makers, planners, and the grass roots sheds light on human-riverine relationships. Chicagoans' interactions with their rivers during the postwar period helped to reshape their ideas about the natural and built environments. As the flood-damaging storms grew worse, the politics of Chicago's waters—its lake and rivers—underwent a fundamental transformation. The politics of water management and public works to stop the flooding came into conflict with the

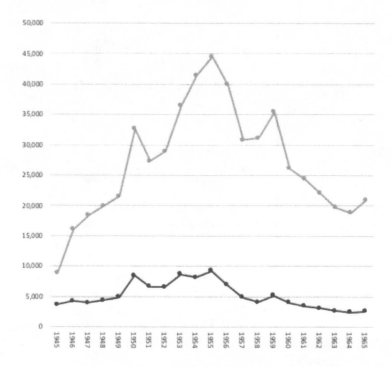

CHICAGO AREA HOME BUILDING

FIG 6.2. Chicago Area Home Building. *Source*: Original chart by author using data from Carl Condit, *Chicago, 1930–1970: Building, Planning, and Urban Technology* (Chicago: University of Chicago Press, 1974), table 3.

politics of outdoor recreation and environmental protection to save the rivers. The city and suburban residents, who were both victims of floods and consumers of nature, brought the two politics together. While demanding solutions from public officials to the problems caused by storm runoff surges, postwar Chicagoans also became defenders of the forest preserves as semisacred places of outdoor recreation and leisure.

This mobilization of the grass roots saved the rivers and forced the engineers to come up with an alternative plan of flood and pollution control. In fighting to protect the rivers, nature conservationists became environmentalists with a modern understanding of ecology. During the 1950s the rising price of damages also forced the sanitary engineers to abandon the incremental approaches of the past, which had worked reasonably well during the previous sixty-year period of below-average rainfall. They were unprepared to deal with either explosive suburban

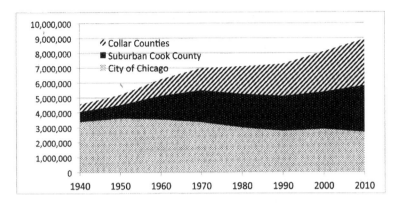

FIG 6.3. Chicago Regional Population Growth, 1940–2010. *Source*: Original chart by author using data from: U. S. Census

sprawl or a wet period of more frequent, extreme events. Frustrated by the protest movement, they would end the policy debate over the fate of the forest preserves in 1965, when they proposed a radical new alternative, the (Deep) Tunnel and Reservoir Plan (TARP).

THE PAVING OVER OF A PRAIRIE WETLAND

In just two decades following war's end, Chicagoans paved over as much land as the previous 150 years of development. Three out of four of the 674,000 new single-family homes they constructed were built in the suburbs.[3] With 3.4 million people in 1940, Chicago maintained a steady state, despite massive suburban dispersal. The city's population edged up by 6.6 percent during the war decade, but lost about an equal amount over the next two decades. The larger metropolitan area continued to spread outward during this period, adding 2.1 million people, a 38 percent increase. It reached 7.6 million people by 1970, when a majority resided outside of the city limits.[4]

After twenty years of building suburbia, Chicagoans doubled the acreage of the built environment. Land-use analysts calculated that the area lost sixteen thousand acres (6,475 ha) a year to development, or the equivalent of a 450-square-mile (1,165-km²) patch of land. In 1945 the size of this area was twice as large as the city borders and equal to the Sanitary District of Chicago's (SDC) boundaries. Its service area would expand to dispose the wastewater of twice as much territory. Chicago's combined sewer and drain pipes were designed to handle a maximum of one inch of rain a day.[5] On one square mile this rainfall amounts to 17.5 million gallons (566.2 million L), or 7.875 billion gallons (29.8 billion L) of additional storm runoff that ultimately found an outlet in the city's river-in-reverse, the Chicago River, and its extension, the Sanitary and Ship Canal. When its Lockport Dam reached capacity to release storm water without flooding communities downstream on the Illinois River, the engineers had to open the gates at the mouth of the Chicago River and contaminate Lake Michigan.

Local public health departments and medical associations were not the only ones worried about the mounting cost of suburban sprawl to the quality of daily life. At the grass roots too Chicagoans sounded the alarm and began mobilizing a movement to save the environment from overdevelopment. The significance of suburbia, Christopher Sellers reminds us, "was one not so much of home buying as home owning." He and other historians posit that the seeds of the environmental movement were sown in the transformative experience of living in a suburban nature. In the case of Chicago at least, the residents of single-family houses within the city were not much different from their counterparts living just across the borderline.[6]

THE EXPANSION OF OUTDOOR RECREATION

In addition to its diverse economy and world-class cultural attractions, Chicago built an environment that institutionalized outdoor sports and leisure activities. Creating an urban nature, the city followed in the Progressive Era footsteps of the city beautiful and recreation movements of Daniel Burnham, Jane Addams, and the landscape architect and park designer Jens Jensen. The Chicago Park District (CPD) managed over 2,700 acres (1,093 ha) of beaches and green spaces along seventeen miles of the lakefront. It also controlled an equal amount of parkland in the neighborhoods. Many of its 137 parks came equipped with year-round field houses and professional staff. Moreover, most of the schools, public and private, now had supervised playgrounds.[7]

During World War II the CPD pumped up civilian morale and bodies by getting the business community to sponsor Sports on the Production Line. Companies mobilized their workers and their children into action on the battlegrounds of baseball diamonds, golf courses, and tennis courts.[8] Civic officials had no need to stoke up use of the city's outdoor spaces; in fact, just the opposite. As the *Chicago Tribune* announced just before the Fourth of July in 1946, "expect record crowds." It predicted at least a million people would flock to the city's beaches and parks, while seats were already sold out on the extra trains and buses added to take them out to the suburbs. In that year alone the CPD counted 26 million participants in supervised recreational activities. It did not attempt to count its less hyperactive visitors. In any case, Chicagoans' enthusiasm for getting outdoors would only grow stronger in the coming years.[9]

The social and spatial limits of open space within the city led its inhabitants to the suburbs. The easing of gas prices and the opening of new superhighways reinforced preexisting patterns of seeking natural green spaces for recreational and leisure activities. By the end of the war, it seemed as if every one of Chicago's ethnic, racial, religious, civic, corporate, and sports groups had marked its favorite spot for gatherings in the Forest Preserve District of Cook County (FPDCC). On a typical weekend in July 1947, several thousand attended the Germania Club's annual picnic in Wolf's Grove, CIO transportation workers headed for athletic contests in Dan Ryan Woods, the Disabled Veteran's Post 849 held an outing in Caldwell Woods, and the 26th Ward Democrats met in Kolze's Park.[10]

Well before the suburban housing boom Chicagoans had made the experience of being out of doors in nature an important part of their everyday lives. They formed a large, if largely unorganized constituency of stakeholders in the environmental protection of the public's existing open spaces. In fact, they supported the plans of local policy makers for the expansion of recreational facilities in the center and watershed/ forestland at the periphery. Suburbanization during the postwar period further reinforced both Chicagoans' personal attachments to experience nature and their political commitments to save it for future generations.[11]

THE UNMAKING OF A FLOOD-PRONE ENVIRONMENT

In fulfilling the American Dream, suburban sprawl had the unintended consequence of turning a prairie wetland into a disaster zone on a regional scale. Of course Chicagoans were no strangers to damaging floods since the earliest days of city building. They were caught unprepared because a sixty-year dry spell had lulled them into forgetting the perils of paving over a marshland. In spite of almost twenty years of record-breaking suburban development, the planners and policy makers were unable to reformulate sanitary strategies that fit the new realities on the ground and in the skies. On the contrary, they remained stuck in a mind-set of incrementalism that had worked reasonably well for the previous generation.

Between 1945 and 1963 the region's inhabitants had to suffer through ten "worst ever" floods and three equally historic corruption scandals before the policy makers at city hall could be persuaded to consider the need for reform. When the rains came, storm runoff surges of bacteria-laden river water had to be released into Lake Michigan, closing the beaches and putting the drinking supply at risk of a deadly epidemic. The growing volumes and speeds of storm water into the rivers resulted in different, but no less destructive, consequences for their aquatic ecosystems.[12]

Chicagoans began organizing to protest against the failures of the powers-that-be to protect the basements of their homes and businesses from repeatedly filling up with contaminated floodwater. They also began mobilizing a proto-environmental movement to save the waters—the lake and the rivers—from ecological disaster. They learned that the SDC's sewage treatment plants' partially treated effluents posed the single greatest threat to this interconnected, underwater world. Policy makers refused to bear the expense of upgrading its technologies from a 70 to a 90 percent level of purification. The contestation of the grass roots against the sanitary engineers' proposal to turn the suburban rivers into sewage channels like the Chicago River would finally force politicians and planners to rethink their strategies of flood control.

After every great inundation of the central business district (CBD), Chicagoans have responded by mounting ever-more heroic, big-technology approaches to turn a marshland into a megacity. After the great flood of 1885, municipal reformers, dreaming of making Chicago the nation's biggest port linking the Atlantic Ocean and the Great Lakes to the Mississippi River and Gulf Coast, put forward an audacious plan to reverse the flow of the Chicago River's contaminated waters away from the lake's drinking supplies. The resultant Sanitary and Ship Canal opened in 1900. Although the city boosters' vision of ocean-cruising ships steaming through the downtown never materialized, their plan for controlling polluted storm water achieved remarkable success over the course of the next half-century. During this period of climate change best known for the Dust Bowl across the Great Plains, Chicago too experienced subnormal amounts of rainfall. When the weather changed after 1945, severe storms came more frequently, punctuated by one record-breaking single rainfall and annual accumulation after another.[13]

These great surges had to go somewhere. Filling up the underground network of drainpipes leading to the treatment plants, the rainwater became contaminated with raw sewage. It gushed into the rivers through hundreds of emergency overflow valves and backflowed into basements and low-lying areas. If the Chicago River were allowed to rise more than five feet above its normal level, the CBD would suffer catastrophic damages. After the Lockport Dam reached its limit, the sanitary engineers had to release billions of gallons of the wastewater into the lake. They had been able to avoid this option of last resort until 1944, despite several notable downpours of between 3 and 4.5 inches (8–11 cm) within a day's time.[14]

At war's end Chicago's sanitary authorities were well aware that they had already fallen behind in meeting their current responsibilities, let alone getting prepared for the dawning era of a suburban housing boom. From the 1920s to the mid-1940s, the SDC engineers had witnessed the volume of runoff pouring into the rivers more than double during heavy rainstorms to reach the maximum amount that could be released at the Lockport Dam. Beside this mounting problem of flood control, the SDC kept pouring more untreated and partially treated human waste into the rivers. Adding to the pollution problem, factories along the riverbanks were allowed to dump their toxic effluents into them. Combined, the 1.2 billion gallons (4.5 billion L) of effluents going downstream every day were loaded with the organic equivalent of the raw sewage of a million people.[15] Over the next decade the SDC would have to grapple with the effects of suburban sprawl, which overwhelmed the infrastructure during a series of historic rainfalls.

In November 1944 sanitary authority experts presented an $80 million improvement plan for the city to the mayor and the aldermen. The Democratic-ruled administration of Mayor Edward J. Kelly quickly passed the proposal of the Engineering Board of Review, and the voters followed in June 1945, when they supported a bond issue to pay the city's share. A closer look at the plan and its creators exposes the strong undertow of tradition on the formulation of flood-control strategies that could meet the twofold challenge of climate change and suburban sprawl. The engineers' blueprint of the future was a laundry list of individual projects rather than an overall strategy of water management and land-use regulation.

The main author of this incremental approach to extend service to the remaining open spaces within city borders was William Trinkaus, chief engineer of the SDC. He was a second-generation employee, joining his father on the job in 1909, and like him, rising to its top professional position. Caught up in a corruption scandal, he would confess to taking between fifteen and twenty bribes from contractors over the years, all the while professing he had done nothing wrong. His successor, Assistant Chief Engineer Horace P. Ramey, had also served a lifetime working for the sanitary district. And he too looked to the remedies of the past to solve current and future problems.[16]

In contrast, one of the city's preeminent planners outlined a much more holistic approach to water management on a regional scale. In July 1947 Chicago commissioner of Public Works Oscar Hewitt suggested, "Chicagoland needs a super water authority." Suburbanization, he reasoned, required expanding the jurisdiction of such an agency to embrace the collar counties. Hewitt admitted that "[the City of] Chicago has lagged behind in building filtration works" and in installing the most up-to-date technologies.[17] The commissioner urged his fellow public officials to enact major administrative and environmental reforms because "it is man, not the lake, that is the great polluter."[18]

But Hewitt's appeal was drowned under the public outcry arising from the disclosure of massive fraud in the very agencies he wanted to entrust with greater, consolidated power. For more than three years beginning in July 1946, stories in the newspapers on the sanitary authorities were more about corruption than construction. In February 1947, just five months after the exposure of the criminal administration of the city's Sewer Department seemed to be fading from public view, an even more outrageous scandal was uncovered involving no less than the president of the board of the SDC, Anton F. "Whitey" Maciejewski.

Just as the boom in public works got under way, Chicago's regime of one-party rule undermined widespread popular support for public works projects to improve the environment. It was replaced by growing skepticism that translated into political demands for structural reforms and budget cuts. The business-minded Civic Federation and the Republican *Chicago Tribune* charged that the SDC was "rotten to the core" and being "headed by a robber."[19]

THE REMAKING OF A WATER-MANAGEMENT POLICY

A hidebound political culture of machine politics meant fewer economic and technical resources, and diminishing popular support to grapple with the double-barreled challenge of suburban sprawl and climate change. Over the next two decades damaging floods and corruption scandals resulted in growing opposition to city hall from the grass roots and shrinking resources from the state and federal governments to fund its public works projects. A closer look at one of these catastrophic events, the great flood of 1954, illustrates the inability of the powers-that-be to respond adequately to these kinds of disasters. Instead, they motivated people to organize protest movements to save not only their basements but also the environment from their own government. When the planners finally proposed a heroic, big-technology approach in 1965, it was too late for the sanitary authorities. They had lost the public trust, turning the political tide against them.[20]

What Chief Engineer Ramey and others called "the most disastrous flood caused by the heaviest rainfall" since 1885 began on October 2, 1954. But unlike that earlier downpour of 6.19 inches (15.72 cm) in a twenty-four-hour period, this weather pattern of record-breaking heat and thunderstorms persisted for the next eight days. It started with the thermometer hitting a high of 91 degrees for the third year in a row, followed overnight by a 4-inch (10-cm) downpour in a five-hour period. While flash floods caused extensive damages in the southern and western areas of the region, the most rain ever in the CBD (6.72 inches

[17.07 cm] in forty-eight hours) began cascading down a week later. The month's total of 12.06 inches (30.63 cm), or four times the average, also surpassed the 1885 figures.[21]

Chicago's city dwellers and suburbanites shared widespread property damages, closed businesses and schools, lost wages, and the breakdown of essential services such as public transportation, telephones, and electricity. In the downtown area the floodwater overflowed into the railroad yards, the post office, the *Daily News* building, and two of Commonwealth Edison's generating stations, knocking out half of the city's power at the height of the emergency. Business came to a standstill as tens of thousands of commuters were cut off from their jobs and equal numbers of other workers were forced to stay home until power could be restored to their places of employment.[22]

The sanitary engineers had to do the "unthinkable," opening the gates of the controlling works at the mouth of the Main Branch to reverse the flow of Chicago's river-in-reverse. Approximately 13 billion gallons (49 billion L) of sewage-contaminated wastewater gushed into the lake. The Health Department increased the amount of chlorine it added to the water supply. In the city the flood caused an estimated $10 million in damages. The suburbs suffered even greater losses because of the maladministration of the public agencies in charge of flood control. Set at $15 million, these damages were primarily the result of sewers back-flowing into basements. Few homeowners carried flood insurance.[23]

At the state level of government, the deluge of 1954 produced political conflict and stalemate rather than concerted, public action. Policy makers in Springfield were unable to decide who should do what. The disaster's effects across the metropolitan area put tremendous pressure on state officials to work out a comprehensive plan. Yet Chicago endured five more catastrophic floods between the great storms of 1954 and 1961, and sanitary authorities failed to prevent damage or even come to an agreement on a plan of action to achieve this goal. On the contrary, they remained deadlocked over the political question of who should control the proposed super-sized agency.[24]

If the outlines of a holistic sanitary strategy of pollution and flood control remained vague after 1954, the political picture was becoming perfectly clear to the residents of this expanding metropolitan area. The following year the Democrats put a rising star within the party, Richard J. Daley, into the mayor's office. A master politician, the last big city boss would impose a one-man rule of power and a one-party regime of government on Chicago until his death in 1976, just after winning his sixth term in office. Behind his shield of protection, the sanitary authorities gained immunity from criticism of their shortcomings in day-to-day

administration. But even this consummate deal maker could stop neither climate change nor its damaging impacts on the built and natural
environments, as well as its human toll of emotional trauma and material loss.[25]

THE ENGENDERING OF AN
ENVIRONMENTAL MOVEMENT

The failure of leadership from within the political establishment opened
up space in the civic arena for professionals and nonexperts alike to offer
alternative flood control strategies. After the defeat of reform in Springfield in 1955, the collar counties', state, and national governments began
implementing their own programs to deal with this ever more frequent
and costly problem. Undertaking similar stopgap projects over the next
decade, the policy makers at city hall were no longer able to contain the
public discourse over water management and land-use regulation. On
the contrary, debate raged among competing factions of consulting engineers, regional planners, and official experts.

Politicians at all levels of government also had to contend with a
groundswell of the grass roots to save the city's waters. Disjointed ges-
tures of eco-protest began to coalesce into an environmental movement
in between the great flood of 1961 and the great corruption scandal a year
later. The ensuing battle of reform to save the lake and the suburban rivers in the forest preserves from the SDC also contributed to the construction of a metropolitan identity. It formed as opposition mounted to the
sanitary authorities, the Democratic Party, and ultimately, Boss Daley.[26]

Even before the next cataclysm hit in September 1961, city dwellers
and suburbanites were already beginning to organize a resistance movement to the proposals of the official agencies in charge of flood control.
Besides destroying the riverbanks to increase flow rates, all of the engineers' designs depended on using Lake Michigan as the sink of last resort
for storm runoff surges. In February 1960 private citizens and regional
planners successfully petitioned the state governor to delay funding of
the engineers' plans to channelize Salt Creek and the Des Plaines River
in the western suburbs. They would have turned these natural streams
into concrete canals. The nonpartisan Northeastern Illinois Metropolitan Area Planning Commission led the lobby campaign to give the
FPDCC time to prepare an alternative approach. Its consultants recommended building five impounding dams on the upper Des Plaines River,
which would add more parkland for recreational use at the same time.
They also announced that the district had worked out a compromise with
the state's engineers to improve Salt Creek in similar ways to minimize
environmental damage.[27]

The triumph of the FPDCC was not surprising given Chicagoans' enduring love affair with its open spaces and facilities. The growth in popularity of spending time in the great outdoors was matched only by the mobilization of political support in favor of increasing open space in the city and the suburbs. In 1961 the city's park district counted fifty million people engaged in supervised recreational activities. Grassroots demands for more forest preserves had already led to an increase of the FPDCC's original charter of thirty-five thousand acres (14,164 ha) to forty-seven thousand acres (19,020 ha) of land. But this was not enough, according to a *Chicago Tribune* May 1961 editorial, because the "forest preserves must grow." The following year the publication of Rachel Carson's *Silent Spring* inspired a national epiphany on the meaning of ecology and the interconnectedness of all living things.[28]

In sharp contrast, the sanitary authorities became the objects of ridicule for their incompetence and failure to design a flood-control strategy for Chicago's expanding metropolitan area. After the great flood of September 1961, this criticism gelled into a unified chorus of demand for reform. Producing over ten inches (25 cm) of rain in ten days, this extreme weather event broke the previous monthly record from October 1954. Given Mayor Daley's success in kick-starting a renaissance of the CBD, the shortcomings of the SDC were no longer tolerable to the city's business community. Its plans for redevelopment of the downtown area depended on turning the river from an open sewer into an urban amenity, lined with walkways and plazas.[29]

After the deluge, the mayor made a major concession to the business community. He agreed to let it conduct a national search to replace the general superintendent of the scandal-ridden SDC. Its blue-ribbon panel was savvy enough to present only a single candidate, Vinton W. Bacon. Mayor Daley got far more than he bargained for in permitting an outsider inside this den of thieves. A highly respected civil engineer, Bacon was not only honest and competent, he was also a tireless reformer. At first he made real progress; a catalog of his disclosures would comprise an encyclopedia of corruption. But after about a year on the job in April 1964, Daley gave a signal to the trustees to begin rolling back reforms.

Although Bacon's political crusade proved ephemeral, his flood-control plan became a permanent legacy.[30] In January 1965 Bacon revealed the outlines of a heroic, big-technology plan to a reporter that "would be one of the great engineering feats in history." But the feasibility study of what would become the TARP seems to have been sidetracked by the trustees. Then, on the eve of the 1967 municipal elections, Mayor Daley made the surprise announcement that the city had a bold new plan and took credit for securing a million-dollar grant from the federal government to conduct the technical studies needed to begin construction. Bacon filled in the details of the multi-billion-dollar "master water plan" that would not only permanently end damaging floods but also clean up polluted rivers within the territory served by the SDC. The Tunnel and Reservoir Plan (TARP) was designed to redirect sewer overflows from storm runoff surges into an underflow system of very large storage tunnels running below the rivers. Later, the contaminated water could be pumped up and properly treated before being released back into the environment. He promised that Chicago's rivers would be turned into natural and recreational amenities clean enough to swim in within ten years.[31]

The year 1965 was pivotal in a related way in the ascendency of the politics of outdoor recreation over the politics of water management. Under the banner Save Our Lake, the city's scattered campaigns of environmental protection became a unified movement. On the heels of Carson's book, the *Chicago Tribune* launched this sustained, front-page crusade, which would contribute significantly over the next seven years to the passage of the landmark Clean Water Act. Coming full circle, the national government would fund the TARP as a make-work project during the following decade of recession.[32]

The suburban rivers were saved during this round of political conflict over a flood-prone environment. During the pivotal year of 1965 the FPDCC reached its fifth anniversary. State lawmakers celebrated by passing a bill to expand its size from fifty-two to sixty-five thousand acres (21,050 to 26,300 ha). Taking pride in these projects to preserve nature and provide the public with more outdoor recreational space, a local newspaper boasted that Chicago was finally entitled to "merit the name of [its motto] 'The Garden City.'"[33]

CHAPTER 7

Water, Oil, and Fish

The Chicago River as a
Technological Matrix of Place

DANIEL MACFARLANE AND LYNNE HEASLEY

ON VISUALIZING THE HIDDEN: IIB AND 6A

Consider the *where* of Figure 7.1. A ragged patch of grass and gravel, a straight line of water slicing through flatland, a row of smokestacks across and parallel to the water whose silhouettes are reinforced by black belches drifting out of the frame on the right. On the far bank, a couple of trucks behind barbed-wire fence. Somewhere industrial.

Dominating the scene is a sign: DANGER: ELECTRIC FISH BARRIER. This was the terminus of a field outing that began with a simple question, *Where is it?* Where was this new electric fish barrier, titled Barrier IIB? IIB was the U.S. Army Corps of Engineers' latest reinforcement to a Great Lakes stronghold against bighead and silver Asian carp, those powerful nonnative fish that arrived from their journey up the Mississippi and encamped in the Chicago Sanitary and Ship Canal behind literal gates barring the way to Lake Michigan.

The location was Romeoville, in Will County, Illinois, about thirty miles southwest of downtown Chicago. While newspapers had announced the barrier, finding it wasn't easy, and once there, nothing barrier-like was visible. So the not-really-so-simple question of *where* morphed into another not-so-simple question of *what* was Barrier IIB.

Sometimes you have to be in a place—you have to look around—in order to ask questions, or to see something unexpected. Go back to

FIG 7.1. Reading an unfamiliar waterscape. Photo by Lynne Heasley.

Figure 7.1 and look past the danger sign. Explore the edges instead—for example, the tree branches poking into the scene from the upper left, or the object below the branches, a metallic fragment arching across the water. *What . . .* is that?

That turned out to be Enbridge pipeline 6A. Figure 7.2 shows it close up. Line 6A was one segment of a binational crude oil trunkline system that began in Fort McMurray, Alberta, heartland of the Canadian tar sands. 6A and IIB intersected at the canal. Tar sands oil and Asian carp, jarring in their togetherness. Even more unexpected was their obscurity.

It may seem strange to call the pipeline obscure, since the arch over the canal is visible from a local overpass. In a way, it's hidden in plain sight. Boats and barges cruise underneath the structure, while cars go over a bridge that offers a distant view. But boaters and drivers must keep moving. They can't get off the boat to look around. They can't be *still* enough to study the place, to know it better. Perhaps the pipeline's operators count on a public moving rapidly past everyday sights. Richard White once wrote about water infrastructure that "boredom works for bureaucracies and corporations as smell works for a skunk. It keeps danger away. . . . The audience is asleep. The modern world is forged amidst our inattention."[1]

Line 6A's brief materialization aside, the landscape hosting 6A was most definitely indistinct. This off-limits property concealed the actual location of 6A and IIB from passers-by. A tree-lined centennial trail lined one outer boundary of the property. But no trail map pointed inward to

FIG 7.2. Intersections in the waterscape. Photo by Lynne Heasley.

pipeline or fish barrier (there was a DANGER sign, after all). Looking for fish barriers meant turning away from the comforting habitual traffic of walkers, runners, and bicyclists. There were discreet entry points—a shallow part of a drainage ditch, a panel of knocked-over chain-link fence. A man walking two tiny dogs hopped the fence, then receded into the tire ruts and wiry gray-green stubble of an industrial barrens. Somewhere belowground the pipeline continued unseen and unimpeded.

Obscurity matters. With water, oil, and fish, critical environmental histories of Chicago and the Great Lakes coexist out of sight, out of mind, or submerged. Pipeline and fish barrier represent the kinds of subterranean networks within which millions of people live unaware. For the remainder of this chapter, we examine three intersecting Great Lakes infrastructures: the Chicago Diversion, Enbridge's Lakehead pipeline system, and the electric fish barrier system. Vast in the scale of their impact, precarious in the intricacy of their design, we try to make these concealed infrastructures visible to readers by applying concepts for "seeing":

- *Technological matrix of place*: Together, diversion, pipeline, and barrier infrastructures intersect at a discrete and bounded location, forming a technological matrix. At the same time each is part of a networked technological system. To understand the three infrastructures requires place-based analysis coupled with notions of larger border-spanning networks.

- *Disguised design*: All three infrastructures are largely unseen despite their scale—underground, underwater, or off limits. They are emblematic of what Daniel Macfarlane has called "disguised design," in that they are meant to be concealed while appearing natural.[2]

- *Environmental risk*: All three contain knowable risks to local communities and the larger Great Lakes region. Whether water, oil, or fish, in each system a resource flows through a conduit. The risk is that something supposedly confined and safe will escape its scripted boundaries.

- We want readers to consider the repercussions of submerging both infrastructure and risk. We want readers to see how these infrastructures undergird not only the natural world and built environments of the Chicago area, but the natural world and built environments of the Great Lakes system of inland seas.

MATRIX #1: THE CHICAGO DIVERSION

The Chicago Sanitary and Ship Canal is a landmark in North American environmental history.[3] By reversing the Chicago River, the canal system marked an audacious start to twentieth-century water engineering exploits. In recognition, it was named one of the century's top ten American public works projects. The canal and its extended river network allowed Chicago to withdraw water from Lake Michigan, making it the first large-scale water diversion out of the Great Lakes. The "Chicago Diversion" was the volume of water the city could legally withdraw and send across the Great Lakes hydrological divide to the Mississippi River basin. From there it would flow downriver to the Gulf of Mexico rather than from Lake Michigan to Lakes Huron, Erie, Ontario, and out the St. Lawrence River to the Atlantic.[4] This east to west Chicago to Mississippi orientation is typical of how people view the canal. By reorienting one's perspective from the canal toward the larger Great Lakes–St. Lawrence basin—that is, from west to east instead of east to west—we can see key but lesser-known relationships.

The canal itself was one example of a water engineering process that transformed the whole of the Great Lakes. Upgrades to the Welland Canal and locks at Sault Ste. Marie connected the lower and upper lakes. Dredged channels in the St. Clair and Detroit Rivers increased the capacity of Great Lakes harbors. Massive hydroelectric installations turned rivers into reservoirs. The St. Lawrence Seaway and Power Project was the most ambitious of these megaprojects; it opened the basin to transoceanic shipping. (The seaway, too, counts among the top ten public works of the century.)[5] These far-flung works formed an integrated tech-no-economic system within which ships and their cargoes of Great Lakes resources circulated from the far western end of Lake Superior to the Atlantic Ocean. At 2,300 miles long, the Great Lakes–St. Lawrence water-way became the world's largest inland maritime system. It also became a transnational technological matrix.

The canal's place within a Great Lakes maritime system helps explain why a chronology of legal disputes now characterizes its history. At odds with Chicago, Canada and the American Great Lakes states had strenu-ously objected to the initial Chicago Diversion. They argued that it low-ered Great Lakes water levels.[6] The diversion became a thorn in U.S.-Ca-nadian diplomacy. Early on it factored into negotiations over the 1909 Boundary Waters Treaty. The treaty provided for joint management of Canadian-American boundary waters. Because of the Chicago Diversion, the final treaty did *not* include Lake Michigan.

Between 1912 and 1924 Canada filed six objections to the diversion. Later the diversion impacted bilateral talks for the joint St. Lawrence navigational seaway and hydropower development. Treaty negotiators for the Great Lakes Waterway Treaty of 1932 had included limitations on the Chicago Diversion. The U.S. Congress voted down the treaty because of these provisions, among other reasons.[7]

U.S.-Canadian wrangling had technological and environmental conse-quences. In 1941 the countries signed an executive agreement that echoed the 1932 treaty. Congress rejected this agreement too. Canada built its own enormous Ogoki and Long Lac diversions *into* the Great Lakes to com-pensate for the Chicago withdrawals.[8] The Ogoki–Long Lac megaprojects channeled water from James Bay and the Albany River into Lake Superior.[9]

In the 1950s the Chicago Diversion reappeared during final treaty negotiations over the joint St. Lawrence Seaway and Power Project. The U.S. Supreme Court had lowered the allowable volume of the diversion to 1,500 cfs (cubic feet per second), but on several occasions Congress temporarily increased the volume, and it twice legislated a permanent increase. Canada formally objected to the legislation, asserting treaty vi-olations. President Eisenhower promptly vetoed the legislation.[10]

Along with Canada, Great Lakes states also fought Chicago and Illi-nois over the Chicago Diversion. In 1967 the U.S. Supreme Court estab-lished an average 3,200-cfs limit for Chicago.[11] In the 1980s the Corps of Engineers studied an increase to 10,000 cfs. This was the maximum flow the canal could sustain. Later Illinois formally, but unsuccessfully, requested the same increase. But Chicago often exceeded its withdrawal limit, sometimes intentionally, sometimes by accident. More recently, the city has stayed within its legal limits.

The diversion itself has continued to weigh down Great Lakes policy and diplomacy. Illinois's refusal to consider any changes nearly sank a decade of interstate negotiations for a Great Lakes–St. Lawrence River Water Resources Basin Compact. The 2008 compact was a landmark framework for governing diversions out of the Great Lakes, as well as a landmark of water policy worldwide.[12]

The issue of water levels has also loomed large. Scientists in the early twentieth century could only hypothesize about fluctuating lake levels; the system-wide impacts of the Chicago Diversion were unclear,[13] but Canada and the Great Lakes states feared it could lower water levels by half a foot, as far as Montreal.[14] Later research on Great Lakes hydrology revealed the complex ways in which the lakes are interconnected.[15] For example, engineering interventions in the Great Lakes–St. Lawrence basin *have* cumulatively lowered water levels, while natural forces have impacted the *scale* of the fluctuations.[16] Likewise the Chicago Diversion itself has slightly lowered water levels throughout the basin, while precipitation and evaporation have determined long-term fluctuations (including the record high water levels of recent years).

Private industry was likewise engaged in the pros and cons of the Chicago Diversion. Shipping and hydropower interests were especially hostile. Far to the east, along the St. Lawrence River, power providers had to modify their hydroelectric works to compensate for lower water levels. At Niagara Falls, lower water levels required remediation work on submerged weirs and control dams, and eventually the whole Niagara waterfall was reengineered.[17]

Industry faced feedback loops. Lower water levels reduced the weight of cargo that Great Lakes freighters could carry. Deeper wing dams and dredged bottoms became essential projects for industrial shipping. In the St. Clair and Detroit Rivers engineers would periodically dredge and channelize the river. But those projects also lowered water levels in Lakes Michigan and Huron.[18] Such iterative actions and reactions, causes and effects, make visible an underwater infrastructure matrix in which a technological "butterfly" flapping its wings in one place had distant hydrological, ecological, and economic impacts elsewhere in the system.

For one-hundred-plus years the Chicago Diversion was a sore infecting interstate politics and international diplomacy. But impacts below the surface were just as important. The brute force of the diversion altered water levels and forged new hydrological networks. By lowering lake levels, the diversion shaped control works: locks, dams, weirs, channels. Each looked like a single infrastructure project initiated in response to local conditions. Together, though, they formed a highly engineered Great Lakes system. Also below the surface were disrupted lake bed and river bottom morphologies and aquatic ecosystems. The Chicago Diversion transformed the waterscapes and water life beneath.

MATRIX #2: THE ENBRIDGE LAKEHEAD PIPELINE SYSTEM

Like water flows from the Chicago Diversion, oil flows through the Chicago region are mostly invisible. At the canal near the Chicago suburb of Romeoville, Enbridge pipeline 6A makes a rare aboveground appearance: a dramatic aerial line featuring a double wishbone arrangement of two thirty-four-inch pipes. The striking arch reaches a height of ninety-three feet (see Figure 7.2). But even this singularity conceals what runs beneath, underground or underwater: a continent-scale network of fossil fuel pipelines.

Enbridge alone has seventeen thousand miles of pipeline in the United States. The Lakehead System accounts for about 1.7 million barrels per day, or 13 percent of U.S. petroleum imports. Enbridge does not disclose how much oil running through its Lakehead System comes from Alberta's huge tar sands deposits. In Chicago the Lakehead System is one of the most important infrastructure matrices that people don't see and know little about.

The first long-distance pipelines in the United States ran near and partially within the Great Lakes basin, especially the southern shore of Lake Erie.[19] The Petrolia-Sarnia region of Ontario, Canada (bordering Michigan), was ground zero for Canadian production; there oil interests in the era of John D. Rockefeller's Standard Oil built Canada's earliest pipelines. By the end of the nineteenth century, Ontario and Michigan had underwater gas pipelines at the bottom of the Great Lakes connecting channels like the St. Clair River. Twenty-eight pipelines now run through the St. Clair.[20] One pair dating to 1918 was long forgotten until 2016, when new owners quietly applied for a permit to reactivate the antique pipes for transporting liquid hydrocarbons, including crude oil. St. Clair communities learned about the extant pipes *and* the permit application only in the last days of the public comment period.[21] Under sudden public scrutiny, the application was withdrawn.

Pipeline infrastructure scaled up after World War II. In 1950 Canadian energy company Inter-Provincial Pipelines—now Enbridge—opened an oil pipeline from Edmonton, Alberta, to Superior, Wisconsin. This was the first leg of a network that includes Enbridge's Lakehead System (Figure 7.3).[22] From Fort McMurray in Alberta, Enbridge's main line runs to Superior, Wisconsin, at the far western tip of Lake Superior. At Superior the system forks into Line 5 and Line 6A (built in 1953 and 1960 respectively).

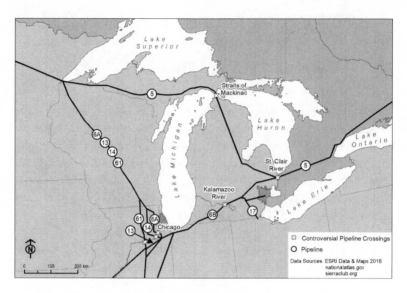

FIG 7.3. The Enbridge Lakehead pipeline system traversing the Great Lakes basin.
Map by Jason Glatz, Lynne Heasley, and Daniel Macfarlane.

Line 6A is Enbridge's largest oil artery supplying its U.S. market.
From Superior Line 6A goes southeast through Wisconsin and Illinois.
Near the Chicago Sanitary and Ship Canal, 6A runs underground, then
soars briefly over the canal as aerial pipes. In April 2010 at Romeoville,
municipal employees, rather than Enbridge itself, discovered a major
spill from Line 6A. All told, six thousand barrels spilled.[23] Authorities
evacuated five hundred people from area businesses when the oil spread
to the retention pond of a nearby wastewater treatment plant. Enbridge
closed three miles of pipeline in order to locate the leak. The closure led
to higher gas prices throughout the Midwest.[24] Tom Kloza, publisher and
chief oil analyst at Oil Price Information Service, said it was like "shutting down the lifeblood that feeds the organs."[25]

Line 6A ends southwest of the canal at Griffith. From there its sister
line, 6B, continues through Indiana and Michigan to Port Huron, and under the St. Clair River to its terminus in Sarnia. Many people in Michigan
know Line 6B even if they don't know its name. Line 6B was the infamous
pipeline of the July 2010 Enbridge oil spill into a tributary of the Kalamazoo River—over 1.2 million gallons (20,000 barrels) of tar sands diluted
bitumen, or heavy crude oil. This was one of the two largest inland oil
spills in U.S. history (both from Enbridge lines).[26] Enbridge's combined
costs for the 2010 Romeoville and Kalamazoo spills exceeded $1 billion.[27]

For its part, Line 5 traverses northern Wisconsin and the Upper Peninsula of Michigan, turns south, runs underwater through the Straits of

Mackinac, and continues to Port Huron, where Lake Huron drains into the St. Clair River. Like 6B, Line 5 terminates in Sarnia. Sarnia serves as a transfer point for oil heading inland, but the city's so-called Chemical Valley also holds 40 percent of Canada's petrochemical industry.

Line 5 has become infamous in its own right. Only in the last decade did Michigan and Ontario communities learn that a major pipeline went through the Straits of Mackinac, a natural channel that connects Lakes Michigan and Huron, making them hydrologically, if not in name, a single lake and the largest freshwater lake in the world by surface area.

With Line 5, a perfect storm of conditions is poised for disaster: the straits' "strong and complex" currents buffet and stress the pipelines and in case of a spill would move oil quickly throughout the lakes.[28] There is no way to slow or remediate a spill if it occurs in icy winter conditions (which last for many months of the year). Sediment covers parts of the pipeline, so its condition is unknown in those places. In other exposed places film footage has shown broken supports and corrosion, unsurprising given that the pipeline is over sixty years old.[29] "If you were to pick the worst possible place for an oil spill in the Great Lakes, this would be it," said hydrodynamic modeler David Schwab.[30] Yet each day, two hundred feet underwater, Line 5 carries 20 million gallons of synthetic crude oil across 4.5 miles of straits. Michigan politicians and policy makers face tremendous public pressure to force Enbridge to shut off the oil through Line 5.[31]

Legally, waters like the Kalamazoo and St. Clair Rivers or the Chicago Sanitary and Ship Canal are held in "public trust." Unlike land, no private "water owner" can buy acreage of river or lake bottom. Parties may only negotiate a conditional right-of-way with the state. With respect to Line 5 in the Straits of Mackinac, perhaps the most shocking discovery was how little anyone knew, including who was responsible for its safety. It turned out that, despite the Kalamazoo and Romeoville spills from the same Lakehead System, the State of Michigan had not revisited its original 1953 easement for the Enbridge pipeline. Nor had the governing federal agency, the Pipeline and Hazardous Materials Safety Administration (PHMSA), reviewed Line 5. Until underwater footage revealed Line 5 to a horrified public, the PHMSA had been nearly as disguised as the pipeline under its watch.

The canal, straits, and rivers are among the thousands of places where oil crosses above or beneath water. Each is a submerged technological matrix, a vulnerable hybrid of engineered and natural hydrological systems. Collectively they form an interconnected network in which leaks and spills are the norm. Since 1999 North American Enbridge pipelines have lost 6.8 million gallons from over eight hundred leaks and spills.[32]

Industry-wide, spills occur daily. Transnational pipeline networks have been called a "dissociating technology." They lie out of sight; they don't need human hands to operate (until oil escapes its boundaries); and they are geographically distant from the original source of the oil.[33] Their disguised design means that communities, and even the pipeline operator, won't immediately know of leaks and spills, so time lags before discovery are also the norm.[34]

By design, a continent-scale pipeline system hides, delays word of, and eventually externalizes the environmental and economic fallouts of oil production to local landowners and communities, and to local landscapes and waterscapes. The system even threatens immense waters like the Gulf of Mexico or, forebodingly, the Great Lakes. Making these infrastructures visible is a precondition for safer policy.

MATRIX #3: ELECTRIC FISH BARRIERS

In 1900 Chicago completed its eight-year quest to flush the city's excrement west, to Lockport and the Des Plaines River and finally to the Mississippi. City leaders celebrated more than Chicago's new stature as a model of modern wastewater treatment. The Chicago Sanitary and Ship Canal symbolized clean, safe water. Hence city leaders could celebrate the blurring and eventual erasure of Chicago's earlier image as a cesspool—a literal cess-river system of human and animal waste oozing through streets, clogging sewers, thickening and chunking the river, rendering the South Fork of the South Branch bloody and bubbling with the methane and hydrogen sulfide gasses of decomposing hog and cow entrails from the meatpacking district. The South Fork of the South Branch had earned its moniker, Bubbly Creek. But the new canal flushed the city clean. The canal also scrubbed away the public's imagination of Chicago as a place of disgust and degradation.

The Chicago Sanitary and Ship Canal is back in the public imagination. This time, though, the canal itself embodies a potential catastrophe: biological degradation, or "biological pollution," of the Great Lakes. The context is this: The unification of the Great Lakes and Mississippi basins created new flows of nuisance aquatic species (NAS in the technical literature). Zebra and quagga mussels and round gobies crossed the Atlantic Ocean to the Great Lakes via the St. Lawrence Seaway, and then migrated to the Mississippi basin via the Chicago Sanitary and Ship Canal. These species fanned southward along the Mississippi River and its tributaries to the Gulf Coast.

Yet the canal was not a one-way-only biological highway. Notre Dame scientist David Lodge called the route "a two-way highway to environmental and economic havoc."[35] Lodge was giving congressional

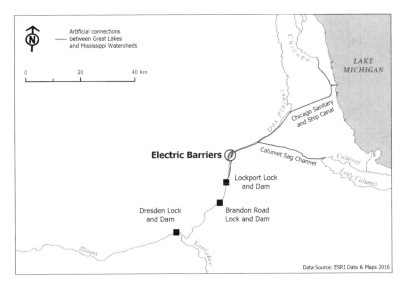

FIG 7.4. Location of the electric dispersal barrier system along the Chicago Sanitary and Ship Canal. Map by Jason Glatz, Lynne Heasley, and Daniel Macfarlane.

testimony about the biogeography of silver and bighead carp, *Hypophthalmichthys molitrix* and *Hypophthalmichthys nobilis*, two of the species colloquially called Asian carp.

From their 1972–1973 introduction to the American south for use as algae cleaners in catfish farms, Asian carp escaped into the wild. They migrated up and down the Mississippi River system, decimating aquatic food webs along the way—"like a school of aquatic bullies," according to an Illinois Department of Natural Resources article.[36] In many parts of the Mississippi basin, silver and bighead carp constitute up to 95 percent of the total aquatic biomass. They reached Illinois rivers in the early 1990s, and in 2002 scientists saw them in the upper Illinois River, which is the conduit between the Chicago Sanitary and Ship Canal and the Mississippi River.

At the same 2010 hearing where Lodge testified, committee chairman and U.S. representative James L. Oberstar likened Asian carp, "this treacherous, dangerous species that we cannot allow into the lakes," to sinister aquatic wolves: "It reminds me of an image in the language of my ancestors, the Slovenes," Oberstar said, "we just think about the wolf, and it is at our doors. And that is what the carp is; it is at our doors."[37]

The Oberstar hearing was focused on plans for the U.S. Army Corps of Engineers Electrical Dispersal Barrier System (Figure 7.4). Technology had assumed a principal role in addressing the continental problem of invasive species like the Asian carp. In Chicago electric fish barriers now

form another disguised infrastructure, a matrix at once powerful and
vulnerable—with both its power and its vulnerabilities determining the
fate of Great Lakes ecosystems.

Today the system consists of barriers placed at intervals in the canal near Romeoville, Illinois. Planning for the barriers got under way
in 1996, when Congress appropriated $750,000 for a "Dispersal Barrier
Demonstration" as part of the National Invasive Species Act of 1996.[38]
The Army Corps of Engineers activated demonstration Barrier I in 2002
to test the technology, IIA in 2009, and IIB in 2011 following the Oberstar hearing. On shore, calibrated generating stations run and monitor
the barriers. In 2017 the Trump administration put on hold the Corps'
most updated study, which recommended another electric barrier farther south, and farther away from Lake Michigan, at the Brandon Road
Lock and Dam, a Des Plaines River choke point for carp.

Smith-Root, Inc. was the Corps' sole contractor to design and build
the dispersal barrier system. The company is an example of historical
happenstance. In 1964 Dave Smith and Lee Root designed an electric
fisher for University of Washington fish biologists. Their invention made
fisheries research more efficient. Smith and Root had stumbled into an
unexpected market. "They got drug into the electrofishing market, not
by design, not by creativity; it just happened," said current Smith-Root
president and CEO Jeff Smith.[39]

Smith-Root soon expanded into other electric devices, including
complex systems for redirecting or containing fish traffic. This was a
technology-science collaboration. Smith-Root worked with fisheries and
other aquatic scientists nationwide, and conversely, scientists assumed
a proactive role in emerging technological approaches to environmental problem solving. Electric dispersal systems like that in the Chicago
Sanitary and Ship Canal marked the culmination of this codependence.

From 1988 through 2016 Smith-Root filed six patents for some version of electric dispersal fish barrier systems. Barrier I used Smith-Root's
original 1988 design for a "fish repelling apparatus using a plurality
of series connected pulse generators to produce an optimized electric
field."[40] A sample patent diagram (Figure 7.5) shows the layout, crudely
stated here, of electrodes along metal strips at the bottom of the canal
supported by electric generation on land, and with cables connecting the
electrodes to the generators. The little fish icons represent the carp. In
subsequent patents Smith-Root expanded their barrier systems.

The fish icons of Smith-Root's patent diagrams are easy to overlook,
yet they might be the most telling part. Buried within the complex energy matrix of generators, circuits, transformers, and electrodes were fish
physiology and behavior. Controlling the fish depended on the precise

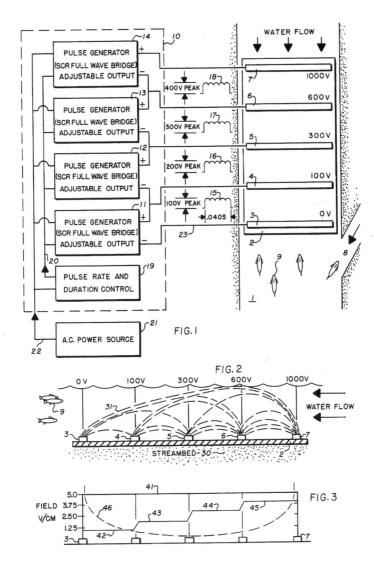

FIG. 1

FIG. 2

FIG. 3

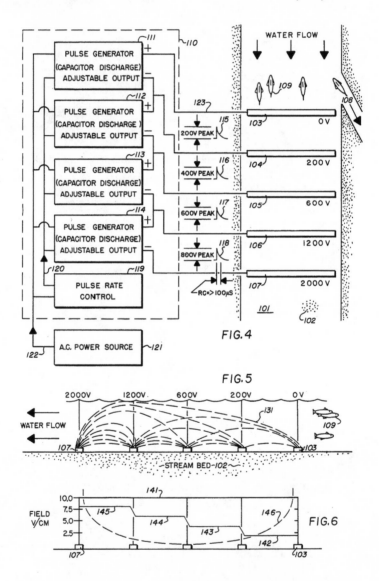

FIG. 4

FIG. 5

FIG. 6

FIG 7.5A & B. Illustrations from U.S. Patent 4,750,451, a Smith-Root electric dispersal fish barrier system.

application of current running through water. This in turn depended on intimate knowledge of how electricity passes through a moving
animal's body and shapes its behavior. For instance, in Smith and colleagues' 2005 patent, Electric Fish Barrier for Water Intakes at Various
Depths, the team detailed such electricity–fish interfaces: "Because a fish
has salts and electrolytes within its body . . . a fish's body acts as a 'voltage divider' when swimming through fresh water."[41] But precision could
never be wholly achieved, because these were real-world uncontrolled
experiments in which two intersecting variables—animals and electricity—were not fully known or predictable.

Smith-Root and their university partners raced to study carp behavior
around electricity. They hoped to adapt the system to newly observed behavior before the fish found their way through the barriers and into Lake
Michigan. Each update of the electric dispersal barrier system became
a new field trial. The most alarming carp behavior involved differences
between large and small animals. It turned out that juvenile fish could
withstand a higher level of pain before turning away from the barrier,
even when the acute pain made them swim abnormally. The initial electric pulses had emitted too low a voltage to repel juvenile carp. Voltage
adjustments were made. Juveniles could also slip through bedrock cracks
in the waterway's surrounding landscape. Regional flooding thereby became high-risk events. Strategic fencing in vulnerable locations followed.

Add in carp tenacity: biologically they were impelled to move up-
stream, so over and over they probed and tested the barriers, finding less
discomfort closer to the surface and toward the canal edges. Acoustic
bubbles or bubble barriers near canal walls might address this particular vulnerability. Barges and other vessels on the canal presented a risk,
altering the direction of the electric current so that fish might draft off
them all the way through a barrier. And then there was safety: Could the
electric current escape the canal and hurt people? Steel "parasitic structures" across the canal dealt with this concern.

The Corps entered all these known risk variables into a risk model.[42]
But the upshot was that carp would adjust, would search for ways to get
through, and their behavior never went entirely according to plans.

The history of the barrier system is inseparable from the history of
late twentieth-century fisheries research. A codependence of science and
technology formed under the intense pressure of fast-spreading invasive
species whose ecological and economic costs were enormous. But this
same interdependence may have foreclosed or delayed other options. The
safer, longer-term solution would be to re-separate the Mississippi and
Great Lakes basins. Hence many of the same Great Lakes stakeholders
who tried to block the canal's opening more than a hundred years ago

are now trying to close the canal. So far, however, an alliance of city leaders, Illinois state leaders, heavy industry (e.g., oil), Mississippi River tow and barge companies, and Chicago tourist operators have prevailed over seven Great Lakes states, two Canadian provinces, a $7 billion Great Lakes fishing industry, and the nearly unanimous ecological anxiety of aquatic ecologists and fish biologists.

Where do matters stand, or drift as it were? Anglers have caught several types of Asian carp in the Great Lakes, but not yet silver or bighead species. Scientists have detected environmental DNA (eDNA) of silver carp in the Chicago River and Great Lakes, though eDNA alone does not mean there are reproducing populations. In 2010 the Illinois Department of Natural Resources reported a nineteen-pound bighead carp in Lake Calumet, past the electric barriers and six miles from Lake Michigan. In June 2017 a commercial fisherman caught an eight-pound silver carp past the barriers and nine miles from Lake Michigan.[43] Such findings mean that carp are finding ways to Lake Michigan, suggesting it's just a matter of time until the fish establish a foothold in the lake.

In the case of oil pipelines, the submerged infrastructure facilitates the dispersal of the environmental risk. In the case of electric fish barriers, the submerged infrastructure is the complex but precarious matrix designed to *contain* the spread of the risk. Celebrated for cleaning up Chicago at the dawn of the twentieth century, the canal is now a threatening conduit for biological pollution. While the Chicago Diversion influences Great Lakes water levels far to the east, and oil pipelines form a techno-geographical relationship with the west and north, invasive carp trace a route from the southern United States through the metropolis and into Lake Michigan. Chicago is connected to most of the continent by concealed technological matrices that intersect at the Sanitary and Ship Canal.

DISGUISED BY DESIGN

In *Nature's Metropolis*, William Cronon's classic study of nineteenth-century Chicago, the city was the gateway through which railways and shipping lanes funneled and dispersed resources cum commodities.[44] The water, oil, and fish of this chapter have a similar metropolitan-hinterland relationship. For example, maps of contemporary pipelines converging on then diverging from the Chicagoland region evoke nineteenth-century railway maps, but instead of pork and lumber, Chicago is now a hub for moving fossil fuels and fish across the continent, and influencing water levels throughout the Great Lakes.

Cronon also posited a "second nature," a city built from nature transformed and commodified. Since *Nature's Metropolis* a robust envirotech scholarship has approached landscapes and waterscapes as intertwined hybrids of nature and technologies.[45] Thus we arrive at ways to understand a twenty-first century Chicago, where boundaries between the natural world and technology are porous; where water, oil, and fish were built into the infrastructures of canal, pipeline, and barrier; and where the infrastructure itself became part of a multidisciplinary scientific enterprise. In present-day Chicago, an electric fish barrier is a technology whose success and vulnerability hinge on the natural phenomena of electricity moving through water and through fish.

Built environment is a common academic term for urban landscapes and waterscapes. The Chicago Sanitary and Ship Canal is also a system of "built ecologies." In the canal natural elements and processes are the dominant aspects of what is nonetheless a human-constructed system. We suggest a unique category of built ecologies and technological megaprojects that are hidden, or at least hidden in plain sight.[46] This category stands in contrast to engineering projects celebrated for their public view-ability, such as epic hydropower dams proudly displayed by the state and admired by awestruck masses.[47] By design, water diversions, buried pipelines, and fish barriers are *unseen*.

Disguised infrastructures have virtues. They are conveniently out of people's way, ingenious in their subtlety. They may even provide peace of mind, because sometimes we don't *want* to see. Disguising infrastructure, however, hides intentional environmental and economic risk taking. The problem is, those taking risks with hydrology, oil spills, and aquatic ecosystems will not bear the full brunt of disasters. Any economic or ecological fallout will be at once highly localized and widely regionalized. The changed hydrology of a reversed Chicago River reverberated throughout the Great Lakes–St. Lawrence waterway. The Chicago Sanitary and Ship Canal became a two-way ecological corridor across two of North America's largest basins, and between North America and Eurasia. Oil pipelines linked Chicago to the Great Lakes in new ways, and the Great Lakes to North America. These disguised infrastructures are implicated in, necessary for, and reliant on continental and global networks.

PART III

THE NATURE OF WORKING-CLASS CHICAGOANS

By 1880 Chicago was the nation's second most important manufacturing center. Forty years later, half of its 400,000 wage earners toiled in heavy industries, including the production of iron and steel, garments, and agricultural and electrical machinery, as well as commercial printing, railroading, and meat packing. Manufacturing employment peaked at 667,407 workers in 1947; the city reached its top population of 3.6 million three years later. Thereafter manufacturing jobs began to decline, falling to only 147,000 by 2000. The city lost more than a million people even as the regional population climbed.

Before World War I European immigrants fueled population growth. African Americans joined Chicago's workforce in the Great Migration from the American South. Mexican-American migrants and Mexican immigrants became part of the metropolitan story after the war. Many of these varied migrants made the transition from rural to urban life with a move to Chicago. Most of these workers lived in the neighborhoods surrounding factories that polluted their workplaces and homes. When one young immigrant arrived in 1893, she longed to see water "but the summer passed and I did not get to see Lake Michigan. The only part of Chicago that I saw that summer was the block on South Halsted Street where we lived, and a few side streets."[1] Over time new outdoor recreation spaces—parks, municipal playgrounds, swimming pools, as well as forest preserves—beckoned workers, but these spaces often became contested ground between racial and ethnic groups, perhaps most dramatically during the July 1919 race riot that began on a hot day at a South Side beach.

Essays in this section explore the different ways Chicago's workers experienced and understood their environments. Colin Fisher, for example, contends that late nineteenth-century anarchists employed the city's green spaces to mobilize workers while offering a proto-environmentalist critique of capitalism and leaving the legacy of May Day. Brian Mc-Cammack analyzes how Chicago's railroads facilitated African Americans' migration and contributed to the ongoing attenuation of "people from the animals and plants that industry turned into food." Yet, he reveals, because African Americans were already deeply embedded in and marginalized by various industrial networks, their foodways changed little in their new urban environment.

Living in the shadow of steel mills on the Far Southeast Side, *Mexicanos* encountered "a polluted industrial zone" but, as Michael Innis-Jiménez explains, used a strong sense of community to respond to political and environmental obstacles that threatened the quality of their lives. Finally, in exploring a distinctive Chicago landscape—the Cook County Forest Preserves—Natalie Bump Vena demonstrates how the government, rather than capitalism, transformed prairies and wetlands into green spaces that workers and others might use, but did so during the Great Depression by employing relief laborers in physically taxing, undercompensated positions.

As the essays in this section reveal, Chicago's working-class environments were shaped by a series of domestic migrations and international immigration. Space does not permit essays on all of the more than 146 racial and ethnic groups whose histories were documented in *The Encyclopedia of Chicago*.[2] Native Americans offer a striking example of resilience. After 1833, the U.S. Census suggests no Native Americans called the city home, although perhaps their small numbers went undetected. Following the Columbian Exposition, more Native Americans migrated to the Windy City, with more "privileged" members of the community working to reshape older cultural perceptions of Indians.[3] The historian James LaGrand picks up the story after World War II when the federal government's efforts to end U.S. tribal relations in various ways contributed to a migration of tribal members from reservations to cities. As their numbers grew in Chicago from a few hundred to more than twenty thousand, Indian residents followed the pan-Indianism that united them along racial lines while still maintaining smaller cultural clubs grounded in individual tribal identities.[4]

CHAPTER 8

May Day

The Green Vision of Chicago's Gilded Age Anarchists

COLIN FISHER

In anticipation of Arbor Day 1889, Illinois governor Joseph Fifer urged students and teachers to leave their classrooms and find places in the newly thawed earth to plant trees, shrubs, and vines. This effort, he maintained, would beautify homes and public grounds, contributing to the wealth and happiness of the Prairie State. The Illinois State Horticultural Society predicted another positive result: trees would serve as a check on "the hordes of socialists and anarchists that crowd our cities and so seriously threaten the life of civil and religious institutions."[1]

The society's claim that planting trees would suppress urban radicalism came three years after significant labor unrest. On May 1, 1886, the world's first May Day, workers brought Chicago to a standstill with a general strike. Three days later at Haymarket Square, an unknown assailant threw a lethal bomb at an advancing police line. The state responded with arrests, a controversial trial, and the execution of four anarchists. Given this history and ongoing labor agitation, middle- and upper-class Chicagoans eagerly sought tools to quiet class tension. The notion that something as simple as tree planting could neutralize radicals must have offered solace. Arbor Day, it appeared, might serve as an antidote to May Day.

Chicago socialists and anarchists obviously saw things differently. "What in the devil's name has the planting of trees, shrubs and vines to do with socialism and Anarchism?" asked the editors of the *Chicagoer*

Arbeiter Zeitung, a radical German-language newspaper. Perhaps "these honorable gentlemen think that we shall multiply so rapidly that the present forests of the state of Illinois will be insufficient for the necessary amount of gallo[w]s?"[2]

That privileged Chicagoans thought that Arbor Day might defuse May Day should come as no surprise to U.S. historians. We know that Anglo-American reformers and landscape architects such as Frederick Law Olmsted believed that public parks would not only improve the health of urban people but also heal class tensions in the body politic. In an increasingly urban and heterogeneous nation wracked by labor conflict, soothing pastoral landscapes would serve as an alternative to saloons and other places of working-class leisure and introduce the masses to the restorative, contemplative nature recreation enjoyed by privileged Anglo-Americans.[3]

Less understood by historians is that urban workers had their own elaborate environmental vision. In this chapter I explore that vision, focusing on the place of nature in the thought and culture of the group targeted by the Illinois State Horticultural Society: Chicago's Gilded Age radicals. After providing historical context, I explore three dimensions of the anarchists' green vision: the radicals' extensive use of urban green space to forge a working-class and, more specifically, anarchist community; the nascent environmental critique of capitalism embedded in anarchist political thought; and the vital importance Chicago anarchists gave nature in arguably their greatest legacy—May Day.

THE RISE OF ANARCHISM IN GILDED AGE CHICAGO

Labor activism in Gilded Age Chicago makes more sense if we understand its context: the dramatic transformation of capitalism following the Civil War. As the historian William Cronon shows, Chicago grew dramatically because industrialists effectively exploited natural resources in the city's vast, ecologically diverse hinterland. From the tallgrass prairies of the Midwest, the white pine forests of northern Minnesota, Wisconsin, and Michigan, and the short-grass prairies of the Great Plains, Chicago merchants and industrialists imported corn, wheat, timber, and animals. Grain was sorted; white pine trees were cut, recut, and then cut again; pigs and cattle were transformed into packaged meat, lard, brushes, and glue. Everything was given a price. Ultimately, Cronon argues, consumers forgot the ecological context of their commodities.[4]

Consumers forgot something else: the workers who did the considerable labor of transforming animals into meat, grain into bread, and ore into steal. After the Civil War industrialists undercut independent craftsmen, relied increasingly on unskilled labor, adopted labor-saving machinery, subdivided tasks, and came to see employees not as distinct human beings, but as faceless, interchangeable resources purchasable in Chicago's labor marketplace. In the absence of a strong regulatory state, managers instituted backbreaking schedules, taxing employees with the same sort of reckless shortsightedness with which they culled northern forests. Most workers toiled sixty hours a week, with only Sunday, the Christian Sabbath, to recuperate.[5]

Many skilled workers responded to changes on the factory floor by joining trade unions. This approach often excluded women, minorities, the unskilled, and the unemployed but marginally improved working conditions for members. Unions also advanced the most pressing and galvanizing issue for millions of workers: limiting the workday to eight hours. In Chicago unions pressed this demand on May 1, 1867, with a dramatic one-week general strike. Although the workers lost, the shutdown was an opening salvo in a decades-long struggle. In 1884 labor organizations called for yet another general strike, this time on May 1, 1886.[6]

While English-speaking skilled workers who dominated trade unions hoped to reform labor markets, German-speaking socialists offered a more radical solution. The socialist movement grew dramatically during the economic depression of 1873. While unions retrenched, socialists organized unskilled workers, supported informal communal strikes, and led the hungry, homeless, and unemployed on marches to City Hall to demand jobs and relief. Unlike trade union leaders, socialists supported the Railroad Strike of 1877, a bloody nationwide upheaval during which large numbers of unskilled workers rose up against their employers. Following the strike, Chicago socialists continued to organize workers but also entered politics, putting forward candidates who called for nationalizing key industries, such as the railroads.[7]

Chicago anarchists were disillusioned socialists. During the 1877 Railroad Strike, police killed thirty workers, wounded two hundred, and violated the civil liberties of thousands more. At the same time, socialist candidates' success provoked a backlash from mainstream politicians who in the late 1870s fraudulently manipulated the voting system to prevent radicals from taking office. State violence at the behest of the city's business class and electoral fraud radicalized some socialists to the point where they believed peaceful change impossible and revolution inevitable. Remembering the violence necessary to end slavery, anarchists organized labor militias and prepared for war between capitalists and "wage slaves."[8]

PARKS AND THE ANARCHIST COMMUNITY

Despite the fact that Anglo-Americans intended English parks on the prairie to quiet class tension, workers quickly appropriated these landscapes and made them central to working-class and ethnic community building. Eager for relief from crowded, smoky, and foul-smelling neighborhoods and repetitive work in factories, tens of thousands of Chicagoans arrived at the city's many public and commercial parks, especially on Sunday, their one day off. They enjoyed fresh air and sunshine in green settings, ate picnic lunches, drank lager, danced to live music, and participated in ethnic athletic activities, such as German gymnastics and Gaelic hurling.[9]

The production and reproduction of working-class and ethnic culture in Chicago parks ran afoul of conservative Anglo-Americans, who believed that Sunday was best reserved for indoor devotion followed by, perhaps, quiet contemplative sightseeing in God's green creation. As early as 1855 Know-Nothing mayor Levi Boone reacted to immigrant enjoyment of the so-called continental Sunday by enforcing the city's Sunday closing law, which stipulated fines for those who dispensed alcohol or disturbed "the peace and good order of society by labor (works of necessity and charity excepted) or by any amusement on Sunday."[10]

German and Irish immigrants responded to the enforcement of the Sunday closing law at the polls and in dramatic protests, such as the 1855 Lager Beer Riot, Chicago's first major civil disturbance. They also lampooned house-bound Anglo-Americans for their inability to fully enjoy the outdoors. One German writer conceded that Anglo-Americans invented the picnic, but only as an indoor affair where participants abstained from drinking and dancing, sang hymns, quoted Bible verses, and searched for joy in pious activities. Those Americans who ventured into nature "bore themselves as though they were being sentenced to death in criminal court instead of relaxing and having fun out-of-doors." The Germans, by contrast, made the picnic an outdoor excursion where "families, far from the four walls of their lodgings, can get away from the toil, problems and strains of day-to-day life" and "hang up the day-to-day mask of constraint and depression and . . . thoroughly enjoy that untranslatable something: Gemütlichkeit." For the Germans and other immigrants, the city's dour "puritans" seemed incapable of fully enjoying and appreciating nature, especially on Sunday.[11]

Ethnic organizations routinely sponsored outings to Chicago green spaces. A Sunday excursion was often an opportunity to meet expatriates and remember distant villages, regions, and nations. As one German worker explained after entering a Chicago urban forest: "Sunshine, woodland green and woodland shade, the sound of horns! On a Sunday afternoon, what more could a German heart possibly wish for? . . . Ha! The Germans like nothing better than a party under the oaks! The life our forefathers had in the woods still clings to us. . . . I forgot I was participating in a party so far away from the homeland in a foreign country." For him and tens of thousands more, urban and commercial parks were vital spaces to escape the city, remember origins, and imagine ethnicity.[12]

On summer Sundays labor organizations also hosted picnics. Regardless of political orientation, the destination was almost always Ogden's Grove, a commercial park opened in 1865 by German entrepreneur Peter Rinderer on the Near North Side. One police-affiliated writer described the wooded park along the banks of the slow-moving Chicago River as the "picnicking ground for labor societies, and particularly for socialists, communists, and anarchists."[13]

At first glance labor and ethnic picnics looked similar. One saw picnic baskets, steins of lager, ice cream, and jovial workers enjoying music, games, and dancing under the trees. The primary difference occurred when speakers climbed the music pavilion. Nationalists prompted attendees to imagine themselves as German, Irish, Swedish, Danish, or Bohemian by pointing backward to origins in a distant homeland. Labor leaders, by contrast, recognized that nostalgia would never convince diverse laborers to identify as a working class, so they forged unity by calling out exploitation and invoking an improved future. Trade unionists pointed to a reformed labor market where workers would have an eight-hour day and more time for leisure. Further to the left, socialists called on the state to seize control of the means of production. Anarchists went even further and celebrated a utopian future free of capitalism, the state, and organized religion.[14]

An anarchist picnic in Ogden's Grove typically began with a parade through city streets. Signs and banners carried slogans such as "Down with Capitalism, Long Live Communism," "Our Civilization—the Bullet and the Policeman's Club," and "Land Belongs to Society." Floats called for worker unity and women's suffrage while lampooning capitalists, politicians, judges, and the church. Most often participants jettisoned the American flag in favor of red and black flags (symbolic of the blood of the workers and the negation of all national flags). Instead of identifying by nationality, anarchists claimed membership in an international proletariat inclusive of women, racial and ethnic minorities, the unskilled, and the unemployed, as well as skilled male workers. After paying a small entrance fee, radicals gathered under the oak trees along the Chicago River, ate their picnics, and heard fiery speeches by luminaries such as the English teamster Samuel Fielden, who urged those gathered to fight for liberty, fraternity, and equality and bring about a future in which "picnics would not be an exception but the rule."[15]

Privileged Chicagoans hoped that public parks would serve as an alternative to working-class and ethnic amusements and introduce workers to genteel Anglo-American outdoor recreation. Workers and immigrants quickly appropriated pastoral landscapes, though, and turned them into important sites for building immigrant and working-class community. Radicals, in particular, used green spaces not only to enjoy themselves but to document grievances, invoke a utopian future free of hierarchy, and forge community across difference by imagining themselves as a multicultural and international proletariat. Far from neutralizing radicals, trees served as the backdrop for anarchist community building and revolutionary action.

NATURE AND ANARCHIST POLITICAL THOUGHT

Beside using green space to build bridges and forge community, anarchists challenged dominant environmental thinking in another way: they called for a revolution that would end private ownership of land, eliminate the unhealthy urban-industrial environments created by capitalism, and make natural resources available to all. In so doing, they anticipated important components of contemporary green anarchist political thought.

Chicago anarchists insisted that there was nothing natural about the ownership of the Earth and its resources. They noted that before the rise of governments, land—like water and light—was free. This world existed in living memory. A writer for the English-language anarchist weekly, The Alarm, observed that in the recent past Native Americans shared the land in common, existing in a state of liberty and anarchy until barbarous white men carrying surveying chains drove them from the soil. "The Indian has been 'civilized' out of existence and exterminated from the continent by the demon of 'personal property.'" Other anarchists recalled common land ownership in Europe. Writing from death row, Michael Schwab, a German-born bookbinder who edited the Chicagoer Arbeiter Zeitung, remembered a lost world of green meadows, shining brooks, green forests of oak and beech, hills planted with vines, and gardens and grain fields that Bavarians had once shared in common, and noted that "the private property system is—to speak historically—but of yesterday."[16]

According to Chicago anarchists, nature, once owned by everyone and no one, had been seized by the few. In England landowners and their obedient servant, the state, employed murder, fraud, and robbery to secure and enclose common lands. This process was repeated in colonies such as Ireland. Despite its republican traditions, vast acreage, and abundant resources, the threat of serfdom even knocked at the door of the United States. Not only had Indians lost their ancestral homelands, but now family farmers were squeezed by bankers and railroad companies, which amassed vast landholdings, some of which they took out of production to drive up food costs. The American anarchist and bookbinder Dyer Lum explained, "Nature has placed us upon the earth, but we are denied its use for productive purposes. . . . Nature's gift has been monopolized." Lum asked readers, "By what title deed has the landlord disinherited you from nature's estate? . . . Have you surrendered your natural claim to a footing on earth?"[17]

In addition to critiquing private property, anarchists addressed the unhealthy effects of the urban-industrial environments created by capitalism. Some identify middle-class Progressive reformers as the first urban environmental justice activists. But at least a half-decade prior to the creation of Jane Addams's Hull House settlement on Chicago's Near West Side, Gilded Age radicals called attention to toxic workplaces, polluted neighborhoods, unsafe housing, and lack of urban green space. While they shared Progressives' environmental concerns, radicals offered a different solution: not mere reform but revolution. Given widespread exploitation, state violence, and electoral fraud, the entire system—capitalism, the state, and organized religion—had to go.

Anarchists chronicled how former cultivators, forced off the land, migrated to industrial cities, where they worked long hours indoors for little pay. Under such unnatural conditions, their health suffered. Schwab noted, "Instead of the melodious tinkling of the bells of his cows," the former independent freeholder "now hears the task bell of the factory, that calls him to work. Instead of the splashing or roaring of the creeks and cataracts, the whispering air of the dark-green forests, he is treated to the clattering of noisy busy machines." Working long hours for some "money-making spinner king," the children of farmers lose their big chests, strong limbs, and rosy cheeks and become a "race of exploited, wretched wage slaves." The German anarchist box maker Johann Most explained that in the name of profit, the industrial worker is "rent apart by machinery, poisoned by chemicals, or slowly suffocated by dirt and disease."[18]

Anarchists noted that meager wages and high rents charged by exploitative landlords meant that urban workers' health continued to deteriorate at home following the end of the workday. The German immigrants in particular were shocked that in a country with so much arable land, urban workers subsisted on partially rotten vegetables scavenged from waste barrels and the occasional scrap of offal from a charitable butcher. Amid opulence, workers paid high rents for cold, unventilated, overcrowded, dark, and fire-prone rooms. "Is it a wonder," asked Schwab, "that diseases of all kinds kill men, women, and children in such places by wholesale, especially children?"[19]

Even worse off were the homeless. Lucy Parsons, an anarchist woman of color and labor organizer, addressed Chicago's thirty-five thousand "tramps" directly, telling them that they had spent years "harnessed to a machine" working ten or more hours a day with only Sunday to escape "unremitting toil." With little pay, they had to scrimp, barely staying ahead of the "wolves of want." And now, they had been let go due to "over production." Winter was coming and they would soon walk icy streets with worn-out shoes, threadbare clothes, and empty bellies. Don't despair, she counseled. *"Learn the use of explosives!"*[20]

While Parsons and other revolutionaries condemned the urban-industrial environments created by capitalism and sometimes romanticized a rural world where land was held in common, they were not antimodernists. Unlike Henry David Thoreau, they did not reject the city for self-sufficient life close to nature. Unlike contemporary anarcho-primitivists such as John Zerzan, they did not call for a return to a hunter-gatherer past where hierarchy was supposedly unknown and all lived lightly off the land. On the contrary, Chicago radicals were forward-looking utopians who embraced cutting-edge steam technology and the liberatory potential of the cosmopolitan city.[21]

Anarchists did not critique technology per se but merely the fact that machines were monopolized by the few and thus unavailable to the many. As the German anarchist, upholsterer, and writer August Spies testified in his Haymarket trial: "Socialism teaches that the machines, the means of transportation and communication are the result of the combined efforts of society, past and present, and that they are therefore the indivisible property of society, just the same as the soil and the mines and all natural gifts should be." In the anarchists' imagined future, industrial workers were no longer "pieces of organic machinery" but unalienated creators who used machines (including factories) for the betterment of humankind. Future farmers, meanwhile, did not retreat to "wasteful practices," but employed science, fertilizers, and state-of-the-art agricultural implements. Science and technology would reduce drudgery, save labor, produce food and material wealth, and create abundant leisure.[22]

The anarchists also embraced the city. For revolutionaries the most instructive examples of real-world anarchism were found not in the rural past but in urban areas. Thousands of Chicago radicals annually commemorated the Paris Commune of 1871, when Parisians seized their city, flew the red flag, created a progressive municipal government, and transformed businesses into cooperatives. For the American anarchist woodworker William Holmes, the Commune was a revolt against the "iniquitous political, industrial, and social systems which then prevailed, and under which we still suffer. It was a complete overthrow, for the time being, of all existing institutions, and an attempt to found a social and industrial republic based upon the inherent rights of man."[23]

The anarchists not only appreciated urban amenities such as good schools, museums, theaters, parks, and cosmopolitan community, they also saw great promise in urban trade unions. For the Chicago group, hierarchical and sometimes exclusionary and conservative industrial unions were in fact the seeds of future communes. As Albert Parsons explained, a union, whether it realized it or not, was "an autonomous commune in the process of incubation." After the revolution, urban communes would exchange goods and services with other communes, including agricultural cooperatives comprised of farmers. Federations of these economic entities would take the place of the state.[24]

Anarchists' embrace of technology and urban life did not preclude a sophisticated environmental ethic. Despite lacking the language of ecology, the French anarchist geographer Élisée Reclus (1830–1905) viewed humanity as "nature becoming self-conscious" and understood the relationship between humans and the environment as dynamic and interactive. Under capitalism and the state, humans had commodified, dominated, and destroyed the natural world. After the revolution and the elimination of hierarchies based on class, race, and gender, free societies carefully calibrated to their regional environmental contexts would emerge. But far from fleeing into the past, anarchists would use science and state-of-the-art agricultural and communication technologies to create deeply interconnected and radically egalitarian bio-regionalist towns and cities.[25]

Chicago's Gilded Age anarchists espoused some elements of utopian green anarchism. They knew Reclus's work well. The *Arbeiter-Zeitung* and *The Alarm* sold his books, *The Alarm* reproduced his essays, and Chicago writers invoked him. Albert and Lucy Parsons included Reclus's 1884 tract, "An Anarchist on Anarchy," in their collection, *Anarchism: Its Philosophy and Scientific Basis as Defined by Some of Its Apostles* (1887). The Chicago anarchists, given their grounding in evolutionary theory, would have surely agreed with Reclus that humans are nature becoming self-conscious. Some even suggested that anarchists would be far better stewards of the Earth's resources. An unknown author in *The Alarm* argued that since under anarchy everyone would benefit from nature's abundance, "all would dread alike a scarcity," and "all would be induced to help avoid it."[26]

Johann Most even anticipated anarchist green cities where diseases disappeared because "bad lodging, murderous workshops, impure food and drink, over-exhaustion have become things unknown." City residents would enjoy healthy homes surrounded by gardens and parks, and children would get instruction in mind and body in the open air. Parks would sometimes serve as sites of political meetings, where women and men arbitrated issues. And rural areas would receive the city's considerable amenities. Isolated homesteads would disappear as farms consolidated into agricultural cooperatives reliant on state-of-the-art machinery. Ultimately, "the former contrast between city and country disappears."[27]

While Gilded Age radicals anticipated some aspects of modernist green anarchism espoused by contemporary figures such as Murray Bookchin, their environmentalism had significant limits. They largely did not envision revolutionaries as artful and careful gardeners of the earth, as did Reclus. They did not link ecological destruction to social inequality, as did Bookchin. Rather, Chicago anarchists typically celebrated humans' progressive domination of nature. The problem for them was not the control and exploitation of nature but rather that the wealthy monopolized land as well as the steam and chemical tools needed to extract the Earth's seemingly infinite wealth. In their postcapitalist future, all would collectively share an endless harvest. As Lum pointedly wrote, under anarchism, "nature alone remained to be exploited."[28]

MAY DAY: SOCIAL JUSTICE
AND THE EARTH'S REAWAKENING

The Chicago anarchists' green vision can be seen in their use of green space and in their political theory, but it is also evident in their holiday: May Day. The day of labor protest originated with an 1884 demand by the predecessor of the American Federation of Labor (AFL) that workers have an eight-hour day by May 1, 1886. Having long struggled for eight hours, Chicago workers enthusiastically supported the drive. Anarchists also joined the eight-hour movement, which they saw as an important first step in the ultimate overthrow of capitalism and the state. On Saturday, May 1, 1886, nearly one hundred thousand Chicago workers from various national backgrounds, industries, and skill levels went on strike. The shutdown effectively paralyzed the rapidly industrializing city. As one reporter noted, "No smoke curled up from the tall chimneys of the factories and mills, and things had assumed a Sabbath-like appearance." While soldiers, Pinkerton private police, and deputized citizens stationed on rooftops trained their Winchester rifles on the streets below and National Guardsmen mustered in nearby armories, tens of thousands of working-class people paraded through Chicago and attended numerous open-air meetings.[29]

The association of May Day with the rebirth of nature is sometimes attributed to the Italian anarchist Pietro Gori, who in 1892 wrote "Inno del primmo magio," a popular song describing workers waiting for the "Green May of humankind." But the association of May Day with nature has a much older history. In ancient Rome plebeians honored the goddess Flora in a weeklong May fertility celebration paid for by fines collected from those who encroached on common lands. May festivals continued throughout medieval and early modern Europe. Carnivalesque gatherings emphasized fertility and sexuality, with dancing around May poles in village greens, crowning of May queens and kings, and gifting of flowers and greenery. Sometimes, Europeans dressed as foliage-covered nature deities called "green men," and in England plays celebrated Robin Hood, who took from the rich and gave to the poor. The Chicago anarchists kept some of these pagan traditions alive. During the 1880s they celebrated nature's reawakening at *Maifest*, a German version of Mayday. In anticipation of the May 1, 1886, labor protest, radicals invoked the rebirth of nature as well as Christ's resurrection. Just days before the general strike, Schwab disrupted a quiet Easter Sunday by speaking to a large crowd at Lake Michigan. He urged workers to stay true to their unions, protect themselves from murderous police, and fight for a shorter day. "The workingmen of Chicago today celebrate their resurrection," Schwab told the crowd along Chicago's great inland sea. "They are resurrected from their laziness, from their indifference in which they have remained for so long. . . . After the 1st of May eight hours, and not a minute more."[30]

On the day of the general strike, radicals and more conservative trade unionists invoked the promise of contact with nature. Tens of thousands of Chicagoans sung "Eight Hours" as they marched through downtown streets:

We mean to make things over, we are tired of toil for naught,

With but bare enough to live upon, and never an hour for thought;
We want to feel the sunshine, and we want to smell the flowers,
We are sure that God has will'd it, and we mean to have eight hours.
We're summoning our forces from the shipyard, shop and mill,

Eight hours for work, eight hours for rest, eight hours for what we will!
Eight hours for work, eight hours for rest, eight hours for what we will!

The song, in its entirety, described nineteenth-century industrial capitalism as a profoundly unnatural, life-destroying, and unholy force that
reduced humans to physically broken slaves who envied "the beasts
that graze the hillside . . . and the birds that wander free." The song also
threatened that if labor was ignored, the movement could assume the
form of "the wild tornado." The song highlighted one important promise
of the campaign: a chance to escape unending hours of work in sunless
factories, mills, and mines, get outdoors, "smell the flowers," and "feel
the sunshine."[31]

On May 4, 1886, just days after the first May Day, the always un-
easy alliance between anarchists and trade unionists disintegrated. That
evening anarchists gathered at Haymarket Square to peacefully protest
police violence. Following the last speech, police moved aggressively to
disperse the crowd. As officers advanced, someone threw a bomb which
exploded among the police, killing one. Confused policemen responded
with their service revolvers, hitting workers and fellow officers. Eight police officers and four protesters died; dozens fell wounded.[32]

The Haymarket bombing sparked the nation's first red scare. Chicago
police arrested hundreds without cause, beat suspects, searched homes
and offices without warrants, opened mail, and shut newspapers. The city
banned the red flag, parades, and open-air meetings in parks and along
Lake Michigan. Despite failing to find the actual bomber and without
direct evidence, the state, in what is now seen as a great miscarriage of
justice, sentenced seven anarchists (Albert Parsons, Spies, Adolph Fischer, George Engel, Louis Lingg, Samuel Fielden, and Schwab) to die by
hanging.[33]

Despite police suppression, Chicago radicals quietly marked May
Day the following year. The Chicago proletarian weekly *Der Verbote* declared May 1, 1887, the first birthday of the labor movement and compared the day with calendar-changing events such as the beginning of
Judaism, Christianity, and Islam and the creation of the French republic.
A year later city anarchists quietly commemorated May 1 at Greif's Hall,
where they organized a picnic in Sheffield, Indiana and made plans to
raise a statue to the martyrs.[34]

In 1888 the AFL tried to reappropriate May Day by organizing another eight-hour general strike, this time for May 1, 1890. Although the
AFL ultimately backed away from the strike, unions launched an unprecedented wave of shutdowns, many of which resulted in shorter hours
and higher pay. In Chicago thirty thousand workers (trade unionists
and radicals alike) marched through the streets. One placard declared
"May 1. This Day Shall Stand in Memory." The procession ended at Lake
Michigan, where workers heard speeches from city labor leaders and

committed a coffin labeled "ten hours" to the cold water. The strike also spread overseas. In solidarity with American workers, the Marxist International Socialist Conference in Paris, which formed the Second International, called on workers around the world to observe the May 1, 1890, protest. Despite widespread opposition and state repression, hundreds of thousands of workers throughout Europe and in Australia, Cuba, Peru, and Chile turned out. The internationalization of May Day and its continued association with radical politics ultimately proved too much for the AFL. The nativist and increasingly conservative labor organization abandoned May Day in favor of Labor Day, which President Grover Cleveland made a national holiday in 1894 following the bloody Pullman strike on Chicago's South Side.[35]

As trade unionists spurned May Day in favor of Labor Day, Chicago radicals, emboldened by events overseas, filled the void, boldly resurrecting the parades and picnics of the 1880s. Thousands of German, Jewish, Czech, and Scandinavian immigrants and native-born Americans marched through the streets with American and red flags to the strains of the revolutionary French anthem, "La Marseillaise." The conservative *Chicago Tribune* noted disapprovingly that this was quite unlike Labor Day, where one heard "The Star Spangled Banner" and saw only American flags. Following May Day parades, marchers listened to speeches at the lakefront and declared solidarity with international workers. Later they regrouped under the willow trees of Kuhn's Park for a May Day picnic, funds from which went to the *Chicagoer Arbeiter Zeitung* and the anarchist memorial fund.[36]

Chicago's establishment found the resurgence of public radical culture intolerable. In anticipation of May Day 1892, the city declared the red flag illegal, banned anarchist speeches, seized the subscription list of the anarchist German American newspaper *Die Freiheit*, and threatened to arrest subscribers if there was any trouble. Nonetheless, six thousand Chicagoans marched through city streets. Hundreds of police officers responded by confiscating flags and harassing marchers. At one parade route bottleneck, officers pivoted into the street, blocked the procession, and attacked the front of the line. At Lake Michigan, traditionally the site of post-parade speeches, the city stationed a team of translators, a legal expert, and 260 officers, with 600 more mustered nearby. Meanwhile gunners stood at the ready behind two rapid-fire Gatling guns. Detectives also patrolled Kuhn's Park, site of the May Day picnic. Asked if she would give a speech at the picnic, the celebrated anarchist Lucy Parsons responded: "We are not here to be murdered or to commit suicide."[37]

This resurgence of police oppression once again drove anarchists indoors and underground. Except for 1921 Chicago would not see another major May Day parade and picnic until the Great Depression. Nonetheless, Chicago radicals continued to associate radical May Day protests with nature's rebirth, linking a modern secular labor holiday with ancient pagan nature worship. As the Chicago socialist William F. Kruse reminded readers of the *Young Socialist Magazine* in 1918, May Day predated the eight-hour movement. Appealing to a nationally, ethnically, and racially diverse working-class population, he explained that "all countries, all religions, all peoples" celebrate May Day since "Nature Worship is the very basis of all religions." He called on labor and socialism to "join hands in great demonstrations, in holy devotion to the day of peace, and plenty, and love, and justice that is to be." [38]

ENVIRONMENTAL JUSTICE DAY?

Illinois Horticultural Society members and Anglo-American park builders and reformers such as Olmsted suggested that Gilded Age immigrant workers had a stunted appreciation for nature. If only native and foreign-born workers would leave their saloons, enter public park landscapes, and enjoy the regenerative powers of nature, they would not only become healthier and more industrious but also assimilate into America's dominant culture. Arbor Day trees would rid the city of anarchists and socialists. Marginalized Chicagoans, though, needed no lectures on nature. Much to the chagrin of Anglo-Americans, workers and immigrants, eager to escape a seemingly unnatural industrial city, appropriated public parks and commercial groves and made them into vital sites for forging working-class community. Anarchists in particular used green spaces such as Ogden's Grove to create a vibrant radical Gilded Age counterculture. Nature also figured prominently in Gilded Age anarchist political thought. Anticipating some aspects of contemporary green anarchism, radicals lambasted the unhealthy urban-industrial environments created by capitalism and called for anarchist garden cities. They also envisioned a world where private property was unknown and all shared the Earth's resources. Chicago radicals also created May Day, a holiday that explicitly linked social justice to nature's rebirth. Far from neutralizing worker unrest, nature was central to anarchist thought and culture.

The vast majority of environmentalists today are unaware of May Day's long association with nature. Typically they defend nature on Earth Day, a holiday that also put ancient spring pagan rites into the service of political activism. That said, growing corporate co-optation and failure to meaningfully address social inequality has prompted some environmentalists—especially those fighting for environmental, food, or climate justice in the Anthropocene (or perhaps better, the Capitalocene, the geologic age not of man but capitalism)—to spurn April 22 in favor of May 1. On May Day environmental activists are increasingly marching alongside workers, women, racialized minorities, immigrants (documented or not), the unemployed, victims of state tyranny, members of LGBTQ+ communities, and others protesting nonconsensual hierarchies. In protests around the world, they are calling attention to how social and ecological problems are intertwined and demanding environmental and climate justice. Such contemporary May Day environmental activism is in many ways novel. But we should not forget that Chicago's Gilded Age radicals bequeathed to us a usable past: a radically democratic, antihierarchical, transnational, capacious holiday that fused social justice activism with a celebration of the Earth.[39]

CHAPTER 9

Black Migrant Foodways in the "Hog Butcher for the World"

BRIAN MCCAMMACK

Many African American migrants who streamed into Chicago between the world wars traveled on the Illinois Central Railroad from southern states like Mississippi, Alabama, Louisiana, and Tennessee, arriving on the city's near South Side. Even shorter journeys took several hours, and many migrants packed food along with all their worldly possessions. Perhaps anxious about how they would feed themselves and their families in a strange land and, more broadly, sensing just how radically altered their connections to nature might be upon arriving in the nation's second largest city, many brought more food than necessary to sustain them for their train ride alone. As Isabel Wilkerson notes in *The Warmth of Other Suns*, migrants carried with them "all kinds of things, live chickens and rabbits, a whole side of a pig . . . jars of fig preserves, pole beans, snap peas, and peaches, whole hams, whatever the folks back home were growing on the farm," clinging to comforting elements of the South they left behind, many for good.[1] Nearly a quarter-million black Chicagoans undertook similar journeys in the interwar era, when the city's black population grew more than sixfold to almost 280,000, more than quadrupling African Americans' share of the population to 8.2 percent.[2]

The very same train lines that brought those thousands of black migrants to Chicago continued to connect them to southern environments, even years after they settled into routines of eating at South Side restaurants and, more commonly, buying food from grocers, peddlers, and butchers for home cooking.[3] Timuel Black, who migrated to Chicago as a small child in 1919, recalled that his uncle who remained in Alabama was convinced that "his poor brother was starving to death" in Chicago, in large part because he "couldn't imagine not having a garden, not having in the back a chicken roost and all that."[4] So Black's uncle sent live chickens by freight train to his brother's family on the South Side, and when he visited them in the city, "he still couldn't believe that we didn't need something to eat. So he brought some chickens with him and raw eggs."[5] Jimmy Ellis, whose parents migrated to Chicago in the 1920s, similarly recalled that his grandfather, a farmer who stayed behind in Alabama, sent "pecans and peanuts and molasses."[6] Far from uncommon, sending southern foods to relatives in Chicago helped maintain cultural and familial ties. But they would not have been nearly enough to sustain a family. What did Black, Ellis, and thousands of other migrants eat in this bustling city far away from the home they knew? How different were those foods from what they had consumed in the South, and what does that tell us about their connections to nature during the Great Migration?

To an extent that may be surprising given how dramatically other aspects of migrants' lives changed—and was no doubt surprising to the migrants who traveled with southern foods as well as the relatives who continued to send food north—migrant foodways actually remained largely unchanged in Chicago. This was true for three reasons, all of which speak to the extent migrants had already been integrated into and exploited by regional, national, and even global industrial networks that produced and transported foodstuffs, collapsing distinctions between very distinct environments. First and most important, whether they lived in the North or the South, low wages combined with high food prices to restrict what working-class African Americans could buy. Women, who generally controlled household budgets, spent the vast majority of their family's income on necessities, and stretching tight budgets meant a heavy reliance on cheaper food staples.[7] Second, modern transportation networks made familiar, affordable foods available to migrants in Chicago. The same rail lines that brought relatives' shipments of southern foods to Black's and Ellis's families transported food to urban markets and groceries on a much grander scale. As Wilkerson notes, migrants "could be assured of finding the same southern peasant food, the same turnip greens, ham hocks, corn bread in Chicago as in Mississippi."[8] In large part that familiarity stemmed from those distribution networks operating both ways: for decades black southerners had consumed foodstuffs from across the country, including meat from Chicago's packinghouses. With many of the same foods available in the North and South, recipes and cooking methods that southern black women developed over generations translated easily to Chicago's kitchens. Third, as those well-developed transportation networks suggest, diets remained fairly constant because women responsible for balancing family budgets and cooking meals purchased most food on the market. In the urban North limited space and cultural taboos worked against keeping small livestock or cultivating a garden, and the women who likely would have tended to such sites of domestic production often took on work in the city's laundries, kitchens, and factories instead. Similarly, while livestock and small gardens were more common in the urban South, they generally did not produce enough to sustain families. In the rural South, meanwhile, the main impediment was a tenant farming system that effectively coerced entire families into producing a cash crop that left little time or autonomy for cultivating substantial food crops or raising livestock.

Like most working-class families throughout the interwar era, black Chicagoans devoted a substantial portion of their income to food purchases. The poorer you were, the higher proportion of your income you needed to spend on food, and most black migrants were among the poorest of the poor.[9] Observing a trend that held throughout the interwar era, the Chicago Commission on Race Relations (CCRR) found in the early 1920s that food was the single largest monthly expenditure for virtually every working-class black family in Chicago, easily outpacing rent and other necessities, and sometimes exceeding half of the family's income.[10] So while black migrants tended to enjoy higher wages in Chicago than in the South, higher prices meant that they still devoted substantial income to food.[11] Many black southerners had hoped to escape these exploitative economies that pushed some families to the brink of starvation—one prospective migrant in rural Mississippi wrote that, "Wages is so low and grocery is so high untill [sic] all I can do is to live," and another in rural Louisiana said similar conditions there meant that "thousands of us can bearly [sic] keep body and soul together."[12] But for many migrants, moving to Chicago did little to resolve that fundamental economic tension. One migrant mother believed that living in Chicago was actually harder than sharecropping in rural Georgia, observing, "In the South you could rest occasionally, but here, where food is so high and one must pay cash, it is hard to come out even."[13] Another black mother faced similar challenges, telling a Chicago interviewer in 1929, "My children would eat between meals if they had something to eat, but I cannot afford any more meals. I have 3 meals a day and I divide what there is between them."[14]

All consumers faced high food prices, but black Chicagoans' food budgets were further constricted by race discrimination that depressed wages, confined families to segregated neighborhoods where landlords

Fig 9.1. Russell Lee, *Kitchen in Crowded Negro Apartment, Chicago, Illinois*, April 1941. Women, young and old alike, were largely responsible for meal preparation in cramped tenement kitchens. *Source*: Library of Congress, Prints & Photographs Division, FSA/OWI Collection, LC-USF34–038642-D.

charged extortionate rents, and forced wives and mothers to work outside the home to supplement their husbands' meager incomes.[15] To cope, black migrants often took in lodgers—whether extended family members, friends, or strangers—who received a room and often home-cooked meals in exchange for paying a portion of the monthly rent.[16] The couple who had been Georgia sharecroppers, for example, lived with their grown daughter and her husband as well as a nephew; all of them contributed to rent and took meals together.[17] Home cooking was also a prominent feature of "rent parties," in which renters opened their homes to all comers for a night, charging a small admission fee for a night of food and entertainment—all in order to make rent that month. *Native Son* author Richard Wright recalled that at rent parties in Chicago he "drank home-brewed beer, ate spaghetti and chitterlings, laughed and talked with black, southern-born girls who worked as domestic servants in white middle-class homes."[18]

Even as their husbands' low wages meant that they increasingly worked outside the home, black women largely adhered to long-standing gender norms and cooked virtually all the food served in their homes, rent parties included. Consistent with Wright's recollections, nearly half of the black mothers surveyed in 1927 worked outside the home, most commonly taking on domestic labor like cooking, cleaning, or sewing for white families.[19] In some cases, women worked explicitly to help feed their families. One migrant mother, the matriarch of a farming family that came from rural Alabama, took up "housekeeping to reduce the food bill," even as her husband and son worked in the stockyards and her daughter worked in a laundry.[20] Women's wage labor outside the home meant that other family members sometimes needed to pick up the slack at mealtime. The CCRR, for instance, found that "when the mother is away all day the food is hastily prepared, which usually means that it is fried. The girl who gets home from school before her mother has finished her day's work usually starts the dinner, or brings something from the delicatessen."[21] Perhaps that was why the *Chicago Defender*, the nation's premier African American newspaper, published a regular column aimed specifically at girls that featured easy-to-make recipes for dishes like spaghetti, chili mac, and grilled cheese.[22] The *Defender* may have catered to that working-class reality, but it also fostered middle-class uplift: the same year the CCRR issued its findings, a column in the "Junior" page for boys and girls stated its aim was "to instill into the minds of our girls to become good cooks first of all. Be better than the average. Let your cooking be such that everyone in your home and in your vicinity can brag about it."[23] By the same token, the *Defender* regularly featured recipes on the Womans Page in the early 1920s, encouraging contributions from "housewives who have dainty or sensible recipes they use [which could] give other women the benefit of discoveries made in your own home."[24] The poverty of the Great Depression only further challenged mothers trying to provide for families, however. At Chicago's first Bud Billiken Day picnic (an annual event created by the *Defender* in 1930 as the Depression deepened), the newspaper wrote that, "Mothers will be asked to bring along an old fashioned basket filled with goodies . . . [but] Because so many of the children will be without the food, Bud is planning to have 'eats' for them," including free hot dogs and ice cream.[25] Free food for children remained a staple of the celebration: in 1935, for example, the *Defender* noted, "The park was filled with picnickers—large baskets were uncovered and huge mounds of food lifted out. There was no selfishness shown and the spirit was beautiful. The few unfortunate children not having parents or guardians able to fix them a lunch, were gathered to the breast of some other good mother who could fully appreciate the little child's feelings, and each was fed until satisfied."[26]

The *Defender*'s annual picnic symbolized the way black Chicagoans came together to support one another amid adversity, but the longer history of reformers urging the working classes to rectify dietary insufficiencies reveals the sometimes tense intraracial class politics underlying

Fig 9.2. Russell Lee, *Corner of Kitchen of Family on Relief, Chicago, Illinois*, April 1941. A mother pours milk for her infant daughter in the kitchen. Milk was among the foods many surveys identified as lacking in African American diets. *Source*: Library of Congress, Prints & Photographs Division, FSA/OWI Collection, LC-USF34–038780-D.

migrant foodways. A doctor who wrote a regular health column in the *Defender*, for instance, exhorted his readers to "buy more fruits and less meats. We should use more fruits in our diet than we do, especially during [the summer], because they contain great nutritive properties."[27] Several years later, upon observing a woman who regularly bought pork neck bones at the butcher, the columnist similarly reflected that, "Our diet should be a well-balanced one. It is not a good thing to eat at all times neck bones or pork chops, potatoes, bread and gravy."[28] These were the most affordable foods, however, particularly in times of economic distress. Reformers rightly believed in the health benefits of a diversified diet, but they also highlighted the marked class divide between the few black Chicagoans who could afford fresh fruits and vegetables and the many who could not. Urging migrants to abandon a diet heavy in meats, fats, and starches was rooted in a desire to distance black Chicagoans from foodways redolent of the South that were often deemed less respectable and even embarrassing by the black middle and upper classes.[29] At other times, however, reformers acknowledged that a balanced diet was an ideal perhaps unattainable—or even inadvisable—for manual laborers who needed to consume a lot of calories on a tight budget. After the influx of migrants began in 1917, the *Defender* columnist maintained that, "People should learn to use more vegetables, cereals, corn meal, cane sugar foods." But he qualified that assertion by acknowledging that, "The fellow digging in the ditch, mauling rails, loading and unloading freight—or, in other words, the fellow doing hard, long, laborious work, requires a more stimulating and proteid or meat diet than the individual doing clerical, indoor or sedentary work. People should buy that kind of food that is within their means, and that has the most food value at the lowest rates."[30]

While restaurants, picnics, rent parties, and other community gatherings reveal that meals could be pleasurable leisure experiences for black migrants, constrained economic circumstances meant that more often than not food was a matter of basic sustenance.[31] Indeed, despite the *Defender*'s uplift aspirations and migrants' substantial expenditures on food, most researchers considered black Chicagoans' diets woefully inadequate.[32] Surveys of black migrants' nutrition in Chicago invariably relied on the self-reporting of women—mostly mothers—because they were the ones almost always primarily responsible for purchasing and preparing food.[33] One such late 1920s survey of African American families in Chicago—nearly 90 percent of whom had been born in the South—found that fewer than one-fifth of those studied "had adequate food."[34] It was a scenario all too familiar for many migrants who found themselves locked into a different iteration of the same discriminatory and exploitative systems in the South that they had tried to escape. In his study of the rural South in the 1920s, for example, the historian Carter Woodson found that African American sharecroppers "live[d] on such food [that was] supplied to them by the plantation commissary and their inadequate income together with the terrorism in vogue makes it impossible for the majority of such persons to improve their daily fare."[35] Long accustomed to meager meals in the South, most migrants transplanted their food culture to urban metropolises like Chicago whether they wanted to or not: low wages inevitably led to diets heavy in low-cost staples like salt pork, potatoes, greens, and cabbage.[36] The wife in the migrant family of sharecroppers from Georgia, for example, prepared meals consisting mainly of "greens, potatoes, and cabbage" along with homemade bread, though families ate milk, eggs, and cereal occasionally and "meat [was] eaten about four times a week."[37]

Meat undoubtedly became more of a luxury during the lean years of the Great Depression when thousands went hungry in Chicago, but in general meat—and especially pork—was the most consistent element of migrants' diets, in large part because, as Carl Sandburg famously wrote, Chicago was the "Hog Butcher for the World."[38] As early as the 1890s, many southern farmers quit producing their own meat because buying it was cheaper, more convenient, and often dictated by labor demands that left them little time or money to raise enough livestock to support a family. Booker T. Washington lamented in 1903, more than a decade before the Great Migration began in earnest, that "in a country where pigs, chickens, ducks, geese, berries, peaches, plums, vegetables, nuts, and other wholesome foods could be produced with little effort," black southerners were "eating salt pork from Chicago and canned chicken and tomatoes sent from Omaha."[39] Washington was right about the southern environment's capacity to produce abundant foodstuffs, but in promoting his brand of race uplift he greatly minimized the constrictions imposed on southern black workers. Whether they worked in Mississippi Delta cotton fields or Birmingham steel mills, black southerners ate meat slaughtered and packed in Chicago not only because the city's packinghouses had extended their reach across the nation, but also because discriminatory and exploitative labor conditions demanded black workers find the most affordable sustenance. Ironically, black southerners' reliance on cheap salt pork from the South Side's stockyards may have helped frame Chicago as the most promising migration destination. The CCRR argued that "Chicago was the logical destination of Negroes from Mississippi, Arkansas, Alabama, Louisiana, and Texas" in part because "the city had become known in these sections through . . . the Stock Yards, and the packing-plants with their numerous storage houses scattered in various towns and cities of the South"; the very same rail lines that shipped meat from the city's South Side brought black migrants there during the interwar years.[40] Low wages continued to restrict African Americans' access to high-quality meat in Chicago, however, having a negative impact on nutrition. As one 1929 study of black Chicagoans concluded, the consumption of "meat increases with the income as does adequacy of the diet."[41]

Nevertheless, black Chicagoans may have consumed more fresh meat than they had down South. Having migrated to the nation's meat processing hub, many gained easier access to fresh meat still subject to spoilage on long journeys despite advances in refrigeration.[42] The high cost of other foods also pushed migrants toward consuming fresh meat: the Georgia sharecropping family, for instance, ate more fresh meat in Chicago "because of the lack of garden space and the high cost of green vegetables."[43] While the stockyards focused their operations on slaughtering and processing pork and beef, fresh-killed chicken was another relatively affordable option. It was not until after World War II that chickens arrived at many Chicago markets as cuts of meat rather than as live animals; as Charles Davis recalled, chicken coops were a "common sight in front of markets" all around the city during the Depression.[44] As a young man, Black worked at a market where he learned how to take a live chicken from one of these coops, wring its neck, defeather it, and break it down into cuts of breast, leg, and back meat.[45] Another South Side resident similarly recalled that during the Depression her mother often sent her to a chicken market where, "The man would put that chicken down, and that chicken would flutter all around and scare you half to death!"[46]

While consuming fresh meat may have become more common in Chicago, families forced to rein in food costs often went without fresh produce, North and South alike.[47] In the Mississippi Delta, fruits were "unaccustomed luxuries" bought at the store with the money gained from a cotton harvest, and one study found that children longed for bananas and apples their parents oftentimes could not afford.[48] Fresh fruit may have been somewhat more affordable in northern cities, but it was often still too expensive for parents to buy for their children.[49] As one Chicago mother told an interviewer, "They all like fruit, but it costs so much to live and fruit is high."[50] When a 1930 study of kindergarten students in one of the poorer sections of Chicago's Black Belt asked what they wished for the most, many children responded that they yearned for toys or candy, but wishes for foodstuffs were also quite common—especially for fruits like oranges, peaches, bananas, grapes, and apples. A boy born in Memphis, for example, whose father was a porter and mother a domestic laborer, wished for milk, cabbage, and an orange; a girl born in Arkansas whose mother was a laundress wanted oranges and peaches.[51] A scene in Lorraine Hansberry's play *A Raisin in the Sun* (based on Hansberry's experiences growing up on the South Side in the 1930s) similarly indicates just how rare fruit was for many black Chicagoans. When Ruth Younger tells her son that they simply do not have the fifty cents he needs for school that day, to Ruth's exasperation her husband almost immediately contradicts her. Ruth's husband gives the boy a whole dollar, telling him to use the extra fifty cents to, "Buy yourself some fruit today—or take a taxicab to school or something!"—both options equally extravagant.[52] Indeed, one study of the Black Belt found that "families whose expenditures were far below standard, for the most part, selected the cheaper cuts of meat, plain breads, and the cheaper green vegetables but omitted fruit," and another noted that, across the board, "The lack of fruit in the diet was outstanding."[53]

Vegetables beyond relatively affordable greens and cabbage were similarly rare, both in the South and Chicago. In the South one study found that "the majority of negro families would serve turnip greens, cabbage or collards daily if they had a convenient supply on hand," but many simply could not even afford those staples; another similarly found that

"A large number of families . . . did not eat vegetables in either winter or summer."[54] Despite long-standing efforts by Washington and others urging sharecroppers to grow more of their own food rather than purchasing a diet high in salt pork and cornmeal, most were locked into the cash crop economy.[55] Woodson observed in 1930, like Washington before him, that "the soil is rich enough to produce vegetables in abundance" in the South, but much more sympathetically—and rightly—went on to note that "the time of the laborers is required in the production of cotton or sugar, and these things must be imported or foregone."[56] While cotton cultivation kept many close to the land, black southerners mostly relied on the same food distribution networks as black Chicagoans. Indeed, when urban migrants ate vegetables, they were often canned: more reasonably priced and sometimes even tastier than supposedly fresh vegetables, they were available year-round in a midwestern climate that could not support much local produce in the winter months.[57]

The occasional fresh fruits and vegetables purchased by Black Chicagoans usually passed through the city's large markets that sold agricultural produce from surrounding midwestern farmlands and beyond, way stations in the broader food aggregation and distribution system that also spread the stockyards' salt pork across the country.[58] These networks meant that in some cases it was easier and more affordable for African Americans to obtain some fresh produce—just like fresh meat—in Chicago than in the South, despite the South's climatic and agricultural advantages. Throughout the interwar era, enterprising black entrepreneurs bought produce from these markets, loaded it in pushcarts, and sold it door-to-door.[59] In the late 1920s Timuel Black and his brother purchased fresh produce from the Seventy-First Street market, transported it several blocks north, and sold it off a pushcart on the South Side's streets and alleys; many other black Chicagoans did the same.[60] The markets supported selling and trading on an even smaller scale, too: one black woman who was clearly adept at stretching her family's budget remarked that "you can always get something to eat at the market like a basket of beans or tomatoes and potatoes for a dime, before they are graded. If you get more than you can use yourself, you can always sell or trade what you don't want."[61] Such entrepreneurial opportunities may have been more limited down South, but if black Chicagoans had migrated from an urban area similarly integrated into these national food networks, odds were that these pushcarts filled with produce would have been a familiar sight.[62] In addition to pushcarts, consumers sought fresh produce at an array of groceries that dotted Chicago's South Side; other than barbershops and beauty parlors, grocery stores were the most numerous black-owned businesses throughout the 1920s and 1930s, offering significant

FIG 9.3. Russell Lee, *Children in Front of Grocery Store, Chicago, Illinois*, April 1941. Children in a buggy outside a grocery store featuring fresh produce often unaffordable to African American migrants. Lemons and other produce grown far from Chicago underscore migrants' embeddedness in national food distribution networks. *Source*: Library of Congress, Prints & Photographs Division, FSA/OWI Collection, LC-USF33–013009-M2. McCammack.

opportunities to entrepreneurs.[63] It may have been easier for migrants to open a grocery in the North, but black-owned groceries were a familiar sight for those who had spent time in the urban South. Indeed, these groceries were often information hubs where prospective migrants gathered news about cities like Chicago.[64] Whether black- or white-owned, Chicago groceries were important because, as the CCRR explained in 1922, they meant that migrants no longer had "to buy groceries at plantation stores where they felt they had been exploited."[65]

Assessing the Great Migration in his introduction to St. Clair Drake and Horace Cayton's *Black Metropolis*, the landmark sociological study of black Chicago, Richard Wright argued that "the advent of machine production altered [humans'] relationship to the earth," and with "their kinship with the soil altered, men became atoms crowding great industrial cities."[66] Later on Drake and Cayton offered a slightly different perspective: in Chicago "over 65 per cent of the Negro adults earn their bread by manual labor in stockyard and steel mill, in factory and kitchen, where they do the essential digging, sweeping, and serving which make metropolitan life tolerable ... still betraying their southern origin, [they] were toilers, working close to the soil, the animals, and the machinery that undergird Chicago's economy."[67] Both close to the soil and far from it: although somewhat paradoxical, both perspectives were accurate, and both speak to how and why migrant foodways remained largely static as thousands settled in Chicago. Although migration to urban centers perhaps represented the final, dramatic act in humans' distancing from nature, machine production and industrialization altered "kinship with the soil" long before migrants left more rural, agricultural lives behind. For most black Chicagoans, separation from the land that provided their food was something to which they had become accustomed in the South. By finding working-class jobs, making homes, and building lives in the city that Sandburg also called a "Player with Railroads and the Nation's Freight Handler," black Chicagoans simply moved nearer the heart of networks into which they had long been integrated.[68] In particular, the railroad transportation networks central to these economies of scale facilitated both migration and the homogenization of foodways, flattening environmental differences between North and South and alienating people from the animals and plants that industry turned into food.[69] Chicago was the epicenter of this alienation from nature, but the ways the food industry increasingly obscured connections between city and country, humans and nature, were already characteristic of both North and South prior to the Great Migration. While migration to a northern city like Chicago undoubtedly forced African Americans to adapt to different environmental conditions and behavioral expectations than what they had become accustomed to in the South, persistent economic marginalization and these already expansive industrial networks meant that migrants' connections to nature through food remained largely unchanged in the interwar years.

CHAPTER 10

"No Cheerful Patches of Green"[1]

Mexican Community and the Industrial Environment on the Far Southeast Side of Chicago

MICHAEL INNIS-JIMÉNEZ

Encarnacion and Juanita Romo, along with their three children, moved from Texas to Chicago around 1923 in search of work and a better life. After about a decade of living near Hull House in the city's Near West Side Mexican enclave, Encarnacion Romo eventually found work in the steel mills of Chicago's Far Southeast Side. He and Juanita, with their now six children, moved near the gates of the steel mill to the 9100 block of Baltimore Ave. In telling his story almost fifty years later, Anthony Romo, the couple's first child to be born in Illinois, remembers the move to South Chicago as a "long cold ride on a streetcar to a cold house with no lights on a broad street." After living the first ten years of his life in the dense Near West Side neighborhood, Anthony was most impressed with South Chicago's streets, "I thought we were going to live on a boulevard because the street was so big and wide." They had moved to a basement-level cold-water flat. Although having been outlawed for over thirty years, basement apartments continued to be available throughout the Far Southeast Side. The flat was a rather large, three-bedroom apartment with its own bathroom, but without hot water. The fact the Romo family moved into an apartment with a bathroom is worth noting because community privies were frequently located in common areas and yards, despite the latter being prohibited.[1]

FIG 10.1. *The Calumet Region. Source*: U.S. Geological Survey.

In a study completed a year after Romo's move to South Chicago, Progressive reformers and University of Chicago social scientists Edith Abbott and Sophonisba Breckinridge called Romo's new neighborhood drab and dilapidated, filled with "dirty shacks and rotten sheds" that were probably "for the fowls which are seen running about in some of the back yards." Although this area of Chicago had large alleyways and many empty lots, Abbot and Breckinridge noted that these spaces were "usually trenches of filth" with "decaying food, dead animals, ashes, and mud."[2] The filth was everywhere. The social worker Mary Faith Adams commented that the area was "always laden, day or night, with the smoke from the steel mills." Although the area wasn't as crowded as other parts of the city, "the vacant spaces present," noted Adams, presented "no cheerful patches of green; alleys and vacant lots are strewn with waste paper, tin cans, and other forms of debris."[3] In an earlier report Abbott and Breckinridge had remarked that with "magnificent enterprise" like the steel mills, came a "hideous waste of human life" where "the men who feed the furnaces and send the products of their toilsome labor to a world market sleep in these miserable overcrowded houses" while having no decent places for relaxation and recreation aside from the "low saloons and dives" near the mills. As social reformers and advocates, these social scientists' concerns fit into their general progressive unease over such a large concentration of young men with no "appropriate" outlet outside of work.[4]

First impressions by Mexican newcomers and visitors were key in determining their long-term perspectives regarding the local environment. This neighborhood was, after all, a robust beacon of American industrialism. With three steel mills on the Illinois side of the Calumet region and another four on the Indiana side, massive buildings and towering smokestacks obscured the horizon and defined the region. Despite this landscape, at least 2,600 Mexicans migrated to the Far Southeast Side by 1928 to work and be part of a vibrant Mexican community. Although some reacted with disgust and dismay as they settled into this polluted industrial zone, many who moved to the area did so with the hope and anticipation of an economic boost and a comfortable living. For a shot at the limited opportunities in the city, within the steel mills and railroads and in the community, *Mexicano* workers and their families endured living in crowded, substandard conditions in one of the most environmentally compromised sections of Chicago. They settled near the gates of the steel mills or, for railroad workers, in boxcar camps. Sharing a culture, resources, and the use of physical space, Mexicanos in the Far Southeast Side considered themselves part of a community that functioned to support its families and their culture. They used a sense of community and camaraderie that came from these shared experiences to help themselves and others around them overcome the social, political, economic, and environmental obstacles that reduced their quality of life. This sense of community also aided Mexicanos in changing their environs to make life safer and more familiar.[5]

Mexican households developed strategies that in time led to the development of physical and cultural Mexican communities on the Far Southeast Side. As a community Mexicanos were able to change their physical environment by, among other things, opening and supporting Mexican-owned businesses and businesses that catered to the community. Mexicanos that entered these "dilapidated" houses were important actors in shaping Mexican South Chicago's physical environment and in improving individual and community welfare. Many Mexicanos chose to live near other Mexicans to take part in the creation of a common culture that could be used collectively to transform their environment and demarcate a Mexican-friendly space. Focusing on the sections of the Far Southeast Side reserved for the newest immigrant laborers in the early to mid-twentieth century—primarily adjacent to steel mills and railroad roundhouses—this chapter examines how Mexicans who settled in the area before 1940 shaped the built environment to make these sections distinctly Mexican. Several elements played a role in creating culturally and physically distinct Mexican communities in this area: the location of Mexican settlement and Mexican owned and operated businesses, the condition of housing that landlords and lodging-house keepers made available to them, growing vegetables and raising pigs or chickens alongside boxcar homes, and the ebb and flow of the size of the Mexican communities.

By analyzing several millgate blocks in the Far Southeast Side neighborhoods of South Chicago and South Deering through contemporary canvasses and the decennial census, we can start to develop a picture of how lives and movement in these millgate communities and boxcar camps were undeniably difficult and unhealthy. Examining the choreography of everyday life in these sections reveals how residents adapted to housing conditions, a compromised environment, and harassment. This was, after all, an area relatively new to Mexicans. It took years of a growing Mexican community before other ethnic groups—and in some cases park staff—allowed Mexicans free and unrestricted access to area parks. The new sights, sounds, and smells; the crowded conditions and limited work opportunities; and the new social rules of an urban, immigrant working-class neighborhood combined to provide a jolt to the senses. Together, argues the historian Nicolas Kenny, the grime, crowded living conditions, dilapidated housing, towering smokestacks, glowing furnaces, and the stench that came along with those many experiences transformed and intensified the "corporeal experience of city living."[6]

From its creation the local steel industry was always dependent on immigrant workers to fill their dangerous, dirty, and physically demanding jobs for low pay. The first European settlers moved into what would become the Far Southeast Side of Chicago in 1830 and in 1839 built their first bridge across the Calumet River in the area that is now Ninety-Second Street. The first railroad crossed the area in 1848. Industrialization and exponential housing and population growth started in the area when work began on Chicago's first steel mill, Brown Iron and Steel Company, less than thirty years later. South Works of the North

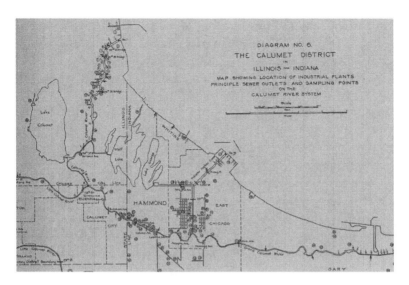

FIG 10.2. Diagram of the Calumet Region in Illinois and Indiana. *Source*: U.S. Public
Health Service.

Chicago Rolling Mill Company began operation in 1880. The mill was
then acquired by U.S. Steel in 1901 and renamed Illinois Steel. The first
foreign-born workers at this mill were mostly northwestern European
from Wales, Scotland, England, and Sweden. By 1897 large numbers
of Polish immigrants were working in the steel mills and living in the
growing neighborhoods surrounding the mills. Within a short fifteen
years, the Polish became the largest ethnic group in South Chicago and
remained such into the Great Depression.[7]

Railroads brought Mexicans to Chicago and the Far Southeast Side in
1916, literally and figuratively. "Imported" to work as *traqueros*, or track
workers, they came in boxcars from the U.S.-Mexico border region to
work on and maintain southwestern rail lines as they expanded into the
Chicago area. Mexicans first entered the steel mills of the Far Southeast
Side in significant numbers in 1919 as nearby race riots and steel strikes
created demand for non–African American workers. The Mexican population of Chicago increased from just over 1,100 in 1920 to almost 20,000
ten years later. After a significant outmigration of Mexicans during the
Great Depression, Mexicans returned to Chicago as World War II in
Europe boosted its economy. Federal census figures enumerated 16,000
Mexicans in Chicago in 1940. With a 1950 postwar economy and labor
force, the Mexican population in the city grew to 62,000.[8]

By the time Mexican immigrants and Mexican Americans started
migrating to work in the steel mills and railroads, the area had already
been home to Serbs, Lithuanians, Greeks, Swedes, Germans, Croats,

Slovenes, Italians, Hungarians, Poles, Slovaks, Jews, and the Irish. African Americans entered the area in large numbers at about the same time as Mexicans. The vast majority of these European immigrants entered the United States as less desirable, unskilled or low-skilled workers with "white" Americans racializing them as less-than-white. This industrial, immigrant neighborhood was, in the eyes of many, a dangerous, polluted zone unfit for human habitation throughout the first half of the twentieth century.

The smokestacks that spewed black smoke and ash, the gray polluted sky, the stench, the rundown houses "smothered in the smudge of the steel mills,"[9] the dirty unpaved alleyways littered with refuse, and the soot that covered everything defined the neighborhood. The sight of rising smoke outlined by the furnaces' glare welcomed those who entered the neighborhood after dark. The smell of slag, raw sewage, and industrial waste dumped by the mills onto the banks of Lake Michigan, the Calumet River, and its tributaries frequently overwhelmed those new to this neighborhood. Although the millgate section adjacent to South Deering's Wisconsin Steel on Torrence Avenue was not "as monotonous in appearance as the blocks in South Chicago," the quality of the housing, the smoke, the dirt, and crime made this area as unhealthy and dangerous as the area around South Works.[10]

Steel's domination of the area went far beyond buildings, paychecks, and the smudge that came from the smokestacks. Its large-scale and long-term water pollution with its cascading health effects further marginalized residents. According to data collected in the fall of 1924 and summer of 1925, Illinois Steel's South Works, located at the mouth of the Calumet River, discharged 223,350,000 gallons of wastewater daily into Lake Michigan or the mouth of the Calumet River through eight separate outlets. Of that, almost 12 million gallons were hydraulic fluid, 113 million gallons came from water used for cooling, 96 million gallons were condenser fluid, and sanitary waste from the plant's employees totaled over 2 million gallons. The much smaller Youngstown Sheet and Tube in the Far Southeast Side dumped 32,727,000 gallons a day into waterways, 80 percent of that into the Calumet River, with the rest directly into Lake Michigan. Also notable was Commonwealth Edison's plant on 100th Street and the west bank of the Calumet River. It delivered 288 million gallons of cooling water into the river through one large outlet. Wisconsin Steel, on the west bank of the Calumet River at 106th Street and Torrence Ave, dumped sanitary sewage, gas-wash water, cooling water, coke waste, and benzol into the river through four outlets. Although data were unavailable, scientists estimated Wisconsin Steel's discharge at approximately 5.5 million gallons a day. Their report lists a total of forty-one industrial sites producing waste on the Illinois side of the Calumet region, with another eighty-two waste-producing industrial plants on the Indiana side of the region.[11]

Raw sewage and other stench-producing organic wastes were also major problems affecting the quality of life of steelworkers and other Calumet region residents. South Works and Youngstown Steel, the two lakefront plants in Illinois, along with five more lakefront plants in Indiana, were significant polluters, dumping "directly into the lake the sanitary sewage of a large number of employees, as well as industrial wastes which, though probably not directly injurious to health, are nevertheless very objectionable as producing extremely disagreeable tastes and odors."[12] Adding to this, the untreated municipal sewage of 21,000 residents in Indiana went into Lake Michigan and the sewage of another 146,400 people in the Calumet region diverted into the Calumet River watersheds. Although some of the latter sewage would not have reached the mouth of the river because of the engineered reversal of the Calumet River to the Calumet-Sag Channel, slow currents and stagnant sections of the river caused septic conditions that came with a stench affecting the quality of life within the region. The region's most polluted lakefront was on the Indiana side of the state line. All eighty-eight samples of lake water taken over two years from the mouth of Indiana Harbor Ship Canal during this study contained "the odor and taste of a diluted mixture of crude oil and phenol compounds." Water samples at the mouth of the Calumet River in South Chicago tested as "generally free" of industrial waste.[13]

City smells were nothing new to urban America and these were not the worst smells in a city that was home to the largest collection of slaughterhouses in the country. That said, the stockyards and city dump stenches stayed relatively close to their origins compared to the wind-swept polluted water that approached Chicago's Loop area and that area's intake stations. An 1873 investigative series by *Chicago Tribune* reporters mapped the city's series of incomplete sewer lines that contributed to lots, alleys, and streets flooded with stagnant water, especially in the poor and industrial areas of the city. This festering landscape was part of the polluted environment that awaited Mexicanos in the Far Southeast Side fifty years later. The Public Health Service study's authors emphasized that winds regularly distributed the smelly waste from the Indiana Harbor Ship Canal up the lakefront as far north as Sixty-Eighth Street in Chicago. This would have undoubtedly gotten the attention of Chicago's political and downtown economic leaders.[14]

Many Mexicans entered the only housing available to them, which in many cases were rundown houses in the mill gate areas. An observer who was a member of the city's Bureau of Social Surveys found that by

1925 Mexicans and African Americans had "found shelter in the most used, most outworn and derelict housing which the city keeps." Mexicans near the steel mills lived in "the old tenement districts [that had] long been experiencing a steady encroachment by industry and commerce" and were residential sections "destined for extinction."[15]

Like Mexican steelworkers, Mexican railroad workers in and around Chicago found the housing adjacent to polluted, industrial workplaces. Around one thousand Mexican railroad workers and their families lived in boxcars in any of twenty-two camps that dotted the Chicago landscape. In 1925 the anthropologist Robert Redfield commented that the Chicago & Western Indiana Railway boxcar colony, located adjacent to railroad tracks between Eighty-Second and Eighty-Third Streets, consisted of around twenty "very dilapidated box cars" scattered around repair sheds and railroad yard buildings. Many had iron-pipe chimneys, while some had porches, potted plants, and even small chicken yards. Workers and their families occupied thirteen of these boxcars. Similar camps existed throughout the South Chicago area.[16]

Unlike workers living in millgate neighborhoods, railroad workers could not mark their neighborhood as Mexican through restaurants, pool halls, or other small businesses. They could, though, personalize their home's exteriors. A 1931 *Christian Science Monitor* article described and praised the personalization of the boxcars, emphasizing that some tenants had created flower and vegetable gardens and those whom the author considered "more prosperous" put up "electric lights," porches, and "tiny farmyards where chickens and perhaps a pig help out with the family's living." Despite the rundown condition of the boxcars and the implied temporary nature of the camps, Mexican men and women living there created a sense of community within the camps by improving the exteriors of their homes. Those living there transformed boxcars from a symbol of how the railroad companies dominated the lives of their workers, controlling their total environment, to an example of personalization.[17]

In changing their industrial landscape by planting and tending vegetable gardens and raising pigs and chickens, railroad workers and their families aimed at making their surroundings more human and humane. Having control over food strengthened their sense of community and provided them with further autonomy in areas far from the Mexican markets in South Chicago's commercial districts or the Near West Side. This is important as Mexican migrants to the Far Southeast Side and greater Chicago linked Mexican food with home and their positive memories of experiences, tastes, textures, and smells that made food central to the development and maintenance of a distinct culture within the community.[18]

Links between Mexican railroad workers and steel workers were clear. Area steel mills profited from the efforts of railroad companies to bring Mexican track workers to the Chicago area. Throughout the United States a significant number of Mexicans contracted to work as traqueros "jumped" their contracts for more lucrative jobs with greater stability. Chicago industry provided these types of jobs, resulting in a large multi-industry influx of Mexicans during World War I. By 1928 southwest transcontinental railroads had a Chicago-area track worker force that was 74 percent Mexican, while other railroads averaged 39 percent. The railroads that had lines in the Southwest had the advantage of recruitment offices which funneled traqueros to Chicago. As Mexican workers left the railroads for better, year-round work, the railroads continued attracting new Mexicans so that the industry served as a point of entry into the industrial Midwest.[19]

Once traqueros jumped their contracts for better-paying jobs in steel mills, many traded their boxcar homes for poor, but non-temporary housing conditions. Probably the most dilapidated section of South Chicago was a roughly four-by-four-block area bordered on the west by railroad tracks running between Baltimore and Brandon Streets, to the east by the fences and gates of South Works, and to the southeast by Harbor Avenue and the Calumet River. Aside from the old and dilapidated housing, the area was described in 1927 as having boarded-up storefronts with "posters declaring violation of the Volstead act" or poolrooms and "soft drink parlors." Calling it a backwater, an observer described South Chicago in 1927 as a place people lived "because they must" in order to be close to their workplace, and where "the new entering elements are of a kind with those already there." In other words, this argument reasons that Mexicans entered these "zones of transition" because the first Mexicans to the neighborhood lived there because these were the only areas where the newest "non-white" workers could live.[20]

The 8900 blocks of The Strand and Green Bay Avenue provide good examples of the living conditions in these newcomer areas set aside for those at the socioeconomic bottom. The 8900 block of The Strand consisted of the mill fence on the east side of the block and twenty-three lots on the west side. Of these lots only fifteen contained dwellings and the northernmost lot housed a restaurant.[21] The census lists 130 residents on the block; of these 76 were listed as Mexicans and 44 as African Americans. The remaining 10 were born in Poland, Yugoslavia, or Austria. Boarders and lodgers made up just over 45 percent of those living on the block. The racial diversity in the neighborhood and the fact that no native-born whites lived there further signaled the area's status as a "zone of transition."[22]

Almost all Mexican families on the street took in boarders to supplement the family's income. The largest boardinghouse, La Casa de
Borde, was comprised of two two-story dwellings and an attached one-
story building that together took up the entire lot at 8920 The Strand. It
housed five family members and twenty-six boarders. All residents were
listed as Mexican with all but one boarder born in Mexico. The building
owner, Anacleto Ramirez, lived on the premises with his wife and three
children and operated a pool hall, possibly the Paseo de La Reforma right
next door. While the majority of his boarders worked at the steel mill,
a few worked for nearby railroads and two worked at a pool hall. This
block was also home to two restaurant/barbershop combinations: La
America and El Porbenir. Additionally, this small block had a small grocery store, a combination restaurant/pool hall, and a barbershop/pool
hall combination.[23]

With a steady increase in Mexican families renting in the area
throughout the 1920s, the number of Mexican boarders increased,
further concentrating the total number of Mexicans in specific sections of the Far Southeast Side. The most common living arrangement
throughout these millgate areas consisted of Mexican families living in
two-floor apartments, reserving the first floor for the family and renting second-floor rooms to boarders. After only a few months of living
in the bunkhouse, the steelworker Alfredo de Avila moved off company property in 1923 and boarded with a Mexican family on Avenue O.
According to de Avila, the standard rate for a boarder in 1923 was $18
every two weeks for room and board. Although many unaccompanied
Mexican men, or *solos*, lived with Mexican families by this point, some
lived in homes and boardinghouses run by ethnic European immigrants
and African Americans. In a 1925 study using a citywide sample that
included 266 Mexican households (as well as 668 African American
households and 590 white immigrant households), Elizabeth A. Hughes
found that Mexicans and African Americans had the highest proportion of one-family households containing lodgers. Forty-three percent
of Mexican one-family households contained lodgers, while 42 percent
of African Americans, 28 percent of native whites, and 17 percent of
foreign-born whites had lodgers. Hughes noted that Mexican men were
the most likely to be living in "non-family cooperative groups." Three
primary factors contributed to the high number of lodgers in Mexican
family homes: first, a large number of unaccompanied men were entering the Far Southeast Side in need of housing; second, many of the workers came to the area with the encouragement of current residents who
then opened their homes to them; and third, rental income provided by
boarders added to family incomes.[24]

The 8900 block of Green Bay Avenue, only one block east of The Strand, was built up on both sides of the street, and only nine of forty-six lots were vacant. Of the 347 residents of this block, 147 were listed as Mexican, 124 white, and 76 African American. Boarders comprised 29 percent of the residents. The largest boardinghouse on this block had 11 boarders and 7 family members, all Mexican. Unlike the largest boardinghouse keeper on The Strand, this keeper was not the building's owner, but did operate a pool hall. Yugoslavians were the majority of non-Mexican immigrants on this block, with Italians, Bulgarians, Russians, Swedes, and Austrians also represented. At the time of the 1930 census this one block was also home to three saloons, one pool hall, and nine storefronts.[25]

Mexicanos made physical changes to their environment by catering to members of the community. A 1928 study lists the 8900 block of Green Bay as home to the El Fenix pool hall (8901); El Azteca restaurant (8905); El Gato Negro restaurant (8904); a combination pool hall, restaurant, and barbershop named Monterrey (8957); and another pool hall, Las Dos Republicas, located just across the street at 9000 Green Bay. This same South Chicago study lists the following Mexican-owned businesses located in the community: a painter, two tailors, five additional pool halls, three more barbershops, two more restaurant/pool halls, a drug store, two moving companies, two bakeries, two newsstands, two grocery stores, and two barbershop/restaurants. Many of these establishments aided in creating a sense of Mexican community through their social gathering spaces and ability to transact business in Spanish.[26]

City parks and park facilities were important outlets for Mexicans in the area. The city's extensive public park system was fundamental to providing Mexicans with amenities such as shower facilities, recreation, and recreational space for adults and families, including organized, supervised activities for the community's youth. That said, turf wars among ethnic groups in the area and discrimination within the park system limited Mexican access to park facilities. A 1928 study on Chicago's park system asserted that parks were not only a character-building necessity for children, but also an important outlet for adults. Children's play was more than just "natural and necessary," Osborn maintained; skillfully supervised and directed play was "one of the most powerful agencies in character building." Access to parks and recreational activities was also important for adults, who needed diversion from the "drudgery and drabness" of the industrial workplace.[27]

Although public parks are best known for open fields and play equipment, Mexicans found South Chicago's Bessemer Park's fieldhouse indispensable. The Romos' cold-water flat provides a great reason for this. The Romo family, consisting of six kids by the mid-1930s, utilized the park's shower facilities as a practical necessity. By the late 1920s, a few years before the Romos moved to Baltimore Avenue, Mexicans were using Bessemer Park shower facilities at a rate of seventy-five people per day in the summer and fifteen in the winter. By 1927, despite interethnic hostility, Mexicans used Bessemer Park for organized Mexican Independence Day celebrations and for the staging of Spanish-language plays in the field house auditorium. Yet records show that only two Mexican children participated in organized park activities that year. The park director reasoned that the lack of youth in park activities was because the facility was too far from their homes. However, the primary reason that young Mexicans avoided Bessemer Park in the 1920s was conflict with the youth of other ethnic groups. South Chicagoan Gilbert Martínez recalled that he and his friends would frequently "get beat up by the guys that would hang around there." As with Bessemer Park, Spanish speakers were not welcome at the area's Russell Square or Calumet Park throughout the 1920s.[28]

Living a racialized existence in an ethnic immigrant environment surrounded by more established European ethnic communities, Mexicans in the mill gate areas were faced with discrimination by their neighbors, primarily Polish, Irish, and Italian. Mexicans created physical and cultural communities that helped them persevere in this hostile environment, which prepared the community, albeit a smaller one, to survive the Great Depression and laid the foundations for one that would survive the eventual closing of the steel mills. Through adaptation of their physical and cultural environment, Mexicans established a community which developed tools to defend against social, political, and economic discrimination and harassment.

Mexicanos migrated to and stayed on the Far Southeast Side despite the lack of park access or the area's somber, polluted industrial landscape. Living near the industrial giants on Chicago's Far Southeast Side meant living with discrimination and ethnic rivalries, with the industrial muck that covered everything, the massive mill buildings that stood between Mexicans and the fresh lake breezes, and the odors of everyday waste mixed with industrial waste, raw sewage, and rotting refuse scattered throughout the neighborhood. Visitors, social service workers, scholars, advocates, and the residents themselves described South Chicago and the rest of the Far Southeast Side as dark, polluted urban and industrial slums. But for the individuals and families who migrated here, this community stood for social and economic opportunity. Not everyone was driven primarily by dreams of economic prosperity; some came to join family and friends and others came to escape political and economic chaos in revolutionary Mexico. Whatever their motives or specific hopes, all came expecting opportunity, security, and a change for the better.

CHAPTER 11

Work Relief Labor in the Cook County Forest Preserves, 1931–1942

NATALIE BUMP VENA

In 1903 the City Council of Chicago hired the architect Dwight Perkins to compile a report for an enlarged park system that would eventually become the Cook County Forest Preserves. The landscape architect Jens Jensen worked with Perkins to recommend properties to include in what they called an "outer belt park," a local vacationland where all Chicagoans could recreate in forests, prairies, and even wetlands. In 1904 the two men published their findings as the *Metropolitan Park Report*, and Chicago's leaders began the legal and political work necessary to create this system of open land for the city. First, they needed state politicians to craft an enabling act establishing a new local government—the Forest Preserve District. Between 1904 and 1913 the Illinois State legislature passed three different Forest Preserve Acts. Illinois's governor did not sign the Act of 1905 into law and the Illinois Supreme Court found the Act of 1909 unconstitutional. Meanwhile, Daniel Burnham included Perkins and Jensen's plans in his key proposals for *The Plan of Chicago* (1909), recommending "'the acquisition of an outer park system, and of parkway circuits.'"[1] Finally, the governor signed the Act of 1913, and the Cook County Board organized the Forest Preserve District in 1916. With jurisdiction encompassing Chicago, the Forest Preserve District of Cook County immediately began acquiring property, and today the system spans seventy thousand acres in noncontiguous parcels that form a C around the city.

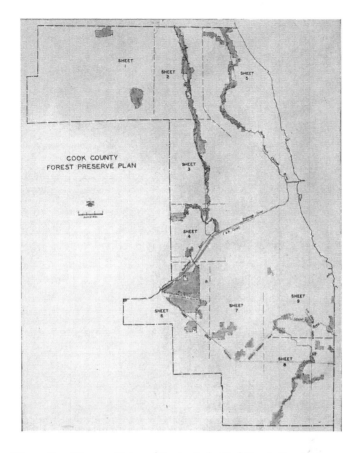

Fig 11.1. *Forest Preserve Maps and Foreign Parks: Cook County Forest Preserve Plan*, undated, FPDCC_00_01_0016_019. *Source*: Forest Preserve District of Cook County Records, Special Collections and University Archives, University of Illinois, Chicago.

For over a century Chicago has had its own preserves originally conceived as retreats from the metropolis, and this chapter explores how Forest Preserve leaders and work relief laborers together produced these seemingly primordial places during the Great Depression. Carl Smith has examined the role of elites, like Daniel Burnham and the Commercial Club, in projects intended to make Chicago more habitable in the early twentieth century.[2] However, the leadership of local government employees and the labor of the poor and the working class in realizing these plans have been relatively underexplored. One exception is Brian McCammack's account of African American relief workers in the Cook

County Forest Preserves.[3] This chapter likewise analyzes the contributions of Depression-era relief labor to the creation of the Forest Preserves as well as the work conditions and the ecological consequences of that transformation. Their physically taxing and undercompensated work sometimes led to the destruction of prairie and wetland ecosystems. Even as they strived to produce a natural landscape independent of Chicago's political and economic history, Forest Preserve leaders used relief labor to create places that reflected institutional priorities as much as they did the nonhuman world.

In the decade just before World War II, thousands of poor and working-class men materially changed land on the outskirts of Chicago, and in the process realized local leaders' plans for an urban retreat capable of absorbing visits from the city's 4 million residents. Even though many of the properties acquired had been truck farms and grazing lands, the district's leaders cast the new holdings as a kind of frontier wilderness. In 1921 Peter Reinberg, the Forest Preserve District's first board president, called the land a "natural playground which is being preserved, with Nature's wonderwork protected and encouraged, from erasure by the grim advance of modern civilization."[4] The historians Stephen Fox and Roderick Nash have both recounted how ongoing urbanization and the closing of the frontier in 1890 precipitated an American preoccupation with wilderness in the early twentieth century, manifested in phenomena like the Boy Scouts (founded in 1910).[5]

In addition to their frontier fantasies, district leaders described the Forest Preserves as a kind of country retreat for Chicago's residents, an escape from the city's stockyards, tenements, and steel plants. In 1920 Reinberg articulated the district's "primal objective" as giving "all our people" the opportunity to "practice the seeking of healthful happiness and life-bettering relaxation in these woodland reaches."[6] He celebrated his namesake camp as a place where "successive groups of wearies and poverty-beset unfortunates enjoyed the pure air and sunshine, gaining health and strength by out-of-doors living and playing among the trees."[7] Fulfilling Reinberg's vision for the district, children and adults flocked to the Forest Preserves to camp and picnic during the 1920s.

Damage wrought from this intensive use undermined promotion of the Forest Preserves as a bucolic wonderland. Concerned about the ongoing integrity of the district's holdings, the Board of Commissioners sought help from an eleven-person Forest Preserve Advisory Committee comprised of Chicago elites in 1927. One member, Rufus C. Dawes, was an energy executive who had led the Commercial Club. Another, D. H. Burnham, was an architect as well as Daniel Burnham's son. And John T. McCutcheon, a Pulitzer Prize–winning cartoonist for the *Chicago*

Tribune, also served as the Chicago Zoological Society's first president.[8] The committee worked with Forest Preserve District staff and the Chicago Regional Planning Association to present wide-ranging recommendations to the Board of Commissioners in January 1929. To mitigate damage by visitors, the committee suggested district staff members provide places for "active recreation . . . so that the remainder of the native woods may be preserved for future citizens." They believed the Forest Preserves needed a "well paved" system of roads as well as "parking places," observing that without them, motorists had "blazed their own muddy and rutted trails into some of the most remote areas to the detriment and destruction of the trees, shrubbery and wild flowers."[9] By relegating intensive use to certain places in the Forest Preserves, the committee recommended the district "strictly segregate nature and culture," as the historian James Feldman has written of the U.S. National Park Service's (NPS) efforts in Wisconsin's Apostle Islands.[10]

The Advisory Committee also believed the district needed to improve the flora and fauna they wished to protect from the public. They directed Forest Preserve personnel to undertake a "rather complete reforestation" of the district's "open lands," recommending that 75 percent of the district's holdings, which included farms and prairies, "be maintained and replanted strictly as wooded or forested areas."[11] Their proposal likely reflected a desire to match the district's land to its name and statutory purpose, which in turn reproduced contemporary ideologies equating forests with American wilderness.[12] The committee also advised the district to turn its numerous wetlands into "lakes and lagoons . . . [that were] stocked with fish, partly to enhance the beauty of the landscape, partly for the recreation of boating and fishing, and also to serve as flood reservoirs and mosquito prevention measures."[13]

To execute these wide-ranging plans, the Board of Commissioners hired Charles "Cap" Sauers to serve as the district's general superintendent.[14] Born in 1893, Sauers grew up in southwestern Indiana and graduated from Purdue University, where he majored in horticulture and landscape architecture. Sauers then attended Harvard University's Graduate School of Landscape Architecture for six months in 1917, at which point he joined the military. During World War I, Sauers served as a U.S. Army captain of the 325th and 326th Field Artillery. Discharged in 1919, he began working for Indiana's Department of Conservation, eventually becoming the assistant director. On April 19, 1929, Sauers started his post at the Forest Preserve District of Cook County.[15]

Sauers must have soon concluded that the Advisory Committee's 1929 plans would be impossible to implement. Just six months after he joined the district, the stock market crashed, devastating Chicagoans and local

governments alike. As the historian Lizabeth Cohen writes, "Only half
the people employed in Chicago manufacturing industries in 1927 were
still working in 1933."[16] The crash left Sauers unable to fund the Advisory
Committee's projects through normal avenues, like tax collection or the
sale of bonds. In one resolution from a 1932 board meeting, the Forest
Preserve commissioners described the fiscal ruin: "Chaos exists in the
matter of collection of taxes in Cook County; thousands of home owners are out of employment and unable to meet their obligations, many
of whom are charges upon the County; delinquency in tax payments is
increasing yearly."[17] In an exchange illustrating both the district's dire financial straits and its intransigent patronage system, Sauers explained to
one commissioner why the district could not afford to hire an Evanston
resident (who was presumably a political supporter) in 1932: "It is utterly
impossible to put him on our regular payroll because on Tuesday of this
week we separated 82 men from our own regular payroll and the rest of
the forces are on half time, in order to meet our budget requirements."[18]

Unable to fund its own initiatives, the Forest Preserve District became an enthusiastic local sponsor of newly developed state and federal
programs designed to put unemployed men to work and to direct money
to families in financial need. Between 1931 and 1942 thousands of work
relief laborers realized the Advisory Committee's vision by planting
forests, digging lagoons, and building comfort stations throughout the
Forest Preserves as part of the Civilian Conservation Corps (CCC), the
Illinois Emergency Relief Commission (IERC), and other government
agencies. In a September 1934 letter Sauers wrote to a friend, "We are up
to our neck in the alphabet soup—FERA, CWA, CCC, etc. We are really
getting a lot of good out of it in preparing our huge holdings for the use
of the millions of people here in the area."[19]

Roberts Mann, who became the Forest Preserve District's superintendent of maintenance in 1932, observed that relief labor had already
"advance[ed] the realization of the Advisory Committee's general plan
by 10 years," noting, "We have made hay while the federal sun shone."[20]
Mann received a degree in engineering from Cornell University and
worked as an engineer until World War I, when he served as a first lieutenant with the U.S. Army Corps of Engineers. He was an engineer in the
Forest Preserve District before his promotion in 1932. During the Great
Depression, he collaborated with state and federal agencies to oversee
work relief efforts.[21]

The district's first relief laborers, who were paid by the Special Work
Division of the IERC, arrived in 1931.[22] The following year IERC increased the number of men assigned to the Forest Preserves from 200
to 1,250 each day.[23] In 1933 the Civil Works Administration (CWA), a

federal agency, employed 9,000 men in the Forest Preserves for four months.[24] Five CCC companies also arrived that year to begin digging lagoons out of the vast Skokie marsh in northeastern Cook County. By the mid-1930s the Skokie Valley had "the largest concentration of CCC men in the country—10 companies and 2400 men," according to their newspaper *Camp Skokie Valley Review*.[25] At that time nine other CCC camps worked throughout the Forest Preserves.[26] Works Progress Administration (WPA) men came in 1935, and in 1938, 2,300 of them labored in the preserves.[27]

The Advisory Committee's written plans for the district made it an attractive partner to state and federal agencies distributing funds during the Great Depression. Robert Kingery, the Advisory Committee's secretary, wrote in 1937, "Having a plan of development and of reforestation, the Commissioners were thoroughly prepared to take advantage of the offer of labor and some materials through the various emergency relief and conservation agencies established by the State and Federal Governments. Early in 1933, when the Federal Civilian Conservation Corps was established, the State Park Authority, knowing Cook County had such comprehensive plans, promptly applied for CCC camps to work in every section of the preserves."[28] Mann corroborated Kingery's recollection, explaining how the district managed to get CCC companies: "In 1933, because we had a general plan and many detailed plans for specific areas, we took five CCC camps."[29] The Advisory Committee's plans then led to a bonanza of work relief labor that could manually recast the Forest Preserves.

Many laborers came from the Chicago area. IERC men commuted daily to the Forest Preserves, traveling by truck from the last stops of Chicago's streetcars.[30] Some could even walk to work from their homes.[31] Federal agencies also paid locals to live and work in the Forest Preserves. When soliciting federal funds from elected representatives, district leaders made sure to mention the benefits of employing Chicago-area residents. In the fall of 1937 the board president Clayton Smith asked U.S. senator James Hamilton Lewis (D-IL) for help bringing more WPA labor to the Forest Preserves, adding, "It is important not only to our physical properties but utilizes hundreds of men from the City of Chicago and will materially assist in the relief picture."[32] When CCC enrollment dropped shortly before the United States entered World War II, Sauers pleaded with his staff to recruit local young men into the corps, partly through the ward offices of Chicago's aldermen: "What I want from all you men is to talk to your foremen and employees in your respective camps and divisions to get boys from 17 to 23 to enroll in the CCC. . . . See your ward offices—take time off or send a ranger out after any likely recruit."[33] Sauers seemed desperate to continue the district's cost-effective reconstruction.

District leaders also believed these programs benefited local men. At a December 1934 meeting of Chicago's Conservation Council, Sauers reported that the district's CCC camps had been "keeping 4,000 men off the streets of Chicago."[34] Mann argued that the work available in the Forest Preserves saved men living in the "crowded wards of Chicago" from the dangers of idleness. In 1933 he concluded that those funded through the IERC "are in great need and desperate, else they could not be here; and left unaided they would drift toward crime, revolution, or suicide." Mann outlined how jobs in the Forest Preserves prevented such perilous ends: "Earned money does something for your self-respect. What's more, it was earned by hard work—hard work in the open air, out in the fresh, wind-swept country—and that does something for your muscles, your lungs, and your blood-stream."[35] Cohen corroborates Mann's insights, writing "observers commonly reported that the unemployed preferred these federal job programs to straight relief payments," and recounts young Chicagoans' excitement at the prospect of joining the CCC: "When a group of teenagers from South Chicago heard a rumor in September 1935 that the minimum age for entrance into CCC camps had been lowered from eighteen to seventeen, they were jubilant."[36]

Despite this teenage hope and Mann's aspirational rhetoric, Forest Preserve labor lacked much financial reward. IERC men received $5 per day of labor, but worked at most one week per month.[37] As one alum from the Chicago area recalled, the CCC paid laborers $30 for a month's worth of work, $25 of which the government automatically forwarded to their families.[38] CCC wages were so low—$1 per day—that leaders of the American Federation of Labor initially opposed the program.[39] Sometimes relief laborers received no compensation at all. Lauding their commitment, Mann reported that IERC men continued working in the Forest Preserves even when the agency had no funds. In 1932 he wrote that many of them had "refused to take the lay-off made necessary by the budget requirements and worked on without pay, preferring to take their time-off at some later, more convenient date rather than cripple the operation of their Division."[40] While these men might have worked for free out of civic duty, they could also have feared losing future paid opportunities with the IERC.

Low pay aside, Forest Preserve assignments could be demeaning. Men who extracted limestone from the district's quarries worked in an occupation that staff members acknowledged carried the stigma of prison camps.[41] And as Forest Preserve personnel entered supervisory roles over relief labor during the 1930s,[42] they had opportunities to abuse their newfound power. In 1937 a concerned resident of River Forest, Illinois, wrote to board president Clayton Smith, "Just a few lines to inform you that many citizens are contemplating to sign a petition to request you to remove Gaynor the Thatcher Woods Supt. from his office." The author listed three reasons why Gaynor was "unfit for the job," including "he is a single man and he uses 2 W.P.A. laborers to make his lunch for him and do house work."[43] The River Forest resident probably found the scheme objectionable, both because Gaynor could take care of himself and the arrangement violated contemporary gender norms that assigned house-keeping to women. The likely exploitation of relief labor occurred in a variety of settings, from mines to homes. McCammack points out that in the segregated CCC, African American laborers were especially vul-nerable as they endured racism while working under "white overseers" who gave them the most arduous and menial tasks in the Cook County Forest Preserves.[44]

In spite of conditions that at best entailed intense physical work at meager wages, Forest Preserve officials complained that relief laborers shirked their responsibilities. A few years after Mann praised IERC men for working without pay, a district staff member wrote, "There is no in-centive, no interest, no pride nor any shame in these men. They feel that the state owes them a living and refuse to go through more than the motions of effort to earn that living."[45] Sauers criticized CWA laborers for working less in anticipation of that program's end. In February 1934 he wrote to Forest Preserve administrators, "There has been a very marked slowing down of our CWA work in the last few weeks. Foremen are not pushing their men, are seen sitting around fires during work hours, are permitting men to bunch up and stand around."[46]

Forest Preserve leaders disdained the "slowing down" of work relief laborers, because they needed them to realize the Advisory Committee's plans that demanded a rather complete overhaul of the landscape. To prepare the Forest Preserves for Chicago's residents, district leaders sent relief laborers to build parking lots, comfort stations, and picnic shelters. They believed such places would attract the majority of visitors and there-fore "protect the interiors and the bulk of our holdings from excessive use, abuse, fire and the automobile."[47] While the CCC only began prioritizing recreational development in the summer of 1935,[48] the Forest Preserve District started using relief labor to accommodate the region's numerous visitors in the early 1930s. By 1934 laborers had built "45 picnic shelters, 9 large log cabins, 2 warming shelters," and "69 trail fireplace shelters."[49] Relief laborers also constructed infrastructure for active winter sports. In 1937 Sauers told his landlady about a CCC project in the Palos Forest Pre-serves, "Some grand ski slides, or toboggan I guess they are, anyway six in a row with sodded lanes at the bottom which are fan shaped and spread out so different toboggans can't run into each other."[50] The Swallow Cliff toboggan slides drew thrill seekers through the early twenty-first century, with the district demolishing them only in 2008.[51]

To produce recreational structures able to withstand "severe usage" and "the destructive efforts of the vandal," relief laborers extracted natural resources from district holdings.[52] Forest Preserve leaders also used these raw materials to match federal funds expended on district projects. Mann explained, "Both CCC and WPA require local (or sponsor's) contributions. Under the WPA this amounts to about twenty-three percent of the federal expenditures. To make up this formidable total we throw in the horns, the hide and the tail. . . . We charge for every item of material we can possibly produce out of our own resources such as sod, black dirt, gravel, building stone, logs and processed or salvaged lumber, all at substantial unit prices."[53] At a district sawmill, relief laborers manufactured "rough sawn lumber" out of "dead or defective trees" collected in the Forest Preserves. Mann estimated that in 1934 the mill generated "over a half million feet, much of it oak worth from $75 to $125 per thousand."[54] As part of their plan to produce recreational areas, Forest Preserve leaders therefore directed relief laborers to remove sod and logs from the property they had a mandate to protect, no doubt disturbing the habitats of plants and animals alike.

Relief laborers also mined the Forest Preserves. Quarries in southwestern Cook County's Sag Valley yielded stone for recreational infrastructure. Men from the IERC, and later the CCC, operated these limestone quarries that Mann said generated "120 to 130 tons of flagstone per day for constructing walks, steps, walls, bridges, fire places and buildings all over the Preserves."[55] One CCC alumnus, Mickie Fisher, recalled using stone from the 111th Street quarry to construct "outlooks" as well as "stone steps and paths" along the I&M Canal.[56] A CCC company composed of World War I veterans ran a stone crusher for driveway and parking lot gravel.[57] In 1934 Mann reported that relief laborers had constructed "18 large parking spaces [lots]" and "124 small parking spaces."[58] The *Camp Skokie Valley Review* described their utility: "Where it was once the custom of visitors and picknickers coming to the Forest Preserves to park their cars wherever opportunity offered and oftentimes causing considerable damage to trees, shrubs and turf, the parking areas in the Forest Preserves have changed all that."[59] Again, protecting some Forest Preserves from the Cook County public justified transfiguring other parcels with relief labor.

While some workers laid parking lots and carved out picnic groves, others planted forests and drained waterways, producing new landscapes and eliminated existing ones. In many ways their goals aligned with those of land managers practicing what Feldman terms the "rewilding" of Wisconsin's Apostle Islands.[60] While "forests regenerated without human interference" after decades of logging in the Apostles,[61] relief laborers deliberately cultivated them in the Cook County Forest Preserves, where the board had acquired farms and prairies alongside woodlands. To fulfill the Advisory Committee's 1929 mandate to turn 75 percent of the preserves into forests, laborers transplanted saplings from wooded places to treeless areas.[62] They also established hundreds of thousands of seedlings. Between 1939 and 1941, for instance, relief laborers planted 1,428,144 trees in the Forest Preserves, one third of which were oaks.[63] Their efforts mirrored those of work relief operations across the country; CCC camps were often called "reforestation camps."[64]

Seeking to hide the human labor behind the new forests, the district's forestry department advised the CCC and other agencies to plant trees in a "staggered" pattern, not in rows, to achieve an unmanufactured aesthetic. The Forest Preserve District also supplied work relief crews with shrub and tree species they believed were native to Cook County.[65] When the district struggled to realize that policy without a sufficient supply of local plants, the CCC built a nursery dedicated to these species.[66] Even as they directed relief laborers to plant tree species characteristic of the Chicago region's flora, district supervisors led them to compromise entire landscapes that contemporaries recognized as native to Cook County. In 1940 an ecologist from the University of Wisconsin observed that tree seedlings were growing on remnant prairies in several Forest Preserves.[67]

To protect young trees, the district directed relief laborers to create firebreaks throughout the district. During the summer of 1935 Mann wrote a memo to all CCC camps, telling them to spend "EVERY AVAILABLE HOUR IN EVERY DAY PLOWING FIRE LANES. THIS WORK TAKES PRECEDENCE OVER ALL OTHER WORK."[68] By 1938 relief laborers in the IERC, CWA, CCC, and WPA had constructed 125 miles of trails that Sauers said served as "the backbone of our fire prevention and fire control system."[69]

In addition to fire prevention, Forest Preserve leaders expected work relief laborers to participate in firefighting. In the district's 1937 fire plan administrators stated, "When a fire is reported, regardless of the time of day or night, whether it is Sunday, a legal holiday or a regular work day, immediate control measures are necessary, and it is the duty of every member of the organization to work without question until the fire is 'black out.'"[70] Stationed with a company in northwestern Cook County, an NPS project superintendent reported that his CCC camp responded to a fire "at the extreme Northwest corner of Deer Grove" on April 25, 1936: "About sixty boys turned out with fire pumps and flappers and fought the blaze for approximately an hour before it was extinguished, with a burnt over area of about three and a half acres."[71] Although district leaders sought to protect what they perceived as natural areas, firefighting further compromised prairie, savanna, and woodland ecosystems that ecologists today understand as fire dependent.

Forest Preserve wetlands also underwent a metamorphosis during the Great Depression, with district leaders instructing work relief laborers to drain some marshes and turn others into lakes. Mosquito abatement motivated many projects, as the Forest Preserve District and partner agencies wanted to protect visitors and neighboring residents from these nuisances and potential vectors of disease. In 1932 Mann reported that the Forest Preserve District lent laborers to the DesPlaines Valley Mosquito Abatement District (DVMAD), where they "lay several thousand feet of tile and dynamite or dig numerous large ditches, effecting the drainage of mosquito breeding places in our Preserves."[72] In 1936 the WPA described extensive mosquito abatement work in Illinois's "low areas": "Hundreds of projects in many sections of the State have been directed to drainage of these swamp areas."[73] Some of these projects happened in the Forest Preserves. In February 1941 Mann relayed, "The WPA is now engaged in digging a deep channel all around the marsh west of Mannheim Road and south of 95th Street in which we once dynamited a big hole." The district planned to stock the channel and others like it with fish that ate mosquito larvae.[74] In addition to controlling pests, these new channels and lakes offered opportunities for fishing and sometimes boating. In their efforts to produce a nature they believed would draw more visitors, district leaders guided relief laborers to alter wetland hydrology, thereby eliminating the increasingly rare places hovering between lakes and dryland.

In the largest New Deal project executed in the Forest Preserves, CCC men recast an eight-hundred-acre sedge meadow, the Skokie, into a series of seven lagoons.[75] In the summer of 1933 Sauers presented the final plans for the Skokie Lagoons to the Advisory Committee, who concluded that "the hydraulics are practical and that quite adequate flood control would be achieved and a fine landscape developed along with it."[76] Developing such a landscape required scores of men, and in 1940 the NPS declared that Camp Skokie Valley was "the largest CCC Camp in the United States."[77] Between 1933 and 1942 CCC laborers removed 4 million cubic yards of earth from the wetland.[78]

As with the new forests, district leaders wanted to make the manmade reservoirs appear as though they had formed without human labor.

By 1937 the district had stocked the Skokie's four completed lagoons with fish species commonly found in Illinois lakes, like bass and pike.[79] Forest Preserve leaders also acquired aquatic plants, such as white pond lily and sago pond weed, from Illinois's Chain O'Lakes State Park and the Fox Lake region to cultivate in the Skokie Lagoons during the summer of 1940.[80]

In the midst of massive projects that rendered places unrecognizable, it is perhaps inevitable that some Forest Preserve leaders worried that work relief projects were harming the district's holdings. Mann, in particular, began to associate relief labor with environmental degradation. In 1938 he expressed concern about New Deal programs across the country: "The past four years of ECW, CWA, FERA and WPA participation have resulted in the creation of new state and county systems, both forest and park, and the development—in many cases the *overdevelopment*—of existing systems."[81] Mann partly attributed the oversized character of Depression-era projects to the federal government's goal of employing as many men as possible. "We are not altogether to blame," Mann told an assembly of park executives. "There has been enormous pressure exerted, in metropolitan areas at least, to coerce sponsors into initiating new projects and into accepting for employment great numbers of men far in excess of what the job or the equipment warranted."[82] With a ready supply of cheap labor from Chicago, and under pressure to use it, district leaders perhaps authorized more tree planting, fire suppression, and hydrological engineering than they would have otherwise.

Nationally known conservationists were also questioning the effects of work relief projects. In 1932, Jens Jensen publicly denounced the Forest Preserve District's plans for the Skokie Lagoons in a *Chicago Sunday Tribune* editorial, "The Beauty of the Skokie." He found the marsh far more valuable than the proposed lagoons, in part because of its flora. "More than ever do I feel that this vast stretch of marsh land should remain as it is forever . . . [a] vastness of waving grass hiding multitudes of beautiful flowers, many of which are very rare in this region."[83] Nearly thirty years earlier, Jensen had celebrated the marsh's beauty when he included it in his *Metropolitan Park Report* recommendations: "The Skokie is a marsh and next to the lake is considered by many Chicago's most beautiful natural feature. The view is unbroken for miles and the coloring from spring to fall is as variable as on the lake."[84] Once Forest Preserve District leaders finally acquired the Skokie through a series of hard-fought eminent domain cases, they decided to transform it, as detailed above. Not even Jensen, the visionary behind the Forest Preserves, could stop them. Beyond Cook County, Bob Marshall believed the CCC's recreational work, which included building roads and picnic areas in national parks and

national forests, ruined America's primitive wilderness.[85] The historian
Paul Sutter has argued that Marshall and his fellow advocates understood New Deal projects, including road construction in parks, as posing
such a threat to wilderness that they founded The Wilderness Society
in 1935. The group subsequently fought for federal protection of "large
expanses of roadless and otherwise unprotected nature," a demand culminating in the Wilderness Act of 1964.[86]

Aldo Leopold, another founding member of The Wilderness Society, criticized drainage projects like the Skokie Lagoons for disturbing
"ecological balance."[87] In a 1938 lecture, "Engineering and Conservation,"
Leopold cited the "drainage of nearly the whole Atlantic tidal marsh
from Maine to Alabama" as exemplifying the ascendancy of engineering, a profession unconscious of "biological equilibria."[88] Leopold made
clear that drainage "was done with relief labor, in the name of mosquito
control, over the protests of wildlife interests. These marshes are the win-
tering ground of many species of migratory waterfowl and the breeding
ground of others."[89] Like Mann, renowned landscape architects and wilderness advocates feared the danger that work relief programs posed to
the natural resources they adored and wished to protect in perpetuity.

In fact, it is likely that Leopold's observations about the environmental consequences of New Deal projects influenced Mann, since the two
men started corresponding in the late 1930s. After Sauers read Leopold's 1938 essay "Conservation Esthetic" in *Bird-Lore* (today *Audubon*
magazine), he urged Mann to ask Leopold for his advice on how to
curb the public's "unintelligent use" of the Forest Preserves.[90] In that essay Leopold had criticized the overdevelopment of protected lands for
tourists: "'Recreational development is a job not of building roads into
lovely country, but of building receptivity into the still unlovely human
mind.'"[91] In the late 1930s Mann recalled visiting Leopold in Madison,
Wisconsin, "That was the first of many pilgrimages to get our thinking
straightened out. We made Professor Leopold very happy. Somebody, at
last, was listening."[92] In 1953 Mann described Leopold, who died in 1948,
as "my friend and mentor."[93]

Of all the Forest Preserve leaders, Mann most clearly documented his
fears about the ramifications of work relief programs. For his part, Sauers was frustrated with the collateral damage that some relief laborers
caused. In 1940 he complained to a mosquito abatement agency about
the "mutilation of the trees and shrubs by your WPA forces because of
improper supervision."[94] Worried about more than a few shrubs, Mann
echoed Leopold as he described how the use of federal relief labor destroyed local biological equilibria. He referred to "exasperating instances" of ill-fated park operations across the country, including "drainage

ditches to eliminate mosquitoes that succeeded largely in eliminating the aquatic vegetation and wildlife."[95] His observations could easily apply to mosquito abatement projects in the Cook County Forest Preserves.

Mann also expressed concerns about the Forest Preserve District's ability to maintain the recreational infrastructure built by Depression-era laborers without a ready supply of workers and federal funding. The CCC and the WPA, the last of the work relief programs, left Cook County in 1942, with the CCC stopping on July 1 and the WPA "dwindl[ing] steadily until its abolishment at the close of the year."[96] In an essay written during World War II, Mann stated, "The past ten have been years of plenty—too fat for those of us who blithely drifted with the flood of federal labor and materials, to find ourselves with more acres, more buildings, more facilities and more frosting on the cake than we possibly can maintain."[97] Mann accurately anticipated the challenge. In 1952 Sauers told the Forest Preserve District's Board of Commissioners, "The federal and local work forces of 1932–1941 expended thirty million dollars on Forest Preserve Development. These have now been under the stress of heavy use and time, and repairs are heavy."[98]

To create a country escape for millions of Chicagoans, Forest Preserve leaders had come to depend on the urban poor and working class who filled the ranks of state and federal relief programs. Still, they remained ambivalent about that labor's legacy on the landscape, as evidenced by Mann's concerns about overdevelopment and the district's efforts to disguise the human hands behind the Forest Preserve's new lakes and woodlands. In spite of their reservations, nature and labor have remained intertwined in the district, with IERC, CCC, and WPA efforts archived everywhere, from the Skokie Lagoons to the stone trails in the Dan Ryan Woods. Chicago's poor and working class not only built the city's downtown and toiled in its steel mills and slaughterhouses. They not only traveled to the Forest Preserves as a reprieve from their tenements and the crowded metropolis. These city dwellers made the Forest Preserves, implementing the Advisory Committee's 1929 plans as participants in state and federal programs.

As the field of ecology and the district's priorities have changed, different kinds of laborers—including volunteers, paid contractors, and Forest Preserve staff members—are remaking the district once more. Practicing ecological restoration, they have reversed some Depression-era projects by extracting planted trees, removing drain tiles, and setting prescription fires. With saws, drip torches, and seeds, stewards of every stripe are, in their view, increasing biodiversity and trying to return the land to its condition during First Nation sovereignty and before Chicago's colossal expansion during the nineteenth century. Those practicing ecological restoration continue the tradition of active land management in the Forest Preserves arguably begun during the Depression. Then as now, district leaders are cultivating places that function both as re- prieves from urban sprawl and as records of our ever-changing relation- ships with nature. Their ongoing interventions make Chicago's preserves no less vital. These places have never been actually distinct from the city, or for that matter, culture. Instead, the Forest Preserves continue to serve as an experiment in how this metropolis can ethically integrate nature.

A CARTOGRAPHIC
INTERLUDE

Commercial printing emerged as an important industry in Chicago, and map production constituted a significant element of this business. Peter Nekola and James Akerman examine the ways that Chicagoans created and used maps to facilitate development, to analyze urban problems and call for reform, and to redesign the city's future. In their essay following the maps, they lead us on a tour that captures essential aspects of the evolution of Chicago's environmental history.

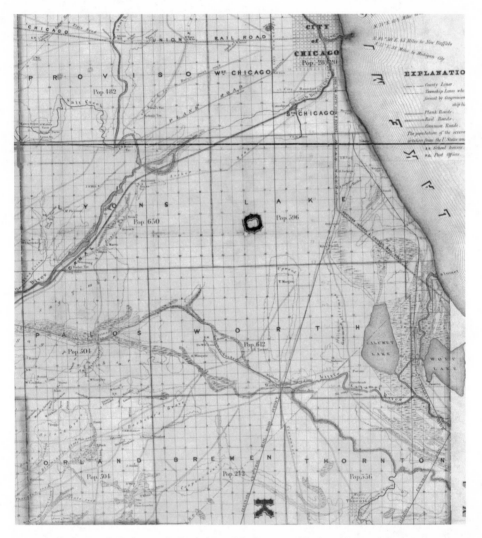

Fig 12.1. Detail, James H. Rees, *Map of the Counties of Cook and DuPage, the East Part of Kane and Kendall, The North Part of Will, State of Illinois* (Chicago, 1851). *Source*: Library of Congress.

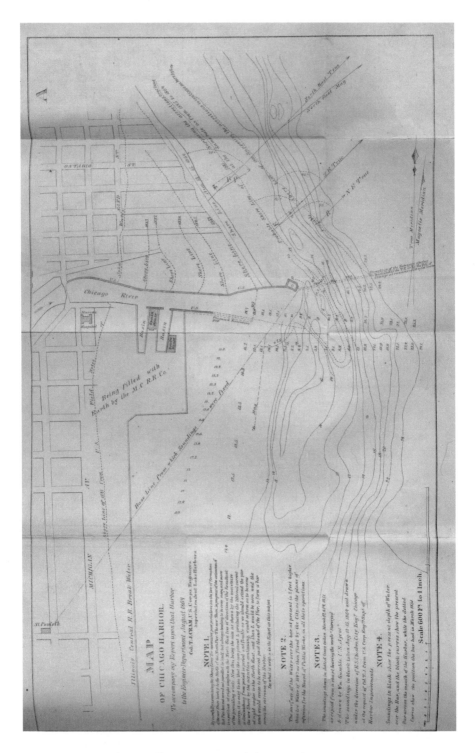

Fig 12.2 . Thomas Jefferson Cram, *Map of Chicago Harbor to Accompany My Report upon That Harbor to the Engineer Department*, August 1864. Courtesy, the Newberry Library.

FIG 12.3. Detail, *Rand McNally & Co.'s United States. Source: Rand McNally & Co.'s Business Atlas* (Chicago: Rand McNally, 1878). Courtesy, the Newberry Library.

FIG 12.4. Van Vechten & Snyder, *Chicago and Environs* (Chicago, 1882). Courtesy, the Newberry Library.

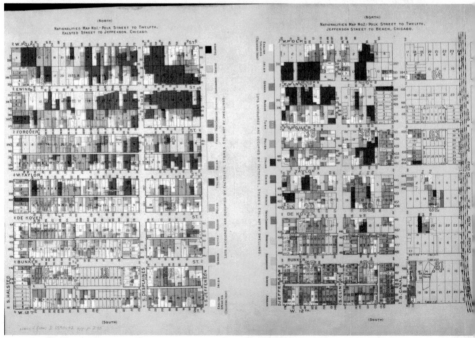

FIG 12.5. Detail, *Insurance Maps of Chicago*, vol. 13 (Pelham, NY: 1901–1950), 44. Courtesy, the Newberry Library.

FIG 12.6. Agnes Sinclair Holbrook, *Nationalities Map. Source*: Samuel Sewell Greeley, ed., *Hull House Maps and Papers: A Presentation of Nationalities and Wages in a Congested District of Chicago, Together with Comments and Essays on Problems Growing Out of the Social Conditions* (New York: Thomas Y. Crowell, 1895), nos. 1–4. Courtesy, the Newberry Library.

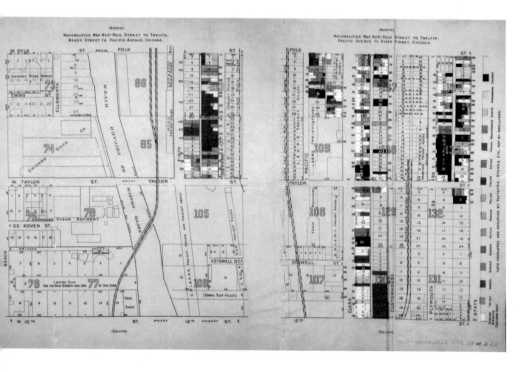

To love and render service to one's city, to have a part in its advancement, to seek to better its conditions and to promote its highest interests,—these are both the duty and the privilege of the patriot of peace.

The thoroughfares of a city may be divided into three classes: the street, by which is meant

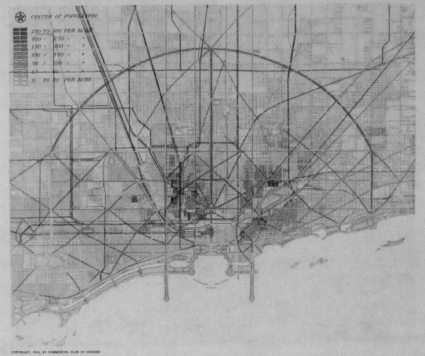

LXXXIX. CHICAGO. DIAGRAM OF GENERAL SCHEME OF STREET CIRCULATION AND PARKS IN RELATION TO THE POPULATION.

The various densities of population, ranging from 0 to 25 persons per acre to 250 to 300 per acre are indicated by different densities of red color. The center of population is indicated by a star. Railroads are shown in blue.

the general type of artery; the avenue, on which tides of traffic and travel surge back and forth; and the boulevard, designed primarily as a combination of park and driveway. The first consideration for all thoroughfares is cleanliness, which is the result of a good roadbed kept in thorough repair, and unremitting care on the part of the city cleaning department. In the congested retail district the desirable street width is from 80 to 100 feet, about equally divided

FIG 12.7. *Diagram of General Scheme of Street Circulation and Parks in Relation to the Population. Source:* Daniel Burnham and Edward Bennett, *Plan of Chicago* (Chicago: The Commercial Club of Chicago, 1909), plate 89. Courtesy, the Newberry Library.

XL. CHICAGO. GENERAL DIAGRAM OF EXTERIOR HIGHWAYS ENCIRCLING, OR RADIATING FROM, THE CITY.
All the arteries composing the system without the city limits exist, except where shown in dotted lines. City limits shown in red tint; rivers
and other waterways in blue.

FIG 12.8. *Chicago: General Diagram of Exterior Highway Encircling, or Radiating from, the City. Source*: Daniel H. Burnham and Edward H. Bennett, *Plan of Chicago* (Chicago: Commercial Club of Chicago, 1909), plate 40. Courtesy, the Newberry Library.

Fɪɢ 12.9. *Chicago and Vicinity Road Map* (Chicago: Chicago Motor Club, 1963).
Courtesy, the Newberry Library Akerman.

CHAPTER 12

Maps and Chicago's Environmental History

PETER NEKOLA AND JAMES R. AKERMAN

As the primary continental nexus of North American industry, transport, and immigration, built on precarious lakeshore and swampland, Chicago needed good maps. Over its two centuries of intensive settlement and development, maps have had a hand both in creating and solving urban problems. This is a history of interacting with and manipulating the local environment in ways mediated by mapping and visualization. Chicago needed good maps, and it produced them. Rather than offer a comprehensive cartographic history of Chicago's built environment, this chapter focuses on four specific themes in mapping of salient significance to the environmental history of the Chicago region: waterways, railroads, living conditions, and urban and regional planning. In the process we hope to demonstrate the value of maps and visualizations as tools for studying urban environmental history.

VISUALIZING A CITY ON SANDS AND SWAMPS

A 1957 report, *Government Problems in the Chicago Metropolitan Area*, devoted its initial chapters to specific problems of water supply, sewage disposal, and surface drainage, stating that "no governmental services are more important to the health and well-being of those who live in cities, villages, and other local communities. No one is wholly unrelated to the others." The report, prepared for an organization then known as the Northeastern Illinois Metropolitan Area Local Governmental Services

Commission, which would evolve eventually into the Northeastern Illinois Planning Commission, came at a time when the city proper's once rapid expansion into surrounding savannah and marshland had begun to be checked by newer, suburban municipalities that, while "functionally" part of the city and its economic, industrial, and population bases, had distinct local governments. While this localist impulse offered a distinct sense of identity to its inhabitants, as well as, for some, a sense of safety and perhaps status, the report concluded that only consolidated or coordinated metropolitan governance could address key urban problems like transportation, education, and public health. By beginning with the latter, the report addressed an environmental context that had been present since before the city existed: the history of Chicago is a story of finding ways to live very near the water table.[1]

That a city on the scale of Chicago would develop at the lowest spot between two of the continent's largest drainage basins, the Mississippi River System and the Great Lakes Basin, made geographical sense. A canoe portage between the Des Plaines and Chicago Rivers served as a major conduit between long-established water-based trade networks in both basins. The fur trade had dominated economic and cultural life in the continent's northern interior for several centuries before intensive settler-colonial agricultural and industrial development supplanted it. Precolonial trades in copper, flint, canoes, and other commodities sustained Indigenous communities along the Chicago's banks for many centuries before the arrival of Europeans, who, beginning with French-speaking explorers, traders, and priests, adopted those routes and began working to integrate them, and the people who lived along them, into their own economic and cultural systems. The low spot used as a portage was ideal for carrying canoes and, later, building barge canals and rail networks. The underlying geology that had made this low spot possible also left a sandy shoreline, constantly modified by what nineteenth-century earth scientists would call the "geological agency" of winds and currents, a hydrology that persistently reminded Chicagoans of both the benefits and complications of living with water, and on new and quickly changing land.

The 1957 report contained seven foldout maps. While the last four represented networks of different forms of rapid transit, all of which required administration to serve a metropolitan population, the first three concerned issues directly related to water: the problem of sanitary drainage in flatland where there is a real risk of contaminating the city's freshwater source and distributing that freshwater to support a large population. Managing this, as well as stabilizing the city's ground, particularly near the lakeshore, required a robust understanding of geological and hydrological processes, creative engineering and adaptation, and the resources and administrative capacity to manage the entire region. All of these in turn demanded the sort of systematic, analytical visualization only specialized maps could offer.

Early maps of Chicago largely took two forms: maps of navigable waterways, which would come to include river modifications and, eventually, canals; and maps of idealized geometric street grids where optimistic developers sought to expand the city's footprint. While the former maps were ultimately practical, working with local conditions, the latter, like their counterparts in imperial European cities and colonial outposts, were designed to erase as much as possible visual evidence of those conditions, offering instead a sense of order and control, of rational, standardized, civilized urban form. Because Chicago, like its downriver cousin New Orleans, was maintaining this grid on land that was constantly being modified by water, that control had additional meaning and demanded constant work.

Real estate plat maps, such as Joshua Hathaway Jr.'s 1834 map *Chicago with the School Section, Wabansia, and Kinzie's Addition*, were the primary purveyors of the grid, the livelihoods of their makers depending on the promotion of the order such views implied. Though illustrations of Chicago's built areas at the time showed little more than a cluster of one-story shacks, Hathaway's plat gave the appearance of a well-ordered city stretching west from the military reservation at Fort Dearborn. This order was interrupted only by a twisting Chicago River, which would be eventually rerouted, reinforced, and channelized in this area to itself become part of the grid. In the 1840s and 1850s Chicago began to experience a rate of growth that overwhelmed this ordering, as well as the city's ability to maintain clean drinking water. These decades saw cholera epidemics so extreme that many found Chicago unfit for habitation. Desperate experiments with underground piping and land-raising were attempted, but problems persisted until the completion of the Chicago Drainage Canal (now the Chicago Sanitary and Ship Canal) which drained sewage down the Des Plaines River to the Mississippi, keeping it out of Lake Michigan.[2]

Chicago's slow growth at the beginning of the nineteenth century can be attributed to a number of factors, among them the limited capacity of the fur trade, already in decline, to support an economic geography constituting much more than small, far-flung outposts. While innovations in agricultural and transportation technology in the 1850s would quickly make Chicago the center of grain, lumber, and livestock markets, a story told well by Cronon's *Nature's Metropolis*, its astronomical growth even before these economic engines took hold tells an additional story of real estate prospecting as an economic driver in itself.[3] And for Chicago's real estate industry in the 1840s, maps were among the most important tools. James H. Rees, whose land agency would eventually become the powerful Chicago Title & Trust Company, was himself a skilled cartographer. His 1851 *Map of the Counties of Cook and Dupage, the East Part of Kane and Kendall, the North Part of Will, State of Illinois* (see Figure 12.1; all figures referenced in this chapter are in the preceding color gallery), which covered an area stretching from the lakefront west to the Fox River Valley and south into the Calumet region, was perhaps the first to visualize in a single map what would become commonly framed in the twentieth century as Metropolitan Chicago.[4]

While the lakefront in Rees's map appears stable and largely featureless, the inland areas contain greater environmental detail than perhaps any previous map of the area. Like these other maps, it synthesized the grids that appeared in city plats like Hathaway's with the township-and-range system grid promoted by the U.S. General Land Office. But it is also unique for its environmental information: nearly every slightly elevated area on the map, with soils that offer sufficient drainage, is represented as tree covered. The surrounding areas are shown dominated by swamp, much of it with little or no natural drainage. These lands would, in Rees's vision, become valuable as urban, suburban, and satellite real estate investments once the swamps could be drained. The Swampland Reclamation Act of 1850, like the acts of Indian removal which had preceded it, effectively provided government subsidies to the real estate industry.

Before the developed insights of ecology and other earth sciences, it was possible to operate on the illusion that "draining the swamp"— that is, eliminating wetlands from a natural system—was possible with minimal effects beyond the economically beneficial.[5] The swampy areas around the southern part of Lake Michigan were characterized as "unsaleable lands" or "wet lands" by local and federal authorities unless they could be drained or otherwise "improved" to support agriculture and settlement.[6] The act effectively transferred such lands to the states, many of which in turn offered subsidies for such "improvement" to promote development. Investors such as the Illinois Central Railroad and George C. Tallman found this buying, draining, and reselling profitable, and a real estate boom commenced in the Chicago area, fueling many industries which would come later.[7]

Efforts to stabilize and develop what was in some ways barely land at all were not limited to the interior, and indeed, what constituted Lake Michigan and its interior was an open question. Spits of sand were constantly forming and reforming lagoons, points, and beaches, as well as the mouth of the Chicago River itself, as the lake's winds and currents ate away at its northern shores and deposited sediment farther south.

Establishing reliable port facilities and lake access in such an area required constant management in order to mitigate the effects of Lake Michigan's "geological agency." A map prepared by the U.S. Army Corps of Engineers for an 1864 report (see Figure 12.2) used field data from 1821 to 1864 to visualize shoreline changes from year to year in order to get a sense of what forces were shaping the lakefront and what efforts could be made to reduce their burden on commerce. To this end it correlated the rate of land formation with bathymetric data to model future shorelines as far distant as 1900. For the Corps and the city, this meant that piers would need to be constantly elongated in order to prevent the channelized river mouth from silting up or closing entirely.

Managing Chicago's lakefront took on new meaning in the twentieth century, when much of it was developed as public parkland, stabilized by jetties and seawalls. While maintaining this shoreline has been a costly, ongoing process, many maps over the years have given the lakeshore a false sense of permanence, which allowed city boosters to promote a sense of the city as a strong, enduring monument of human accomplishment. As Paul Petraitis of the Chicago History Museum has noted, "Historians, especially Chicago historians, have proved to be the slowest to change their concepts, preferring to picture the pre–Fort Dearborn landscape as roughly the same as today's."[8] The work of maintaining such facades on precarious foundations is part of Chicago's complicated relationship to an environment that has been impossible to erase, even if it can easily be put out of mind by residents and visitors.

These two maps illustrate not just the rough transition of Chicago from a settler-colonial outpost to one of the world's major centers of population and industry; their tentative, nonstandard techniques also illustrate the rough beginnings of today's earth sciences. The lake's currents finally began to be understood when the U.S. Weather Bureau in the 1890s set adrift hundreds of bottles containing identifying papers and then pursued and collected them.[9] Theories of the glacial processes that left Lake Michigan in the wake of a retreating ice sheet were in their infancy in the nineteenth century, and civil engineers employed techniques that had changed little since the Byzantine Empire. Cartography, similarly, had no standardized way to represent landforms for systematic study until the 1890s, and Chicagoans had to wait until Fryxell's *The Physiography of the Region of Chicago* to find a systematic account of Chicago's landforms and their formation.[10] By this time other compelling visualizations had inspired successive waves of urban fabric, with the city and its residents, planners, and builders stumbling blindly on, attending to problems after they had become acute, with layers upon layers of makeshift solutions.

VISUALIZING RAIL NETWORKS

Chicago's advantageous position had been remarked on by the earliest French travelers through the region. Louis Joliet suggested that "we could go with facility to Florida in a bark [canoe]. . . . It would only be necessary to make a canal by cutting through but half a league of prairie," not only providing a nearly direct link between eastern North America's two great waterways, but also opening the way to the prairies beyond.[11] By the early decades of the eighteenth century the location of the Chicago River and portage was carefully and consistently indicated by French mapmakers. Yet Joliet's prescient advice to construct a canal was not followed until Congress authorized the construction of the Illinois and Michigan Canal in 1822. The portage and canal may account for Chicago's emergence as a transportation node, but it was the railroad that opened the way to the prairies beyond. Mapping played a crucial role in this transition from canoes to canals to railroads and, eventually, to highways.

The much delayed opening of the canal in 1848 was nearly simultaneous with the arrival of the city's first railroad, the Galena and Chicago Union, which was originally intended to link Chicago with the lead mining region of northwestern Illinois and southeastern Wisconsin. The economic importance of the canal was further diminished by the construction of additional railroads in the 1850s. "By the eve of the Civil War," Roger Grant writes, "11 railroads served the expanding metropolis. Boosters claimed that a hundred trains daily served their city. More significant, was Chicago's role as the principal transshipment point between eastern and western rail networks. No city before or since assumed such a strategic position."[12] This favorable position for rail traffic depended in part on the same advantageous position exploited by the canal. The southern tip of Lake Michigan was both a natural terminus for lake traffic and an obstacle to be circumvented by land-based traffic from the populous and industrial Northeast and the mid-Atlantic states.

A Rand McNally map of the United States from 1878 (see Figure 12.3) shows how railroads heading west from Indiana, Michigan, Ontario, Ohio, and points east are constricted by the lake and into a narrow corridor in northwestern Indiana. To the west, in contrast, the railroads fan out, seemingly unimpeded, in all directions. The next major physical barrier, the Mississippi River, was already bridged in fourteen places by the time this map was published. The first of these bridges, linking Rock Island, Illinois, and Davenport, Iowa, was built by the predecessor of the Chicago, Rock Island, and Pacific Railroad, almost due west of Chicago. The next thirteen were situated along the river between St. Paul and St.

Louis. Each made its crossing point into railroad hubs in its own right. Chicago, however, sits two hundred miles east of the midpoint between the northernmost and southernmost crossings. Virtually all of the rail traffic between the northwest, the southern shore of the Great Lakes, and beyond must pass through the city. Though Indianapolis and St. Louis appear as secondary hubs, a pronounced pattern of lines radiating southwest from Chicago reflects the attraction of the lake city's connections with northern lines.

This map also reflects Chicago's rail and commercial dominance of the Midwest in a subtler, but no less significant way. It is a product of Rand, McNally & Company, established by William H. Rand and Andrew McNally (migrants from Boston and Scotland) ten years earlier in 1868, when they bought out the job-printing arm of the *Chicago Tribune*. Rand McNally followed the model of many American newspaper printers, which sought to recoup the costs of expensive equipment by serving the printing needs of other local businesses. In Rand McNally's case, one of the fastest growing "local" industries in Chicago in the 1860s and 1870s was, of course, the railroad. The company found a niche printing tickets, timetables, and handbills for railroad companies based or terminating in Chicago.[13] Many of these were land-grant railroads, which were then aggressively engaged in building transcontinental and regional lines. Through these client relationships Rand McNally became engaged in publications, including maps, that promoted the development of the "Great West." By the early 1870s the company advertised map engraving as a specialty. It published its first comprehensive guide to western railroads in 1871. This compendium of maps, general timetables, and ticketing information became national in scope in 1872 and dominated the market to the end of the century. The company's embrace of wax engraving (cerography) enabled the production of maps in large press runs, maps that could easily be combined with printed text, updated and customized to the specifications of clients, and which were receptive to the use of machine coloring.[14] The facility with which Rand McNally could customize its maps meant that even small local clients could afford to incorporate maps into modest promotional pamphlets and handbills.

Other Chicago printers, such as Poole Brothers, adopted similar business models, but Rand McNally ascended to national prominence as a map publisher, aggressively branching out into general mapping, including atlases, school geographies, and a line of popular pocket maps. Here too customization helped expand their market. General atlases were sold in bulk or by subscription through local newspapers with custom covers and title pages, and often the insertion of some local maps or advertising content. By the turn of the century it was the largest map publisher in the country. In this respect too it was able to exploit its rail position, employing mass-marketing techniques and access to cheap and rapid transportation in much the same manner as the famous Chicago mail order houses, Montgomery Ward and Sears, Roebuck & Company.

The 1882 map *Chicago and Environs* by the firm of Van Vechten and Snyder (see Figure 12.4) shows the influence of Chicago's extensive inter-regional rail network on metropolitan growth. The constriction of main lines arriving from the east in the Calumet region of Illinois and Indiana is visible at the far southeastern margin of the map. Several lines parallel each other hard by the shoreline of Lake Michigan, and to the south is a tangle of four additional main lines. Across the state boundary these and other lines from the south intertwine with the Calumet waterways serving the emerging heavy industries of the city's South Side and southern suburbs.

To the west and north, in contrast, the railroads, the same main lines that reach far to the west, fan out at intervals of four to eight miles. In DuPage and northwestern Cook Counties the railroads cross higher agricultural ground situated on moraines and glacial till and drained by south-flowing rivers. A string of suburbs has developed along these lines, often with the collusion of the railroads, which realized revenue from commuter services. These include suburbs such as Riverside and Norwood Park, which developed expressly as commuter railroad communities, and older towns such as Naperville, Downers Grove, Hinsdale (Fullersburg), and Elmhurst, whose historic cores along highways have expanded or migrated to the vicinity of new rail depots. An older fan of regional roads parallels the railroads. The old highways also still support independent communities and market centers, such as Addison, Bloomingdale, and Niles Center, but those that lie at some distance from the railroads are smaller and more compact. Concentric circles measure the distance in miles from Chicago's central courthouse, helping readers calculate the expense and time required for the commute to the city center.

The dominant radial pattern of the rail network is countervailed only by the newly constructed Chicago Belt Railroad, an inner circumferential route developed by a consortium of main lines to move heavy freight between lines at a distance from the congestion of the city center. Substantial suburban rail yards, focal points for industrial development in the late nineteenth and early twentieth centuries, are developing near the intersections of the Belt Line and the main lines. These patterns were more than lines on a map; where they concentrated, so too did people.

VISUALIZING LIVING CONDITIONS

Chicago's development at the heart of a nexus of freight lines, raw materials, commodities, and other manufactured goods made it among the world's fastest-growing population centers in the industrial era, with a growing demand for any emigrants willing to accept menial jobs in its factories, stockyards, steel mills, and freight yards, and work their way toward distant promises of education, home and business ownership, and other marks of prosperity. Experiences like those dramatized in Upton Sinclair's novel *The Jungle* were common for many new Chicagoans from central, eastern, and southern Europe as well as from the American South. Migrants found themselves living in cramped, often unsafe quarters with unreliable water and sewer service, working long hours for scant wages, with few opportunities for education or acculturation to the urban environment.

Despite several cholera outbreaks and multi-block fires from the 1830s on, Chicago before the Great Fire of 1871 had few safety or building codes; it had few health codes or zoning requirements well into the twentieth century. Many residential neighborhoods were intermingled with industrial areas, and though sewer and drainage infrastructure alleviated risks of such diseases as cholera, these neighborhoods faced constant pollution and stench from slaughterhouses, smelting operations, tanneries, and other industries. While Chicago offered opportunity, particularly to second- and third-generation arrivals, this came hand in hand with navigating regular and ongoing public health crises and environmental disasters.

Housing and factory building codes in the wake of the fire, which destroyed much of the city core, are perhaps most easily visualized in a relatively new genre of maps arising from these concerns. Fire insurance atlases color-coded buildings according to their construction materials and the apparent degree of safety those materials implied. Chicago's factories, warehouses, offices, and residences turned largely to brick and eventually steel-frame construction after the fire. These maps also included building dimensions, predominant uses, and the locations of water towers and, eventually, sprinkler systems. They were compiled into large, bound volumes which could be updated over decades by pasting new data over old, chronicling changes to urban built environments.

These atlases offer windows into Chicago's often overlapping industrial and residential areas. A view of the 4200 blocks of South Ashland and Marshfield Avenues (see Figure 12.5), updated as recently as the 1970s for the People's Gas Company of Chicago, shows a mix of houses and industry near the end of a blocked-up branch of the Chicago River

known infamously as Bubbly Creek, due to methane and hydrogen sulfide gas bubbles resulting from the decomposition of animal matter
dumped into the river by surrounding slaughterhouses. Visible on the
left is the remaining facility of Thomas E. Wilson and Company, originally a slaughtering, meatpacking, and hide-tanning interest which
became better known in the twentieth century for its contributions to
sporting goods, including the development of now-standard designs for
American footballs, basketballs, and baseball gloves.[15] Here the Wilson
facility's operations are noted, from a pickling room and hide storage
cellars to hide washing platforms, the runoff from which was likely to
flow into the river and wash into local streets, as residents recalled.[16]

The streets on this map contained other industries, such as the Darling and Company Garage and Repair Shop at upper right, with a handful of houses at the southern end of the same block. While the Wilson
and Darling buildings were brick, seven of the nine houses on the block
that remained into the 1970s were made of wood, represented in yellow, which was characteristic of Chicago's poorest working-class housing stock. Clear pasteovers and a proliferation of empty lots suggest
the block was in the process of residential abandonment when the last
updates to this map were made. In 2017 a few of these structures remained, converted into small businesses, including a tire repair shop, a
sandwich shop, and a beauty supply shop, with the Wilson facility now a
yard stacked high with shipping containers. Ironically, these containers
rest empty in many such yards in Chicago, after having delivered goods
from manufacturers across the world, possibly including the sources of
today's Wilson goods.

In the 1890s activists from Hull House, among the most famous of the
settlement houses established to study and work to improve living conditions in industrial cities, employed similar maps in their studies of "a
congested district of Chicago . . . [and] problems growing out of its social
conditions."[17] The mission of Hull House, articulated by its founder Jane
Addams in 1892, involved "scientific study of the causes of poverty and
dependence, communication of these facts to the public, and persistent
pressure for [legislative and social] reform."[18] Its employees surveyed the
local population and mapped their living conditions, in the hope that the
public, and government, would come to understand the relevance of a social science perspective on problems in urban life—namely, that disease,
crime, violence, and persistent poverty result from poor living conditions,
often imposed on people with few options for escape. Settlement houses
sought to provide options locally, through classes in literacy, trades, and
the arts, even as they advocated for government, industry, and charity to
recognize the nature of such problems and cooperate in working toward

structural solutions. Fieldwork in social science research, seen here in its disciplinary infancy, employed visual, cartographic strategies to improve living conditions in the urban environment.

An 1895 map of overlapping residential and industrial land use along another section of the Chicago River, one of many published in Hull House reports, illustrated some of those conditions with bright colors (see Figure 12.6). This section, bounded by Beach and State, Polk and Twelfth Streets, showed one of the "congested" areas of 1890s Chicago occupied by recent migrants, densely but sporadically distributed among rails, docks, a sugar refinery, stoneworks, freight houses, abandoned canal slips, and a train station. Inhabitants are labeled and color-coded in the American English terms of the time as "English Speaking (Excluding Irish), Irish, German, Dutch, Russian, Italian, Swiss, French Canadian, Arabian, Greek, Syrian, Polish, French, Chinese, Scandinavian, [and] Colored." In some places as many as nine distinct ethnicities are recorded as occupying a single house-sized residential property, attesting to the degree of crowding in the neighborhood, and it was common for houses to be divided multiple times, or for many haphazard dwellings to occupy a single residential city lot.

Simply providing adequate clean water, air, and light to local residents, to say nothing of sewage and waste disposal, would have been a challenge for city, state, and federal governments even if they had actively invested in making such provisions. That independent organizations such as Hull House were needed to study them suggests that little or no such active interest existed, at least insofar as this neighborhood was concerned. It was this image of Chicago that other sorts of reformers and, we might say, revisionaries sought to change by showcasing a gleaming, aesthetically conceived "White City" of the future at the 1893 Columbian Exposition, and by working to realize that vision for the entire metropole.

VISUALIZING PAST FUTURES

The urban planner and architect Daniel Burnham was not an environmentalist in the sense we understand the term today. But the character of the urban environment and its influence on the quality of urban life lay at the heart of the *Plan of Chicago* that Burnham published in 1909 with his protégé Edward H. Bennett for the Commercial Club of Chicago. The City Beautiful Movement, of which the Burnham and Bennett plan was the most complete expression, was ultimately an aesthetic movement, devoted to recasting cities as works of art, but such abstract and seemingly superficial goals were driven by practical environmental concerns that Burnham's progressive contemporaries would have understood. For

Burnham, Chicago's problems were the consequences of the city's rapid expansion, the despoiling of its lakefront, unsightly and unclean slums, consistent problems of pollution, and traffic congestion:

> The growth of the city has been so rapid that it is impossible to plan for the economical disposition of the great influx of people, surging like a human tide to spread itself wherever opportunity for profitable labor offered place. Thoughtful people are appalled at the results of progress; at the waste of time, strength, and money which congestion in the city streets begets; at the toll of lives taken by disease when sanitary precautions are neglected; and the frequent outbreaks against law and order which result from narrow and pleasureless lives. So that while the keynote of the nineteenth century was expansion, we of the twentieth century find that our dominant idea is conservation.[19]

This is a remarkable choice of words. John Muir and Burnham were contemporaries, though Burnham's "conservation" was more aligned with Gifford Pinchot's utilitarian concerns. Parks, forest preserves, and the beautification of Chicago's lakefront were major elements of the *Plan of Chicago*. So too were highway construction, the development of industrial infrastructure, and reshaping neighborhoods in the interest of efficiency and beauty.

Most of all, Burnham wished to conserve—that is, to sustain and support—Chicago's industrial and commercial growth. Ironically, the plan strives to hide the city's industrial and commercial landscapes, either by moving them outward from the city center or masking them. Accelerating a trend already well under way, heavy industry was to be moved west and south, encouraged and accommodated by new port facilities, belt railways, and a massive freight-handling facility on the city's perimeter. Central railroads were to be hidden underground and their passenger facilities consolidated. The main harbor was to be remade for pleasure craft and light freight. Railroads and industries along Chicago's south shore were to be screened by a shoreline of parkland and lagoons. Double-decked roadways and viaducts would relegate commercial traffic to a lower level, while upper decks would handle lighter traffic and promenades.

The plan's extensive visual component of maps and views reflects its ambivalence about the commercial environment it sought to conserve. The famously dreamy views and elevations by Frederik Janin and Jules Guerin diminish the presence of people and vehicles, except where needed to demonstrate how to design segregated traffic. The color illustration "View Looking North on the South Branch of the Chicago River"[20] shows the activity of teamsters and dockworkers handling freight from barges and flatboats from water-level roadways, while light traffic and

pedestrians pass unaffected by and perhaps unaware of the activity below. The *Diagram of General Scheme of Street Circulation and Parks in
Relation to the Population* (see Figure 12.7) articulates how a new web
of wide diagonal avenues and boulevards was intended to overcome the
inefficiency of the dominant street grid and shift the burden of through
street traffic away from a recreational and ceremonial city center and
densely populated inner neighborhoods. A plainer and larger-scale plan
of roughly the same area shows "the city as a complete organism in which
all its functions are related one to another in such a manner that it will
become a unit."[21] However, while the proposed new rail stations and
commercial piers can be discerned, the railroads themselves that are the
commercial foundation of this "unit" are invisible.

Burnham and Bennett's plan for a suburban highway system is one
of its most radical and far-reaching proposals, and among the first to envision a coherent highway system for an entire metropolitan area.[22] The
scheme (see Figure 12.8) extends the idea of crisscrossing, radial, and arching circumferential highways proposed for the city proper's thoroughfares.
These highways were to be constructed at state and county expense, and
would, the plan argued, offer both economic, health, and communication
benefits to those in city and suburb alike, and also greater access to the
countryside. Burnham and Bennett saw the periphery of Chicago in largely pastoral and agricultural terms. While Burnham himself commuted to
his downtown office from suburban Evanston, he saw the automobile's
recreational value as its primary contribution to the economic life of the
metropolis. Motorized trucks and affordable vehicles for the middle class
like the Model-T, introduced in 1908, were new on the market. While
Burnham and Bennett's metropolitan highway scheme might resemble
urban bypasses constructed on a massive scale in the 1950s and 1960s, the
Plan of Chicago did not fully anticipate the urban automobile age to the
extent that it might, had it been written even ten years later.

A map published by the Chicago Motor Club in 1963 (see Figure 12.9),
midway through the construction of Chicago's superhighway system,
suggests that Burnham and Bennett's plan for new diagonal avenues in
the city experienced similar difficulties with established transportation
networks, development, and private property claims. Though highway
officials endorsed the idea of cutting across the existing pattern of radial
highways and farm roads set on the Public Land Survey grid, few of the
links that would integrate local roads into this scheme were ever built.
Illinois Highway Route 83, the one circumferential route constructed before the interstate highway era, connected southern, western, and northern suburbs from the Indiana to Wisconsin state lines in an L-shaped
rather than semicircular route. Established radial highways were the

most important east–west suburban routes, and the most important cross-highways followed township, sectional, and subsectional boundaries. To this day Interstate 294 (the Tri-State Tollway) is the only recognizably complete circumferential route. A planned crosstown expressway near the alignment of Cicero Avenue met fierce resistance in the 1960s and 1970s and was never built. Efforts to extend the north–south Interstate 355 and other plans for major cross-pattern superhighways have encountered similar local opposition.

The motor club map shows that Chicago area expressway construction followed, wherever it could, the path of least resistance, paralleling railroad lines or routing through forest preserves. The Stevenson Expressway (I-55), still under construction on this map, was built over the original, obsolete alignment of the Illinois and Michigan Canal. In suburban DuPage and northwestern Cook Counties, expressways were built miles away from commuter railroads and towns on less-developed agricultural land. Already in 1963 it is evident that new suburbs like Hoffman Estates and Barrington Hills were developing along these new or newly widened highways, and older Naperville appears to extend tentacles of automobile-oriented housing tracts northward. By the end of the twentieth century similar maps would show that almost all of DuPage County's rural land had been incorporated into one town or another. The isolated small towns, oriented to the Chicago market and linked by improved rural highways envisioned by Burnham and Bennett, had evolved instead into today's sprawling landscape of residences, retail centers, light industry, and dispersed service companies.

CITY OF SWAMPS, RAILS, CROWDS, AND PLANS

Sprawl; rail, water, and sewer infrastructure; public health and safety; and aesthetic considerations are all, in the end, environmental concerns. Chicago, an industrial city built on sands and swamps, with unparalleled rail infrastructure and an impressive commitment to parklands and beautification, has been a unique site of both ecological catastrophe and reform. It is a built environment where nothing has been left unconditioned by social and economic activity. Its richness as a case study is made richer by the complex layers of cartographic visualizations of the city and its problems and solutions across two centuries, and through eras of unregulated development, deindustrialization and abandonment, crisis and poverty, as well as social reform, pollution mitigation, reinvestment, and sustainable redevelopment. Reading this environmental history in its vast archives of maps can illuminate these topics. We have sought to provide only a few of many possible examples, and we invite the reader to discover more.

PART IV

MANAGING (OR NOT) URBAN-INDUSTRIAL COMPLEXITY

After visiting Chicago in 1906, Rudyard Kipling wrote, "I urgently desire never to see it again. . . . Its water is the water of the Hughli and its air is dirt."[1] Despite this characterization, some officials and individuals attempted to ameliorate pollution almost as soon as factories started spewing waste. Local governments addressed the threats to public health by developing physical infrastructures as well as ordinances reining in the worst excesses of rapid industrial urbanization. In his essay, for example, Joshua Salzmann observes that Chicagoans frequently risked life and limb in the built environment that undergirded Chicagoans' commodification of nature. Accidental deaths prompted a "safety first movement" to improve the industrial economy's efficiency and lessen its mortality.

While Chicago's local governments made real gains, challenges continued over the twentieth century. With the 1906 publication of *The Jungle* by Upton Sinclair, conditions in Chicago's stockyards drew national attention. Progressive researchers and reformers turned their attention to ameliorating the intransigent environmental problems found especially in the city's working-class neighborhoods. Residents at settlement houses pushed for city, state, and national legislation to protect workers, their children, and their neighborhoods from industrial pollution. Sophonisba P. Breckinridge and Edith Abbott, drawn to Hull House by its founder Jane Addams, found around them "the presence of uncared-for-waste, the sight of blood, the carcasses naked of flesh and skin, the suggestion of death and disintegration—all of which must react in a demoralizing way, not only upon the character of the people, but the conditions under which they live."[2] Local pollution was class-bound and increasingly race-bound over the course of the twentieth century. Steven Corey's essay explores this geography of industrial pollution and racial inequality. Emphasizing the long-standing reliance on coal, he charts the lengthy, circuitous path to the clean water and air regulations.

Garbage collection and disposal remained a challenge. In the 1890s, when Jane Addams reported inadequate garbage collection in her Hull House neighborhood, the city made her a garbage inspector. She found this new role to be deeply enmeshed in the tangle of ward boss politics that made lasting reform difficult. Looking at the whole city, Craig Colten argues that Chicagoans handled refuse by displacing it, first as landfill along the lakefront and then as fill for streets or reclaimed wetlands and quarries. While the city continued this latter practice after World War II, it pushed its efforts far beyond its borders, creating exurban garbage landscapes.

Across the twentieth century industrial pollution most affected working-class neighborhoods. As hundreds of thousands of African Americans migrated to Chicago, they often found themselves relegated to areas in close proximity to noxious enterprises. Sylvia Hood Washington examines the citizen activism that emerged from racially segregated neighborhoods well before the 1970s when it took the name of "environmental justice." She identifies "Mrs. Block Beautiful," a campaign launched during World War I and continued into the 1950s, as an early urban conservation movement led by African American women to address the environmental consequences of this marginalization.

CHAPTER 13

Blood on the Tracks

Accidental Death and the Built Environment

JOSHUA A. T. SALZMANN

A few minutes after 5 p.m. on July 17, 1893, the Halsted streetcar driver Charles Stalnecker came to a pair of railroad tracks that cut across the path of his horse-drawn car, which was carrying about sixty people. Stalnecker's conductor, Frederick Barnett, jumped off to scout the intersection. The first set of tracks looked clear. Barnett's view of the second was blocked by several idle train cars, but the crossing gates were raised, and he waved Stalnecker onward. As Stalnecker rolled over the first set of tracks, Barnett saw a puff of train smoke rise over the second. Mistaking Barnett's sudden command to stop for a go-ahead sign, Stalnecker whipped the horses and drove the streetcar in front of an oncoming freight train.[1]

The collision sent passengers flying through the air. A dozen people suffered terrible injuries. Several died. The body of a plumber named Finn was cut in two, his head crushed. Frank Vandenberg, just three years old, died from internal wounds. Witnesses saw Stalnecker "wandering about crazed by the calamity to the people in his charge."[2] His overwhelming sense of guilt is understandable, but not entirely warranted. If the collision on Halsted resulted in part from miscommunication, new conditions in the urban environment—which put Chicago's inhabitants in perpetual peril—were equally at work.

Progressive Era American cities were becoming healthier in certain respects and, in others, more deadly. They were made healthier by civic reformers such as Robert Woods, Caroline Bartlett Crane, and Jane Addams who pressed city governments to install sewers, pick up trash, and remove animals from streets. Their efforts—dubbed by the historian Theodore Steinberg the "great cleanup"—led to sharp declines in deaths from infectious diseases.[3] Yet, even as the modern urban landscape was becoming healthier, new technologies were making the built environment hazardous to human life.[4]

Building great cities like Chicago involved the construction of tall buildings, mechanized worksites, and technological systems for lighting and transportation. City dwellers were frequently crushed by machines, suffocated by illuminating gas, burned in fires, and mauled by trains and streetcars.[5] Their deaths were meticulously documented by the Cook County Coroner's office. When a person died in an accident, the coroner summoned an "inquest jury" of six men to determine the cause of death and who, if anyone, bore responsibility. Cook County inquest juries investigated over seventy-four thousand deaths between December 1872 and November 1911.[6] The records of these investigations reveal how Chicagoans thought about bodily risk and safety in an increasingly mechanized urban landscape.

Oftentimes juries found accidental deaths to be cruel twists of fate, especially when they did not occur with great frequency. Over time, though, juries—as well as journalists, judges, and doctors—began to regard accidental deaths differently. They viewed them less as freaks of fate and more as symptoms of individual, corporate, and governmental failure. Coroner's inquest juries, journalists, and doctors thus implored public officials to enforce existing safety rules and add new ones on an ad hoc basis. They also called for better engineering in particularly deadly parts of the city and urged people to become more safety conscious. Their efforts helped fuel the rise of a national "safety first" movement, which sought to make the built environment more modern—orderly, efficient, and less deadly.[7]

DEADLY WORKSITES

Chicago's phenomenal growth assured that the city ranked among the world's largest construction sites. Located at the confluence of the Great Lakes and Mississippi River watersheds, Chicago became a hub for the transshipment of grain, lumber, and livestock.[8] Butchers built enormous meatpacking houses surrounding the Chicago Union Stock Yard, which opened in 1865. Grain elevator operators erected towering warehouses on the banks of the Chicago and Calumet Rivers. The riverbanks also became sites for manufacturing steel and iron, brewing beer, and shipping dry goods. Over two hundred lumber dealers crowded the banks of the South Branch of the Chicago River, stacking piles of timber as tall as buildings.[9] The lumberyards supplied wood to builders on the midwestern prairie; they also supplied builders in Chicago's booming residential construction industry, who were trying to keep pace with the city's astounding population growth from 350 people in 1833 to over a million in 1890.[10] The residential construction boom, meanwhile, was paralleled by the growth in commercial buildings. Using steel frames, Chicago's architects pushed downtown buildings to new heights.[11]

Many Americans understood Chicago and its skyline's growth as an embodiment of the force of nature. In an 1880 lecture to the Chicago Historical Society, former Illinois lieutenant governor William Bross maintained that "He who is the Author of Nature selected the site of this great city."[12] While Bross naturalized the city's growth, other observers acknowledged a greater human role in city building. The architect Louis Sullivan claimed that one could find in Chicago "the primal power [of man] assuming self-expression amid nature's impelling urge."[13] Chicagoans like Sullivan reworked the raw elements of nature into new forms—commodities and buildings—that the geographer Matthew Gandy calls "metropolitan nature."[14] It was perilous work.

For the laborers who forged the city's tall buildings, death was just a misstep away. As buildings grew in height, the coroner documented more deadly falls. A worker tumbled from the roof of the Rock Island Grain Elevator in January 1874, for example, and on June 22, 1884, George Gunderson fell to his death while working on the construction of the new Board of Trade Building, a ten-story, 320-foot-tall structure.[15] Sometimes pieces of buildings fell onto people. A loose brick fell off the Newberry Apartment Building at the corner of Oak and State on March 8, 1894, cracking open the skull of Carl Erick. Another man died when he was hit in the head by a coal bucket that fell from the roof of the Chicago Commercial Club Building on January 23, 1894.[16] The skyscraper was a symbol of the modern city; it was also a deadly worksite.

Lethal hazards abounded in many of Chicago's sites of production, which were increasingly mechanized and largely unregulated. According to the historian David Von Drehle, an average of one hundred people died in workplace accidents every day in America.[17] Chicagoans died in myriad ways. They succumbed to blunt force trauma: One man suffered, an 1874 inquest jury reported, mortal "injuries . . . while having a cake of ice fall on him at the brewery of Bartholomew and Roesing."[18] Another died later that year from "injuries received by the caving in of a coal pile in the yard of A. L. Zbedstram."[19] Some workers suffocated: James Sckudrna choked to death under a pile of grain at the Chase Elevator in January 1894.[20] Burns also claimed lives: a man died in March 1894 "from scalds and burns received by falling in a catch basin at Armour's No. 2 Beef House in Union Stock Yards."[21] Sometimes machines snagged laborers, tearing limbs from their bodies: William Thunes died in February 1884 at the Chicago Steel Works when he caught his arm in a machine as he tried to repair it while it was running. The coroner's inquest jury suggested, "While repairs of the belting are being made . . . the machinery should be stopped." The jury added, "If it is done while in motion experienced men should do it."[22]

Employers often blamed accidents on workers, even children. When Hull House founder Jane Addams confronted several factory owners about the accidental death of a child in 1890, she assumed they "would share our horror and remorse, and . . . do everything possible to prevent the recurrence of such a tragedy." To her surprise, "they did nothing whatever" to address the issue. Instead, the owners showed Addams a standard contract "signed by the parents of working children, [stating] that they will make no claim for damages resulting from 'carelessness.'"[23] With few incentives to protect workers, on-the-job safety was an afterthought and death an ever-present possibility.

DEATH BY GAS

Late nineteenth-century cities were shaped by what the historian Thomas Hughes called "technological systems," or the interplay between new material artifacts, organizational forms, and human users.[24] When these systems failed, technologies designed to provide human comfort destroyed lives. Illuminating gas offers a case in point.

Before electric lighting was widely adopted in the twentieth century, many buildings were equipped with indoor gas lamps, and Chicagoans sometimes suffered death by gas asphyxiation. Gas flowed to the lamp through pipes, which could be opened and closed with the turn of a cock, or valve. When the cock was open, a person could ignite the lamp. The grave danger lay in shutting the light off. The proper way was to turn the cock and stop the flow of gas. Sometimes, however, people simply blew out the flame and left the gas flowing. In those instances, the room would fill with odorless, colorless gas, suffocating its inhabitants, often as they slept. Mary Casey of 1330 Wabash Avenue made this error in April 1884 after, the inquest jury noted, "having neglected to turn gas off properly."[25] Hotel guests unaccustomed to gas lighting frequently made this mistake too. A guest at the Palmer House Hotel in March 1884 died from "asphyxia caused by inhaling gas, having left the gas cock accidentally turned on when retiring for the night."[26] After four guests died of asphyxiation in one downtown hotel, the manager posted a placard in each room with bold print that bore this warning:

Caution!

Turn off the Gas. Never Blow It Out.

Sure Death

To those who do so.

If you do not understand how to turn it off ring your bell and a servant will
    come and show you.[27]

In other words, the operator of the lamp could prevent death by following proper procedures.

In some instances, however, deaths were caused by the gas company.
Glitches in the gas distribution system killed six people in Hyde Park in
October 1892, for example. The victims included Curtis Goddard and
John F. Giasnier, teenage boys whom the *Chicago Tribune* described as
"well educated, intelligent, and sober." The description suggested they
knew how to operate the gas lamp in their bedroom. When Goddard
and Giasnier went to bed, they attempted to light a gas lamp, but the gas
did not flow. Supply shortages were dogging the Mutual Fuel Gas Company that supplied Hyde Park. The boys fell asleep. During the night, the
gas began to flow, suffocating them as they slept. Customers outraged by
the deadly, erratic flow of gas berated company officials, who maintained
the problems were beyond their control. The company had purchased a
massive new steel gas holding tank to be built by the Carnegie Works in
Homestead, Pennsylvania, but "the protracted strike there is the cause
of the delay."[28]

DEATH BY FIRE

Fires claimed the lives and property of Chicagoans with terrible frequency. The Great Chicago Fire of 1871 stood out for its destructive power;
in thirty-six hours, the blaze engulfed the city's downtown and much of
the North Side, laying waste to over 18,000 buildings and leaving ninety
thousand people, about one third of the population, homeless.[29] But it
was not unprecedented. Just a day before the Great Chicago Fire, a blaze
destroyed property valued at over a million dollars on the west side of
the city.[30] Three years after the Great Chicago Fire, in July 1874 another
fire burned forty-seven acres on the South Side, destroying 812 buildings
and killing twenty people.[31]

The frequency of deadly blazes inspired Chicagoans to make their city
"fireproof" through engineering and regulation. After the 1871 fire architects increasingly used steel, brick, limestone, and marble for Chicago's
newest buildings.[32] In 1875 the Chicago Department of Public Works,
the Board of Underwriters, and the Citizen's Association sponsored a
contest with a $1,000 cash prize for the person who could design the

best fireproof home.[33] The City Council, meanwhile, passed ordinances calling for fire alarms, light shields, fire exits, occupancy limits, and fire extinguishers.[34]

Despite this "fireproofing," the most deadly fire in the city's history occurred thirty-two years after the Great Fire, and in a new building which had been advertised as being "absolutely fireproof"—the Iroquois Theater. The 1,602-seat structure opened in November 1903. During a sold-out performance on December 30, 1903, a spark from an arc light onstage set the curtain ablaze. Flames and smoke enveloped the theater, killing 602 people.[35]

Why did so many people lose their lives in a new building that had been recently inspected? Political corruption likely made the fire deadlier than it might have been. Even though the theater had passed safety inspections just before opening in November, postfire inspections revealed numerous code violations. The theater lacked a fire alarm, exit signs, shields for arc lights, and an onstage fire extinguisher.[36]

How did these violations go unnoted in the recent inspection? One possible explanation is that the theater managers had bribed the building commissioner and inspector with free tickets. Chicago theaters routinely created a "free list" of city officials who could attend shows at no cost if they overlooked code violations.[37] After the Iroquois fire, a grand jury indicted the theater manager, assistant manager, and the stage carpenter for manslaughter. It also indicted the building commissioner and the inspector for professional malfeasance. The Chicago City Council passed a series of laws updating the city's fire safety codes.[38] It aimed to send a message: government could—and, next time, would—save people from fire.

DEATH BY TRAIN AND STREETCAR

Railroad tracks ensnared Chicago in a web of steel, and trains caused the greatest number of accidental deaths in the city. Railroad crossing deaths were the indirect result of government policies enacted between the 1850s and 1870s. Those accidents inspired people to demand that city and state government change how trains and streetcars moved through town.

The Galena and Chicago Union laid the first tracks in Chicago in 1848. In less than a decade, the city became the nation's biggest hub for rail traffic. By 1856 ten trunk lines with three thousand miles of track ran through Chicago. Fifty-eight passenger trains and thirty-eight freight trains rolled in and out of town each day.[39] The City Council helped lure these railroads to town with right-of-way agreements that made streets horribly dangerous.

During the 1850s the City Council passed numerous ordinances offering railroads generous rights of way across and along streets. This
delighted railroad executives, since it meant they could avoid slow, costly
negotiations with property owners. The city, meanwhile, benefited from
a new source of tax revenue. Rather than taxing all residents to maintain
Chicago's streets, the City Council levied special assessments, taxes, on
the people and businesses that owned property on the street in need of
improvement. If a railroad ran along a street, it had to pay any special assessments. Thus many alderman and some residents welcomed railroads
on their streets as a source of tax revenue.[40] It was a dangerous bargain.
Trains ran on the same level as pedestrians, turning streets into what the
novelist Upton Sinclair described as "a deathtrap for the unwary."[41]

The death toll grew with Chicago's population and railroad traffic.
According to the Chicago Department of Public Health, trains and
streetcars killed 31 people in 1870; 51 in 1875; and 124 in 1881. The death
toll in 1886 was 152. It rose to 299 in 1890; and in 1901 it reached 326.[42]
Sometimes people fell from trains; in most cases, though, people died
crossing the tracks. Chicagoans often struggled to spot trains snaking
their way through a landscape cluttered with buildings, and they sometimes failed to hear the sound of an approaching train over the din of
the city.[43]

The City of Chicago took steps to protect people, but its regulations
were not always enforced. The city occasionally required a specific railroad to post a flagman at a particular crossing, but oftentimes it made
no such demands. The City Council began to augment its piecemeal railroad safety requirements in the 1860s by passing citywide rules. In 1867,
for example, the City Council required railroads to post a flagman "on
all streets . . . on which horse railroad cars run."[44] Theoretically this law
would have prevented the dreadful collision on Halsted Street in July
1893, but at that time there was no flagman in the intersection. The city's
laxity in enforcing the law likely owed, in part, to corruption and apathy,
but it also had to do with nineteenth-century Americans' evolving understanding of fault in railroad accidents.

The Cook County Coroner's office records show that Chicagoans
blamed humans for crossing accidents with greater frequency over the
course of the late nineteenth century. According to the coroner's records,
twenty-eight people died from train accidents in the first six months of
1874.[45] The coroner judged the company or its employees at fault in just
four of those cases. In most instances the coroner's inquest juries described the accidents in terms that denied human agency. A report from
January 20, 1874, followed a standard linguistic construction: the victim
died "from injuries received by being run over by cars [of the C&NW]

and the jury from the evidence find that said death was accidental."[46] The jury absolved the city, the railroad company, and its employees alike with grammar. The injuries "were received" by the victim; nobody doled them out. Another report on a May 27 accident revealed another, common construction. The jury reported the victim died "from injuries received by being run over by engine No. 105."[47] Thus engine No. 105 had killed the victim—not the engineer, the railroad, or the City Council, which had invited trains into the streets.

Coroner's inquest jury reports occasionally assigned blame to people and railroads. On February 5, 1874, one report noted, a mother saw her daughter Josephine standing in front of an oncoming train "and ran to save her and in so doing stumbled and fell on the rail and was killed along with the child."[48] The mother made a tragic misstep. In other instances victims acted less sympathetically. On April 29, 1874, a fifteen-year-old-boy "went to the [C&NW] to steal a ride to the stock yard" and "slipped and was run over." His fault lay in trying to "steal a ride."[49] Railroad companies earned rebuke from coroner's inquest juries only in cases where someone had violated a clearly stated safety procedure, as when the Chicago, Burlington, and Quincy killed a pedestrian at an unguarded crossing on Union Street. The inquest jury determined that the death was "produced by gross carelessness on the part of the directors of said . . . rail road . . . in not keeping a flagman on said crossing."[50] The cases in which inquest juries held railroads accountable remained rare, though, until the 1880s and 1890s.

As the number of railroad deaths increased, inquest juries began to find railroad companies and employees at fault in more accidents. During the first six months of 1874 the juries found railroads and their employees at fault in just 4 of 28, or 14 percent, of railroad deaths.[51] Railroad death rates increased substantially in the next decade. In the first six months of 1884, 77 people died from railroad accidents. Inquest juries judged railroads or their employees at fault in 19 of those accidents, or 25 percent of the cases.[52] A similar pattern held true for 1894. During the first six months of that year 113 people died in railroad accidents, and inquest juries held the company or an employee responsible in 32 cases, or 28 percent.[53] In holding the railroad companies liable, inquest juries were suggesting that something could be done to avoid deaths.[54]

Coroner's inquest juries began to tell railroad and government officials what steps they should take to save lives. In the 1870s jury reports seldom made safety recommendations, but by the 1880s and 1890s they did so routinely. For example, when an Illinois Central train cut down a pedestrian at a crossing in March 1884, the jury noted that the railroad "should have better protections in place in the way of a fence or watchman to keep the people from getting on the tracks."[55] Following a crossing accident on Clifton Avenue in January 1894, the jury recommended that the "Chicago Burlington and Quincy R.R. Company employ competent watchmen and erect gates at said crossing."[56] After a train on the Erie Railroad smashed into a pedestrian in the winter of 1894, the jury noted: "We the Jury favor track elevation and would recommend to said R. R. Cos that they employ labor at street crossing that understand why bells are put in tower houses at said crossings."[57] Juries often called for railroads and municipal leaders to transform the built environment, making it safer. Who, though, would pay to erect gates and fences, build warning bell towers, and elevate tracks?

Chicagoans debated whether railroads or taxpayers should pay to protect people. As the death toll mounted, even pro-business newspapers like the *Chicago Tribune* denounced railroads for "murdering" pedestrians.[58] The railroad companies perpetrated so many "murders," the *Tribune* opined in 1889, they should be compelled to build viaducts, safety gates, and even elevate tracks at their own expense. It was a matter of restoring the streets to their previous condition. "Before the railroads entered Chicago," the paper recalled in 1891, "the crossings were safe." Then the "big locomotive steam horse came in and the crossings ceased to be safe." The paper concluded: "It is the duty of the companies to restore the crossings to their former safe condition."[59]

Railroad companies often wanted to elevate tracks and build viaducts, which eliminated costly delays at crossings; they just opposed footing the whole bill for such improvements. The railroad attorney George Willard told the *Tribune* in 1891, "Railroad companies, with few exceptions have been and are willing to bear a just proportion of the cost of building viaducts where they are really needed to lessen danger and save time." Willard simply wanted construction to be financed in the same manner as the city streets had been built—through special assessments on *all* the property owners and businesses located near the tracks.[60] Willard's arguments seemed logical, given that the City Council had forced railroads to contribute to the cost of paving streets. The State Supreme Court thought otherwise, though.

The court held railroad companies liable for the cost of making crossings safe. In January 1892, the court ruled that the City of Chicago had the right to extend a section of West Taylor Street across the tracks of the Chicago and Northwestern Railway. "Every railway company," the court said, "takes its right of way subject to the right of the public to extend the public highways and streets across such right of way."[61] Not only did the city have a right to run a street across the tracks, the railroad had to pay the price of making the crossing safe. Where the

TABLE 13.1: Leading Causes of Violent Death in Chicago, 1900 and 1905

Year	Rail road	Falls	Drowning	Burns	Street Railway	Suffocation	Homicide	Suicide	All Violent Deaths
1900	258	219	131	126	73	84	102	356	1,652
1905	319	286	145	140	147	120	185	430	2,107

TABLE 13.1. Leading Causes of Violent Death in Chicago, 1900 and 1905. *Source*: Thomas Grant Allen, "Chicago Takes Greater Toll of Human Life: Chicago's Loss from Violence," *Chicago Tribune*, April 15, 1906, F2.

tracks intersected with streets, the court noted, railroad corporations were "subject to the police power of the State." The state, in turn, could use that power to protect "life and liberty" by passing safety rules. Even if those regulations cost the company money, they did not constitute an unlawful taking of property without due process. The Illinois Supreme Court concluded, "The railroad must make crossings safe and it must stand the expense."[62]

PREVENTING ACCIDENTAL DEATHS

Dr. Thomas Grant Allen articulated an idea that had been incubating in the city for decades in a feature article in the April 15, 1906, edition of the *Chicago Tribune*. Allen argued that humans could prevent accidental deaths through engineering, regulation, and by being careful. The doctor's article reflected a growing emphasis on safety in America. As the historian Barbara Welke has noted, American Progressives regarded accidents as "products of habits inconsistent with the modern world."[63] Allen, like other Progressives, lamented the inefficiency of accidental death. Chicago offered a worst case. The city led America in "violent deaths," including those resulting from murder, suicide, falling, burning, drowning, suffocating, and train and streetcar accidents. With over two thousand such deaths in 1905, Allen noted, Chicago lost four times as many people "as perished in the Iroquois theater fire," costing its residents a staggering $2 million in lost wages. The ultimate causes of these deaths, Allen held, were lax enforcement of safety laws and the city's "spirit of hurry and rush."[64]

Allen's analysis of violent deaths underscored the perils of Chicago's built environment. Not all of the doctor's categories spoke to environmental hazards, to be sure. Suicide and murder ranked first and fourth among the causes of violent death. Deaths by drowning might or might not have been caused by conditions in the built environment. In some cases people fell off city bridges or drowned in privy vaults. In others

TABLE 13.2: Comparison of Violent Deaths to Total Mortality in Chicago

Period	Population	Average Deaths, All Causes	Average Deaths, Violence	Deaths, All Causes per 1,000 Pop.	Violent Deaths per 1,000 Deaths	Deaths, All Causes to 1 Violent Death
1856 to 1865	109,000	2,800	120	23.39	43	23
1866 to 1875	307,000	7,400	300	23.82	41	21
1876 to 1886	503,000	10,700	460	20.30	46	22
1886 to 1895	1,209,000	21,300	1,300	20.09	60	12
1896 to 1905	1,699,000	26,200	1,800	14.71	72	14

TABLE 13.2. Comparison of Violent Deaths to Total Mortality in Chicago. *Source*: Thomas Grant Allen, "Chicago Takes Greater Toll of Human Life: Chicago's Loss from Violence," *Chicago Tribune*, April 15, 1906, F2.

they perished while swimming, fishing, or boating. The other causes outlined by Allen were linked to the built environment, though. Railroad deaths were connected to transportation infrastructure, suffocation to systems of illumination, and burning and falling to buildings.[65]

Allen understood these accidental deaths in the context of advances in health care and the great cleanup of the Progressive Era. During the 1870s and 1880s scientists began to locate the origins of disease not in "miasmas" but in microscopic entities.[66] Acting on this knowledge, sanitarians pushed urban politicians to take measures to cleanse cities by building sewers, picking up trash, and inspecting sanitary conditions in homes and workplaces.[67] These efforts made the city healthier, Allen noted. The "progress in saving lives from the attacks of preventable diseases," Allen's data suggested, led to falling rates of death by nonviolent causes. In spite of that progress, violent deaths were on the rise in Chicago.[68]

Allen invited Chicagoans to think about accidental deaths in similar terms to those caused by disease. He asked readers to imagine an epidemic had occurred in which doctors could have prevented "50 per cent of the deaths" but failed to do so because they were "incompetent or wicked." That, Allen insisted, was the case with accidental death. Instead of incompetent doctors, though, residents were dying because "our administrators of law and order . . . have failed to enforce the ordinances and laws regarding railway crossings, the speed of vehicles, the safety of buildings, the closing of saloons at prescribed hours and the safety of our public streets."[69] Accidental deaths could and should be prevented. Allen's sentiment was shared by a growing number of business and civic leaders; it was, moreover, the premise of America's burgeoning Safety First Movement.

By 1915 the walls of America's public buildings and factories were cluttered with posters urging people to "Think Safety First." The slogan of
America's Safety First Movement was so ubiquitous that it was the title
of a satirical musical written by the Princeton student F. Scott Fitzger-
ald. The movement sought to decrease accidents and promote efficiency through education and engineering; its acolytes founded the Safety
First Federation of America in 1915. The historian Arwen Mohun has
shown that the movement had its roots in "lots of places" where people
confronted the mortal hazards of the built environment, including "railroads, streetcars, insurance companies," and coroner's offices.[70]

The story of Chicago's deadly built environment sheds light on how
and why those Americans came to believe that they had the ability—and
the responsibility—to prevent accidental deaths. Chicago led America
in accidental deaths because its denizens had so radically, and quickly, transformed the built environment. They had constructed factories,
grain elevators, and stockyards; erected soaring skyscrapers and hotels
with creature comforts like gas lighting; and spun a dense web of steel
rails ensnaring the city. Chicago's built environment shielded people
from the elements and facilitated transportation and production. It was
also a deadly hazard. Chicagoans fell off skyscrapers, got maimed by machinery, choked on gas, burned in fires, and were mauled by trains and
streetcars.

When Charles Stalnecker drove the Halsted streetcar in front of a
freight train in July 1893, he blamed himself for the ensuing calamity. He
was partly right. The actions of individuals did play a role in such tragic events, but their frequency demonstrated that conditions in the built
environment were also to blame. The city and railroad corporations had,
after all, sent freight trains barreling through busy streets. As the death
toll from railroad and other accidents mounted, it became increasingly
apparent—to Dr. Thomas Allen, the *Chicago Tribune*, and the members
of Chicago's coroner's inquest juries—that lawmakers, engineers, and
citizens could take steps to make the built environment safer. That realization, though, was hard won. It took time and the repetition of horrible
scenes like the one on Halsted Street on July 17, 1893. It took blood on
the tracks.

CHAPTER 14

Air and Water Pollution in the Urban-Industrial Nexus

Chicago, 1840s–1970s

STEVEN H. COREY

MACHINE CITY

In stark contrast to the present, smoky skies and dirty water dominated much of Chicago's physical environment from the 1840s to 1970s. This essay provides a framework for how and why visible air and water contamination went from being a ubiquitous part of Chicago life to a barely noticed occurrence today. Central to this transition is seeing Chicago as a nexus point where two processes, industrialization and urbanization, converge to generate an ever-changing landscape. In order to historicize the human manipulation of air and water, the geography of industrial pollution and social inequality, transformations in the city's overall economic base, and debates over coal consumption and other select responses of citizens to threats from environmental deterioration will be examined.

Most salient in the long-term improvement of Chicago's skies and waterways were the actions of people from all backgrounds who engaged for nearly a full century through formal and informal political means to curb pollution. Such engagement itself changed over time and reflected the city's history of domination by political machines and commercial enterprises. During the nineteenth and early twentieth centuries, social and economic elites sought voluntary cooperation and public policies that favored business practices and often exempted the largest sources of pollution. By the 1960s and early 1970s, though, widespread environmental activism at the grassroots and ad hoc levels demanded change and accountability from industries and public officials that transcended the political and economic paradigms which allowed pollution to go largely unabated for the better part of a century and a half.

The nineteenth century witnessed a period of unprecedented population growth and urbanization in the United States (U.S.). Chicago exemplified the astonishing rise of cities, growing from 4,470 residents in 1840 to 503,000 in 1880. Incredibly, just ten years later the population had doubled to over 1 million people, with Chicago becoming the nation's second largest city behind New York. By 1910 Chicago's population had doubled again to more than 2 million people, and by 1950 reached its peak at 3.6 million residents.[1]

Like population growth, industrialization in Chicago and the surrounding region occurred at exceptional speed and scale. In 1860 there were 750 manufacturing concerns in the city employing almost 6,700 workers. By 1880 Chicago was the preeminent industrial city outside the East Coast, with 3,519 firms employing almost 80,000 workers. In 1900 there were over 19,000 separate firms with more than 260,000 workers, making Chicago the second largest manufacturing city behind New York. Twenty years later Chicago produced more food products, furniture, iron and steel, railroad supplies, and telephone equipment than anywhere else in the country. Throughout the 1920s to the late 1930s, over 500,000 people throughout the greater Chicago metropolitan area worked in manufacturing.[2]

Chicago typified the "Machine Age" city at the center of industrial metropolitan areas throughout North and South America, Europe, and Asia by the early twentieth century. Such cities shared a common set of features that embodied a mixed legacy of social and economic progress, as well as technological limitations in attempting to harness and control the natural environment as part of urban and industrial processes. Machine Age cities depended on and generated an almost insatiable demand for energy, primarily fossil fuels, and in turn produced extensive pollution through the release of solid, liquid, and gaseous wastes, often without any treatment, into the land, air, and water.[3]

THE GEOGRAPHY OF INDUSTRIAL POLLUTION AND RESIDENTIAL INEQUALITY

The distribution of manufacturing facilities in Chicago formed a compact pattern with a central core, today's so-called Loop and the banks of

the Chicago River, and suburban industrial districts anchored by waterways and railroads radiating out from the center. As a result, the Chicago
River, and later the Calumet River, became heavily industrialized and
polluted, with business establishments located on or near riverbanks, using water as a means of transportation and waste disposal.

Adjacent to factory zones along the rivers stood the homes of industrial laborers and other working-class families, while the more affluent
lived farther away from manufacturing pollution and close to Lake
Michigan. Cheap labor helped fuel industrial growth. The migration of
thousands of people each year into Chicago from other U.S. cities or rural areas, as well as immigrants from abroad, provided a steady workforce for factory owners. Laborers endured grueling conditions on the
shop floor with low pay and little or no guarantee of steady employment.
Nor could the working class escape the ever-present side effects of urbanization and industrialization. Smoke, soot, and filth of all kinds lingered
in the air they breathed, the clothes they wore, and the water they drank,
cooked with, and, when so fortunate, bathed in.[4]

In terms of specific industries, clothing, printing, and leather works
dominated the Loop, while furniture, carriage, and metal shops abounded in the immediately surrounding districts. Along the North Branch
of the Chicago River were lumberyards; tanneries; distilleries; factories
such as the McCormick Reaper Works, fully operational by 1850; and
woodworking, meatpacking, and iron and steel firms, such as the North
Chicago Rolling Mill Company which opened in 1865. The Near North
Side contained a heavy concentration of breweries. Along the South
Branch of the Chicago River were stone cutting and brickyards, lumber
and woodworking mills, meatpacking plants, and iron and steel factories, such as the Union Rolling Mill Company built in 1863.

During the 1870s and 1880s industrialists further expanded to the
south and west. In 1872 a new and larger McCormick Works (known
as the McCormick Works of International Harvester from 1902 until its
closure in the 1950s) arose at Blue Island and Western Avenues. Directly
south, down the South Fork of the South Branch of the Chicago River,
stood the meatpacking district centered around the Union Stock Yards,
which opened in 1865. Residential areas in and around the slaughterhouses included Packingtown, later known as Back of the Yards, whose
hardscrabble way of life became infamous through Upton Sinclair's book
The Jungle (1906). Even in the 1890s, neighborhoods like Back of the
Yards had the character of a stereotypical western frontier, with wooden
sidewalks, dirt streets, and ramshackle wooden buildings. A network of
railroads, manufacturing plants, contaminated water, and putrid odors
from rotting animal flesh and huge dumps along Robey Street (now

Damen Avenue) underscored the isolation of this region from the rest of the city.[5]

In 1880 the North Chicago Rolling Mill Company began construction on a new facility at Lake Michigan and the Calumet River (the "South Works"). Other factories opened in the Calumet region, forming the steel communities of Southeast Chicago, specifically Hegewisch, South Deering, East Side, and South Chicago, as well as Chicago Heights, Illinois, East Chicago, and Gary, Indiana. As with the area adjacent to the stockyards, the industrial infrastructure of steel mills and accompanying scrap yards, slag heaps, and dumping grounds hemmed in residents within land-, air-, and waterscapes of waste. However, for some residents dirty air itself was a sign of prosperity. As South Chicago resident Estelle Uzelac Latkovich reminisced in 1983 about the area in the 1930s, when "orange ore dust belched from the enormous smokestacks, we knew times were good—the men were working."[6]

The cumulative impact of manufacturing in the Chicago metropolitan area on the physical environment was the release of an especially noxious and often deadly brew of water and air pollution that permeated daily life for every person—indeed, almost every form of flora and fauna—in the city. Slaughterhouses, tanneries, and breweries released a slurry of wastewater, blood, entrails, manure, and decomposing plant-based materials. These wastes mingled with untreated human sewage, drainage water from streets and buildings, as well as toxic dyes and chemicals from printing, textile, iron, steel, and machine production. Exacerbating the discharge of these materials into waterways was the city's "first nature" typography as a swampy prairie, with shifting shores and a low water table, that absorbed water in its myriad states of purity and released it beyond the control of human technology. In terms of air pollution, almost every building in the city had a heating system that released smoke into the air. Industrial flues and open furnaces spewed forth every conceivable hue of smoke and variety of particulate matter such as soot, ash dust, and smoldering cinders, as well as chemical substances like iron oxide that turned skies a red glow at night.[7]

ANNIHILATING CLEAN AIR: COAL

Smoke and soot from the combustion of bituminous or soft coal represented the largest single source of air pollution in Chicago. Beginning in the mid-1840s, the city's skies became darker and more odiferous from the increased use of bituminous coal, given its high sulfur, ash, moisture, and volatile matter composition. Soft coal was—and still is—an abundant source of energy found throughout much of Illinois south of

Chicago, western Indiana, and western Kentucky, in the so-called Illinois Basin or the Eastern Interior Coal Field.

Coal's impact came in three principal areas: as a form of energy, as a commodity, and as effluvium or a by-product of combustion. Coal was essential to steam generation for manufacturing and the railroad locomotives that traversed the region. Coal was both cheap and plentiful, with dozens of storage and distribution yards lining the waterfront and railroad corridors. Across the United States coal consumption sparked a maelstrom of controversy over how much smoke residents could tolerate given the dominant political economy that allowed for generally unfettered business practices. After the Civil War public health and urban beautification movements took aim at reducing coal emissions to improve the quality of life in cities.[8]

Chicago's first smoke ordinance came in 1881 after several years of lobbying by an elite group of residents and business executives called the Citizens Association. Although the regulation, one of the earliest of its kind in the United States, did not apply to residential buildings, it did declare dense emissions from the smokestack of any boat, locomotive, or chimney associated with a commercial establishment to be a public nuisance. Compliance with the 1881 measure proved difficult, though. Enforcement fell to a small cadre of sanitary inspectors in the Department of Health already busy enforcing numerous other ordinances related to sewage and garbage. Even when inspectors issued citations, city attorneys were often too busy with other matters to take action, or politicians used their influence to delay or dismiss cases. The ravenous demand for more energy also undermined antismoke efforts.[9]

Not that the city or its newspapers gave up trying to document and shed light on the smoke problem. For example, in July 1890 the *Chicago Daily Tribune* reported on the air quality in one industrial neighborhood of the Near North Side along the river. A collection of distilleries and light industry produced soot so foul that a postal worker's face became black and his eyes sore every time he walked down Kingsbury Street, prompting multiple baths each day. Coal furnaces at the Shufeldt Distillery produced a plume of smoke twenty feet high before strong winds dissipated it onto rooftops and the streets. The largest offenders, though, were the Edison power plant and the city's own electrical light works which burned soft Illinois coal from sunset to sunrise, forcing residents downwind to keep their windows closed to protect their linens. Paradoxically, as the city generated large volumes of smoke, it also tried to keep emissions from locomotives in check. As the *Tribune* reported, the city's law department charged the Illinois Central Railroad for multiple smoke violations. Unfortunately, such litigation typically dragged on for years, even decades, and ultimately had little impact on reducing air pollution.[10]

The selection of Chicago as the site for the World's Columbian Exposition in 1893 brought a renewed round of interest in cleaner skies. In late 1891 a public meeting of civic organizations held at the Union League Club brought about the formation of the Society for the Prevention of Smoke (SPS). Leaders of SPS included numerous Citizens Association members, directors of the Columbian Exposition, and leading newspaper publishers, bankers, real estate developers, manufacturers, and meatpackers.[11]

The immediate target of SPS was the dense and dark smoke from buildings located in, and tugboats that operated around, the downtown area. An initial campaign to seek the voluntary cooperation of building and tug owners to reduce emissions met with mixed success. The SPS hired engineers who lent technical expertise to those interested in modifying their equipment. By July 1892 SPS announced that 40 percent of building owners downtown, and a few tugboat owners, had "practically cured their smoke nuisance."[12] Another 20 percent unsuccessfully attempted smoke abatement, leaving 40 percent making no efforts.

In December 1892 the SPS announced that smoke was under control downtown. Such a claim, though, was premature. As the national depression of the early 1890s began to take its toll in Chicago, compliance began to slack. Since participation in SPS initiatives was purely voluntary, personal values determined commitment levels. Absentee landlords were a particularly difficult group to negotiate with. Some building owners could not afford upgrades to their boilers and furnaces, while others remained unwilling to reduce their profits, and still others could not see how compliance would make a difference. Chicago's dark industrial and coal-ridden skies did not clear in time for the Columbian Exposition.[13]

THE EVOLUTION OF WASTEWATER TREATMENT

Prior to the widespread acceptance of the germ theory by the early twentieth century, citizens and public health reformers often linked water and air cleanliness to the assumption that, under certain conditions, wastewater, smoke, effluvium, and putrescent materials caused disease. Given this understanding, the best way to prevent sickness and death was to halt the generation of noxious matter altogether—not a serious option in a booming industrial economy—or to dispose of it in a timely and efficient fashion. Given the ability of waterways to dilute and carry wastes far from the point of production, they became de facto sinks.[14]

Relying on waterways as sinks, though, threatened their simultaneous status as sources of potable water. Prior to World War I Chicagoans were at the mercy of a disease chain wherein raw or untreated wastewater emptied into rivers that flowed into and contaminated the city's main drinking supply, Lake Michigan, bringing about episodes of cholera, typhoid, and dysentery. In 1889 the Illinois legislature created the Sanitary District of Chicago (SDC), which in 1900 finished the Sanitary and Ship Canal, or Main Channel, that permanently reversed the flow of the Chicago River to send wastes away from Lake Michigan. However, a 1902 outbreak of typhoid underscored the fact that changing the river's course did not bring an end to the water-borne disease chain.

Nor did the Chicago River cease being a dump. In fact, reversing the flow eliminated the natural drainage of Bubbly Creek—a 1.25-mile portion of the South Fork of the South Branch of the Chicago River named after the release of bubbles from the decomposition of animal processing wastes—and turned it stagnant. Though imperfect, substantial progress in eliminating disease did begin in the 1910s and 1920s with the construction of large intercepting sewers, filtration plants, and chemical treatment facilities to neutralize harmful bacteria. Chemical augmentation of the SDC canal system, though, also allowed for the continued use of rivers as waste sinks.[15]

ELECTRIFICATION AND THE POLITICS OF SMOKE ABATEMENT

Smoke returned to the forefront of city politics in 1907 with the creation of a new municipal agency dedicated to smoke abatement, the Department of Smoke Inspection (DSI), and calls for the conversion of locomotives from coal to electrical power. In September 1907 Paul P. Bird assumed control of the DSI and soon after called for the electrification of railroads to reduce pollution, traffic delays, and railroad operational costs. Despite the success of electrification in New York City, the Chicago City Council failed to act.[16]

During the summer of 1908 a coalition of mostly middle-class women from the South Side, led by Annie Sergel, president of the city's Anti-Smoke League, circulated a petition that called for the city to require electrification. With the support of forty-two other civic organizations, such as the Chicago Women's Club and the South Side Business Men's Club, two hundred members of the Anti-Smoke League presented the electrification petition, with forty thousand signatures, before the City Council in October 1908.[17]

Once again, the City Council failed to act. Over the next eight years, railroad companies responded to calls for electrification by using their

TABLE 14.1. Chicago Sources of Visible Smoke Generation

Source Classification	Percentage of Visible Smoke 1911 Report	Percentage of Visible Smoke 1915 Report	Difference in Reports (1915 from 1911)
Steam Locomotives	43	22	−21
High-Pressure Steam Power & Heating Plants	36	44.5	+8.5
Metallurgical Furnaces, Gas & Coke Plants	12.5	28.75	+16.25
Low-Pressure Steam Power & Heating Plants	4.5	4	−0.5
Steamboats	4	0.75	−3.25
Total	100	100	

Note: "Source Classification" is a consolidation of categories used in both the 1911 and 1915 reports. Source: Paul P. Bird, Report of the Department of Smoke Inspection City of Chicago (Chicago, February 1911), 39; Chicago Association of Commerce, Committee of Investigation on Smoke Abatement and Electrification of Railway Terminals, Smoke Abatement and Electrification of Railway Terminals in Chicago (Chicago: Committee of Investigation, 1915; reprinted Elmsford, NY: Maxwell Reprint Company, 1971), facing p. 136.

vast economic, social, and political capital to refute or trivialize concerns that their business practices undermined the quality of life in Chicago. Two key reports illustrate the issues.

In 1911 the DSI issued the first comprehensive overview of Chicago's smoke nuisance and concluded that, at minimum, the financial loss to the city from smoke ran at least $17.6 million annually, or about $8 per capita. In making this calculation, Bird and the DSI considered property loss for all people living and working in the city, as well as the impact of smoke on the health of people, animals, and plants. Most interestingly, given statements from the American Medical Association that smoke was not a danger to public health, the 1911 DSI report noted that medical doctors all over the world were directing a fight against smoke, which they knew to be extremely harmful to human membranes.[18] Using the Ringelmann Method that compared smoke and soot density levels against a grayish grid of black lines on a white surface, the DSI found that railroad locomotives generated more visible smoke than any other source, at 43 percent of the total (Table 14.1).

In contrast the Chicago Association of Commerce released in 1915 a massive $100,000, multiyear study, Smoke Abatement and Electrification of Railway Terminals in Chicago, which used the same Ringelmann Method but found that locomotives accounted for only 22 percent of visible city smoke. Rather, *Smoke Abatement* found that high-pressure steam power and heating plants (used throughout industry and large commercial buildings) produced almost 45 percent of the visible smoke. As Table 14.1 illustrates, the two reports assigned responsibility for visible smoke very differently.

Smoke Abatement shifted the blame for visible smoke from locomotives to industrial furnaces and larger power plants. In a rather remarkable, and to a degree disingenuous, line of reasoning, *Smoke Abatement* argued that approximately a third of atmospheric pollution in Chicago came not from smoke, but from so-called allies of smoke which included: the bare ground which produced dust when dry; building roofs and backyards where people beat their rugs (here the study failed to mention that much of this dust came from dirty coal soot); building demolition and construction; street construction and repairs; normal wear and tear or "abrasion" of streets and sidewalks; and street cleaning and municipal solid waste disposal.[19]

Smoke Abatement then, in its own logical way, argued that electrification of locomotives would not make an immediate or dramatic difference to the city's skies. In fact, the report asserted that electrification would result in a reduction in visible smoke by no more than 20 percent; a reduction in solid constituents of smoke, soot, and other particulate matter by only 5 percent, and dust and dirt from all sources by no more than 4 percent. Instead, *Smoke Abatement* called for the creation of a permanent Pure Air Commission, made up of persons with technical qualifications, to oversee the installation of boilers and furnaces, investigate all sources of air pollution, supervise street paving, and clean thoroughfares, roofs, and all other sources of smoke allies.[20]

Even though the American Medical Association finally caught up with their European peers and acknowledged public health risks from air pollution in 1915, the energy of the Anti-Smoke League campaign had waned, and another reform moment had passed in Chicago. In just a few years World War I ushered in an economy that relied heavily on the very sources of pollution that made the skies darker and more dangerous than ever.

Electrification became a hot political topic again in the mid-1920s, ironically because the Illinois Central finally opened an electrified south shore suburban line. Once again citizens and newspapers pressured remaining railroads to electrify, and once more these companies balked and the city declined to act further. In fact, the only visible reductions in smoke production between the two world wars occurred during the Great Depression, as factories closed or precipitously reduced their output.[21]

POSTWAR DECLENSION AND
CLEAN AIR REGULATION

Beginning in the late 1940s and intensifying in the 1960s and early 1970s, Chicagoans renewed their efforts to improve city skies, which by federal survey and reputation, ranked second only to New York in terms of poor quality. While many of the same themes and conditions from antismoke campaigns of the 1880s–1910s remained, three factors account for eventual success by the early 1980s. First, a heightened national awareness of the dangers of environmental degradation, and specifically air pollution, brought about new sensibilities on acceptable levels of smoke in daily life and opportunities for effective regulation. Second, a flurry of citizen activism through community coalition building cut across class and ethnic cleavages to unite public health, environmental, and local neighborhood organizations. Finally, urban and industrial realignments in the 1950s–1980s diminished the political and economic clout of principal pollution sources.

Academic and social commentators commonly characterize postwar urbanization in the United States as a period of declension for large cities, specifically in terms of losses in population, quality housing stock, industrial production and employment, the commercial tax base, and the ability to provide adequate public services. Of particular note was the era's residential and community instability, as large numbers of middle- and working-class white ethnics left for the suburbs while lower-income whites and African Americans from the rural South moved in, along with Latino, Caribbean, and Asian immigrants.

Even before the suburban exodus, though, Chicago homeowners organized to fight the noxious impact of factory emissions. On October 1, 1948, residents of Wolfe Park in the city's South Side woke up to find their homes covered in a dust comprised of copper, carbon, iron, iron oxide, zinc, and other minerals. While such material had been falling on their property for years, this occurrence etched windows and peeled paint off homes. In response the Wolfe Park Civic Association met with nearby manufacturers to hold them accountable for property damages. Only Republic Steel indicated a willingness to monitor emissions to keep the neighborhood clean and eliminate such waste in their production process. Without serious enforcement of antismoke ordinances by the city, Republic Steel's actions were the best any community organization could hope for. Indeed, it could have been much worse. The very same month Wolfe Park residents were dusting off their homes, a five-day smog from an air temperature inversion in Donora, Pennsylvania, trapped emissions from a zinc works and killed at least twenty people.[22]

Nationally the late 1940s saw public health authorities articulate the need for greater regulation by linking environmental quality to the future of cities. At the 1949 meeting of the American Medical Association's Congress of Industrial Health held in Chicago, experts argued that, left unchecked, smoke would lead to declining residential property values, an exodus of citizens to the suburbs, and the creation of slums and other "rotten centers."[23] The 1950s and 1960s saw a proliferation of state and local air pollution control programs in response to poor urban air quality and assistance from the federal government in the form of several so-called Clean Air Acts.

The first federal measure, the Air Pollution Control Act of 1955, authorized the U.S. Public Health Service to research and disseminate information on air quality. The Clean Air Act of 1963 went a step further by encouraging citizen participation in government hearings, policy making, and enforcement. With the 1967 Air Quality Act, the federal government expanded its approach to empowering local agencies and citizens through the establishment of metropolitan air quality regions throughout the country and requiring states to establish air quality standards.[24]

Chicago eventually enacted more stringent smoke measures in 1953, 1958, 1963, and 1965—many modeled after those in Los Angeles—which, at least on paper, required more sophisticated incineration technologies. However, as in the past, provisions allowed large polluters to delay or even forgo compliance. By the mid-1960s the scope and scale of air pollution in Chicago was nothing short of daunting. In a city of 3.5 million people, there were approximately seven thousand industrial enterprises and literally hundreds of thousands of chimneys. Although most chimneys were residential and discharged little visible gas or smoke, in the aggregate they laid a baseline of air pollution exacerbated by emissions from thousands of industrial and metallurgical smokestacks and more than twelve thousand large coal-fired boilers.

Smoke, of course, respected no geographical boundaries. Southeasterly breezes brought heavy doses of industrial pollution from the Calumet region in northeastern Illinois and northwestern Indiana. On those days, as much as 70 percent of all air pollution came from steel manufacturing. By 1966 the average Chicago resident inhaled an average of three pounds of dust annually. In contrast, the city's Department of Air Pollution Control, organized in 1958, had only a dozen inspectors who used monitor stations and helicopters to issue violations and fines. In 1965 alone inspectors reported 6,192 violations, issued 2,635 tickets (at a fee of $10), and received 6,085 complaints from the public, a number that almost doubled by 1970.[25]

"IT'S GETTING SO BAD THAT EVEN PEOPLE ARE COMPLAINING."

FIG 14.1. Bill Mauldin, "It's Getting So Bad That Even People Are Complaining," 1965, *Chicago Sun-Times*. Copyrights held by and image courtesy of the Pritzker Military Museum & Library.

The stakes for air pollution control became higher with each passing year. After Donora, several large cities experienced even deadlier temperature inversions that trapped pollution under clouds for days at a time. Among the most notable were the five-day "Big Smoke" in December 1952 that killed upward of 4,000 people in London, England, and New York City's three inversions in 1953, 1963, and 1966, the latter killing more than 170 people. While Chicago experienced inversions, there were no mass deaths, and the city proclaimed itself a national leader with tougher standards. On days when northerly winds prevailed, Chicago was a showcase for improvement, while southerly winds brought dense, filthy smog.[26]

"NO PLACE TO HIDE": GRASSROOTS ACTIVISTS AND COMMUNITY ORGANIZING

Despite heightened expectations for improved federal, state, and local regulation, little had changed in the skies over the nation's largest cities. Into the void of serious enforcement, Chicago's community and environmental organizations united to exert political pressure on Mayor Richard J. Daley and other government officials. In the summer of 1969 the Tuberculosis Institute of Chicago and Cook County helped organized the Clean Air Coordinating Committee (CACC). The CACC brought together twenty-five neighborhood, environmental conservation, and public health groups.[27]

The CACC immediately mobilized membership to attend public hearings of the Illinois and Indiana Air Pollution Control Boards to testify on the need for stringent emissions standards to comply with provisions of the federal 1967 Clean Air Act. Other activities included condemning the City Council and Mayor Daley for granting coal users a one-year delay on a 1969 deadline to limit the sulfur content in coal to only 2.5 percent (most coal burned in Chicago had a content of 6–7 percent). Dr. William R. Barclay, Tuberculosis Institute president, specifically criticized the city's extension for economically benefiting a few at the discomfort and danger to millions of citizens.[28]

Just a few months later in November 1969 Chicago experienced a "super smog" that killed at least one hundred people and forced children and those with respiratory ailments indoors or to hospitals. Almost immediately the underpinnings of the social, political, and economic paradigm that tolerated visible air pollution shifted. Over the following year media coverage on the deadly effects of pollution fueled an unprecedented level of political activity from community groups that forced Mayor Daley to switch course and adopt the most salient features of his environmental critics' agenda.[29]

On January 13, 1970, WBBM-TV (CBS) aired an evening special on air pollution titled "No Place to Hide" that pulled no punches in blaming Commonwealth Edison (Com Ed) power plants as the primary source of deadly sulfur dioxide emissions. The WBBM-TV team created an *Action Guide on Air Pollution* that provided an overview of health risks and resources to help citizens demand the enactment of stricter air pollution controls. The *Action Guide* included a directory of citywide environmental organizations, including CACC, the Campaign Against Environmental Violence, and the Industrial Areas Foundation, along with eighteen other neighborhood and nature-specific groups such as the Hyde Park Clean Air Committee, Southwest Air Pollution Committee, Peoples Group of Garfield Ridge, Citizens Revolt Against Pollution, Rainbow Neighbors, Save the Dunes Council (Indiana), and Illinois Federated Sportsman Clubs.[30]

A second special called "Up in Smoke" aired in March 1970 on WTTW-TV (PBS). Produced by the city's Better Government Association and the *Chicago Today* newspaper, "Up in Smoke" attacked the incompetence of pollution inspectors and the entire system of fines and violations. Of note was the ineffectiveness of the city's Air Pollution Appeals Board, which "Up in Smoke" ridiculed as nepotistic and incompetent in allowing large polluters like U.S. Steel and Republic Steel to operate immune from prosecution.[31]

The biggest target for environmental and community activists was Com Ed. In January 1970 the newly formed Citizen's Action Program (CAP) began a campaign called Proxies for People that encouraged citizens to buy stock in the company and then assign proxies for them to CAP leaders, who advocated for changes in corporate policies and practice. Proxies for People—modeled after the community organizer Saul Alinsky's Rochester, New York, campaign to force the Kodak corporation to hire more African Americans—was part of a larger series of confrontations between CAP members and political and corporate leaders. CAP drew its strength from working-class, white ethnic Chicagoans who were more interested in saving their neighborhood than the planet, and wished to improve Chicago rather than moving to the suburbs. Led by the activist priest Len Dubi, CAP chalked up a series of early successes, including a March 1970 announcement from Com Ed that they would switch to burning coal with less than 2 percent sulfur.[32]

Intense pressure from CAP, CACC, and other organizations also paid off the following month when Mayor Daley announced even tougher standards than the ones he had delayed the previous summer. Alderman William S. Singer (44th Ward), a consistent critic of the mayor on environmental issues, praised the new measures and noted that the "result should be much cleaner air for Chicagoans to breathe" (Figure 14.2).[33] At the federal level, passage of the Clean Air Act Amendments of 1970 lent further support for clean air activists and heightened their expectations for more accountable government. Under the measure federal officials

FIG 14.2. Flier for a community meeting on air pollution at the Lake View Library led by 44th Ward Alderman William S. Singer, April 14, 1970. *Source*: Hull-House Association records, Box 106, Folders 1295–1308, Special Collections, University of Illinois at Chicago.

were finally given ultimate authority over air quality and the ability to set air quality standards for states.[34]

At the grassroots level in Chicago, constant agitation and enforcement by environmental and community groups helped prevent backsliding to old practices. For example, starting in February 1971 CACC and the Tuberculosis Institute began a Court Observers Program to monitor the proceedings of the Circuit Court overseeing air pollution cases. Volunteers found that of the 1,100 cases brought to the court, judges fined all but two offenders well below the minimum level. Judges were not only lax with enforcement, they often flaunted pollution control laws, and officers from the Department of Environmental Control (which replaced the Department of Air Pollution Control in 1970) exhibited a lack of knowledge of specific cases. After the CACC went public with their findings, the Circuit Court began to toughen fines and crack down on repeat offenders.[35]

In addition to switching to lower-sulfur coal, Com Ed reduced its visible smoke emissions throughout the 1970s by phasing out old power generating facilities, installing new air pollution control equipment, and constructing nuclear power plants outside the city. In the 1950s and 1960s nuclear power held great promise for smokeless electricity and low-cost energy. The construction of nuclear facilities around Chicago began in 1956 with Dresden 1, the nation's first privately built plant, located fifty miles southwest of Chicago in Grundy County, Illinois (in operation between 1960 and 1978). However, cost overruns associated with plant construction and reactor safety, along with a host of other environmental concerns, severely limited the appeal of atomic energy by the late 1970s. From 1965 to 1973 Com Ed announced plans for a dozen more nuclear plants, although only six were completed by 1988, almost a decade behind schedule and more than $11 billion over budget. Despite a decade of heavy financial burdens for Com Ed and an all but stagnant demand for electricity, in 1980 nuclear power accounted for 40 percent of its electrical generation.[36]

AIR AND WATER POLLUTION IN A DE- AND REINDUSTRIALIZED LANDSCAPE

By the late twentieth century the political and economic conditions that had generated massive environmental pollution shifted dramatically from their earlier foundations. Politically, in stark contrast to the nineteenth and early twentieth centuries, people of all ages and socioeconomic backgrounds from across the entire metropolitan region engaged in various forms of environmental activism aimed squarely at eliminating sources of air and water contamination and holding corporations and public officials accountable for their role in generating and sustaining foul air and water. Such activism played out at the neighborhood organization and regulatory policy levels in ways that transcended loopholes, exemptions, and other mechanisms traditionally used by industrialists and political machines to thwart change. Economically, a long period of industrial realignment across the United States, known as deindustrialization, resulted in widespread factory closures in Chicago and de facto reductions in—although by no mean the elimination of—noxious emissions.

By 1979 visible smoke and water pollution from industrial discharge and coal combustion had largely subsided in Chicago through stiff enforcement of clean air regulations. In fact, controls targeting sulfur fumes and dust were so effective that only seventy-five large coal boilers remained in Chicago. Symbolically, the same year saw the city's Department of Environmental Control replaced by the Chicago Department of Energy and Environmental Protection, with an expanded role in monitoring building demolition, watercraft, noise pollution, and auto emissions.

Water quality also improved with upgrades to sewage treatment facilities and the modernization of manufacturing facilities. For example, five major steel plants implemented industrial recycling processes that discontinued the discharge of 546 million gallons of wastes per day into the Calumet River and Lake Michigan. Cleaner air in the form of reduced coal emissions also resulted in the overall improvement of water quality throughout the Great Lakes.

Another significant reason for the decline in air and water pollution came from ongoing deindustrialization, which resulted in the loss of approximately 25 percent of the city's factories during the 1970s. The sudden collapse of the American steel industry in the early 1980s brought even more closures and high rates of unemployment as thousands of workers in South Chicago lost their jobs. Other manufacturers followed, and as a result between 1982 and 1992 the city lost another 18 percent of its manufacturing jobs.[37]

Despite dramatic declines in manufacturing, air and water pollution remain an integral, though less discernible, part of Chicago's urban environmental landscape. The smoke problem has given way to concerns about excessive levels of automobile emissions, heavy metals and other toxins released from commercial smokestacks, and industrial waste processing facilities located near residential neighborhoods. With the elimination of visible smoke, though, most city residents now rate the air quality in their neighborhood as good or even excellent. Dramatic improvement in water quality has again become a top priority for the SDC's current successor, the Metropolitan Water Reclamation District of Greater Chicago, as it seeks to comply with recent federal mandates to make all the rivers in and around Chicago "fishable and swimmable." Most especially troubling are the handful of days each year when massive rainstorms overwhelm the drainage capacity of sewage outflows and release millions of gallons of untreated raw sewage and other forms of wastewater into the Chicago River.[38]

Finally, and perhaps most unexpectedly, manufacturing is now returning to Chicago in the form of more environmentally friendly enterprises. The former Union Stock Yards is currently experiencing a revival as the Chicago Stockyards Industrial Park, which by the early 2010s had over seventy companies and fifteen thousand employees. In stark contrast to the past, though, these businesses are built around green technologies that are by design more environmentally sustainable and release as little waste as possible with the expectation of avoiding the contamination and ecological calamities of the previous industrial era.[39]

CHAPTER 15

Chicago's Wastelands

Refuse Disposal and Urban Growth, 1840–1990

CRAIG E. COLTEN

The disposal of waste has long been an important aspect of city build-ing. In America prevailing concepts of environmental management have guided urban expansion within a context of municipal government and public health. In metropolitan regions zones of active disposal preced-ed urban land uses, creating a series of outwardly progressing waste frontiers. Throughout the nineteenth century modification of water-fronts, using general urban refuse as fill, conformed to transportation and commercial demands by enlarging valuable harbor facilities. Use of garbage and general refuse to reclaim wetlands and exhausted quarries addressed late nineteenth- and early twentieth-century concerns with public health and safety and altered urban fringe topography sufficiently to permit new uses of former disposal grounds. Post–World War II waste disposal policy reflects many refinements in waste management technol-ogy and scientific understanding of environmental processes, although practice has not always followed the methods advocated by experts since municipal authorities were unable to control the practices used beyond their boundaries. As state governments asserted legal authority during the 1960s, the effects of a generation of exporting urban wastes to fringe sites had already left an imprint on the physical form of rural areas and on unseen subterranean waters.

This chapter will explore the role of general urban wastes[1] in the devel-opment of a major American metropolis. Poised originally on the soggy

plain left by ancient Lake Michigan, Chicago occupied a site ill-suited to
large-scale urbanization in the early nineteenth century. Hoisting it up
from the mire and creating a well-drained foundation was a multifaceted
process in which the waste of the city played a vital role. Chicago stands
as an example of a widespread process involving massive reworking of
the land to accommodate a street network and support the built environment. Between 1840 and 1990 it underwent three distinct phases of waste
disposal. The first revamped the shoreline by using vast quantities of refuse to create the foundations for a major park and other cultural amenities and lasted until about 1900. Following the 1871 fire, reconstruction
efforts called for clay and limestone, excavated beyond the built-up area
of the city. The pock-marked landscape of the turn-of-the-century urban
fringe provided convenient and acceptable repositories for refuse during
the second phase, and once refilled these sites served other purposes. In
addition to reclaiming pits and wetlands, a variety of urban wastes, including garbage, provided fill for raising streets and other low areas. This
period lasted through the 1950s. The third generation of disposal also
used exhausted quarries, but represented a major leap outward to a new
garbage frontier in exurban sites and defied contemporary public health
standards. Displacement of disposal activities to rural sites represented
the city's rejection of its own wastes and introduced massive amounts of
refuse to suburban regions beyond municipal authority.

The process of land reclamation and reconfiguration in Chicago represents a practice followed to some extent in most cities.[2] A series of se-
quentially and outwardly displaced zones of waste disposal initially provided a remedy to perceived waterfront shortcomings, followed by the
conversion of suburban excavations to other urban land uses as industrial, commercial, and residential activity surrounded the spent quarries. Reclaiming what was considered at the time useless land served to
prepare a highly altered topography for new uses. Finally, in recent years
regulation and public perception of wastes, along with the demand for
larger repositories, has forced this urban-based activity to rural areas.
This has displaced urban problems to suburban areas, created permanent
mounds that deter subsequent reuse, and introduced contaminants to
rural groundwater supplies. Thus waste disposal was part and parcel of
a broader pattern of urban expansion associated with the transportation
revolution and changing land values.[3]

SHAPING THE URBAN ENVIRONMENT

In a discussion of the squares of London, Henry Lawrence proclaimed,
"A city is a built place, is often seen as the antithesis of nature, since
buildings and pavements displace forests and fields."[4] Although he

acknowledged that society selectively maintains portions of "living nature" in its "hard fabric," his statement suggests a highly naive view of the
physical geography of the city. The process of city building almost inevitably involves several stages of landform modification between field and
forest and the urban fabric. Within this transition, waste disposal was a
critical stage in preparing pre-urbanized land for subsequent occupants.
Only tantalizingly brief allusions to reclamation of areas such as Boston's
Back Bay indicate any attention to this process.[5] There are also many
natural processes, both seen and unseen, that continue to operate within the city, sometimes causing great consternation to the city builders.
Public responses to the "urban heat island" effect and subsidence due to
groundwater extraction represent human accommodations to persistent
natural processes within the city.[6] Additionally, the creation of new land
forms, such as massive heaps of residue, has presented long-term land-use
dilemmas for urban administrators.[7] Geographers, like early town planners who ignored topography in laying out rigid grid street patterns, have
commonly presented analyses devoid of physical geography.[8]

The three stages of waste disposal and land modification in Chicago
have parallels in the general process of urban growth.[9] Each fits within a period of urban growth that was influenced by public health policy, technological capabilities, and local environmental conditions. The
first phase centers on the central business district before 1900 and represents efforts to remedy the physical shortcomings of an urban setting.
Nineteenth-century commercial cities required adequate wharfage, and
refuse provided material to build up waterfronts.[10] Increased congestion
and waste production forced Gilded Age and Progressive Era urban administrators to centralize the management of waste removal, both sewage
and garbage. In response to public displeasure with offensive conditions
created by existing disposal methods, they made headway with sanitation
problems largely by exporting the wastes to suburban or downstream locations or by experimenting with new technologies, mostly after 1900.[11] Historical examinations of environmental development tend to focus on internal elements chronicled by municipal records, although recent analyses
have shown that the impacts are not circumscribed by city boundaries.[12]

Suburban development often followed new transportation technologies to areas unsuited to urbanization. Land modification was a fundamental accompaniment to this outward movement of people and institutions. Encroaching urban land uses, infrastructure, and city boundaries
hemmed in extractive activities and presented new options for the reuse
of former quarries once enclosed within the administrative reach of the
city.[13] Industrial development in satellite cities also created a demand
for land expansion or reclamation, particularly wetlands that had been

avoided during the initial phases of city building.[14] Zones devoted to industrial and waste disposal activity diverted residential land uses and contributed to the formation of sectorial growth patterns. Expansion and infilling reworked the fringe, ultimately displacing extractive activities, and thereby fostered another generation of fringe modification.

Post–World War II urban expansion again encroached on existing land disposal sites.[15] The outward press of urban land uses and increasing volumes of wastes forced municipal authorities to turn to more highly dispersed and remote disposal grounds beyond their control. This third phase of waste disposal represented a departure from previous waste disposal activities. No longer did garbage and other wastes provide fill for transitional land uses. Rather, sanitary landfills mounded the garbage high above the previous land surface and precluded most subsequent land uses. This occurred despite widespread recognition that land disposal could contribute to groundwater contamination. Thus the exportation of huge quantities of waste has not only limited land-use options at the disposal sites, but has introduced environmental problems to rural areas.

The municipal management of Chicago's wastes and the selection of disposal locations are the central objects of this investigation. Several primary records provide the essential documentation of the outward expansion of disposal activities. The earliest phase of disposal activity is recounted in the recently rediscovered City Council minutes, long thought victim of the 1871 fire. Various annual reports from the city agency charged with waste removal responsibility document evolving policies, disposal practices, and quantities of refuse. Other government reports and municipal ordinances provide further primary evidence of the progression from waterfront to exurban waste disposal.

LAKEFRONT MODIFICATIONS IN CHICAGO

Much of Chicago's early land reclamation was an ad hoc process that reflected little careful planning and required constant administrative adjustments to the emerging problems created by growing waste heaps. Pre-1850 disposal relied on the open expanse of the lakefront, a location convenient to the source of refuse. Construction of jetties to protect the mouth of the Chicago River commenced in 1833 and caused natural accretion of sediments on the north side of this man-made obstruction (see Figure 15.1). The beginnings of Streeterville, a lakefront subdivision, arose on the newly formed tract of land, while sediment-poor currents cut into the shoreline south of the jetty. This erosion, which threatened Michigan Avenue by the 1840s, prompted the City Council to take defensive actions. In 1844 the council "ascertained by the estimate of competent Engineers, that the Public Grounds East of Michigan Avenue can be protected from

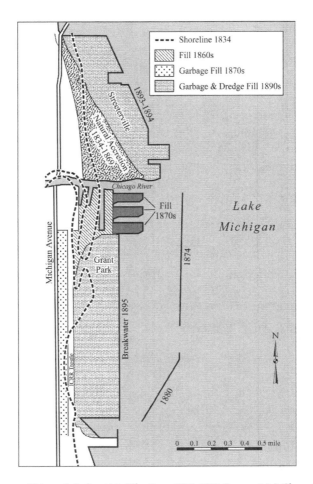

FIG 15.1. Chicago Lakefront Modifications, 1834–1910. *Sources*: M. J. Chrzastowski, "The Building, Deterioration and Proposed Rebuilding of the Chicago Lakefront," *Shore and Beach* 59 (1991): 2–10; and U.S. Army Corps of Engineers, Chicago District, *Annual Report*, 1903.

the encroachment of the Lake; and whereas without Protection, the whole will be shortly destroyed, and the City deprived of one of the most extensive Public Parks, than can be found in any American city."[16] Based on this sentiment, the council approved a series of plans to stabilize the shore. Repeated efforts between 1840 and 1852, however, failed to halt the erosive forces of the lake's currents. Nevertheless, they set in motion a process of using the lakefront for the disposal of urban wastes.

During this same period, the city began systematic garbage collection and disposal. In 1849 the City Council passed an ordinance "to preserve the public health" by collecting garbage and refuse and using it to fill areas perceived to present public health problems—namely sites with standing water or the lakefront.[17] The City Council acknowledged that the use of the lakefront for private dumping of slaughterhouse refuse was an emerging nuisance in 1852 and prohibited such nuisance-causing behavior.[18] Despite a city ordinance outlawing the dumping of refuse into the lake within the city, problems persisted. Lakefront residents complained that unknown offenders were in the habit of "depositing and leaving upon the Lake Shore . . . large quantities and heaps of offal, stable manure, decaying vegetable . . . and various other offensive substances" and that these deposits constituted "a serious annoyance and nuisance."[19] Furthermore, the health officer petitioned the City Council in 1855 to extend a road to the lakeshore north of the city limits so that scavengers could deposit and bury offal and night soil along the lakefront.[20] Thus, within and immediately beyond the city limits, waste disposal contributed to the extension of land into the lake.

An agreement with the Illinois Central Railroad provided the city with another convenient repository for waste disposal during the 1870s. The company built a masonry wall to protect the waterfront and its trestle east of Michigan Avenue, creating a lagoon between the tracks and the old shoreline.[21] The enclosed basin served as a bin for the city's ash, garbage, paving blocks, and other refuse.[22] In 1895 the Army Engineers constructed a breakwater farther out into the lake and granted permission for the city to fill in this basin. Municipal refuse, along with dredge spoil taken from the outer harbor, reclaimed what eventually became Grant Park.[23] Extension of the shoreline north of North Pier took place during the same period. General municipal refuse and ash heaped on the shore extended the Streeterville neighborhood over five hundred feet into the lake.[24] Citizen complaints revealed industrial dumping near the Grand Haven Slip where breweries and a soap factory used a pond as a "receptacle" for offal and other wastes. When confronted with this situation, the City Council called for a special assessment on the businesses surrounding the pond to pay to fill it and thereby prevent the nuisance.[25] Refuse disposal behind protective bulwarks along the lakefront continued into the early twentieth century.[26]

Filling also took place in industrial districts. Along the lower reaches of the Calumet River and on either side of its mouth, steel mills purchased the right to the lake bed and created massive areas of "made land." Slag was the principal source of fill, although liquid wastes and sludges were poured over the slag as the manufacturers pushed their operations out into Lake Michigan between 1870 and 1910. In addition, dredge spoil from the river accounted for much of the land created by industries.[27] Although initially the dredged material was naturally occurring sediments, by 1895 the Army Corps of Engineers voiced concern at the navigational

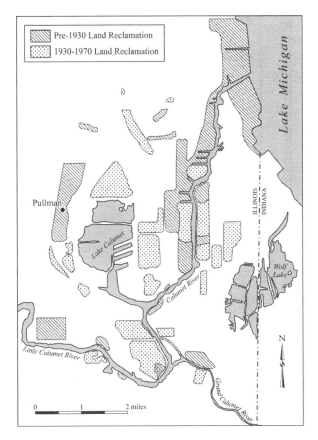

FIG 15.2. Land Reclamation in Southeast Chicago. *Source*: Craig E. Colten, "Industrial Waste in Southeast Chicago: Production and Disposal," *Environmental Review* 10 (1986): 95–105.

obstructions caused by industrial wastes dumped into the waterway. In reference to the section of the Calumet River west of the state boundary, the district engineer reported, "What goes into it stays there for lack of current to carry it off. Several towns and some great filth-producing manufacturing establishments have filled the dredge channels with filth as fast as excavated."[28] The army waged a continual campaign to maintain a navigable channel along the Calumet River, and this meant transferring waste dumped into the river back onto its banks. Thus industrial wastes contributed significantly to the "made land" upon which the southeast Chicago industrial complex was built.

Dumping did not stop at the water's edge. The city barged municipal garbage out into Lake Michigan in 1882, and industrial waste disposal in the lake persisted until at least 1910.[29] Realignment of the shorefront near Chicago's central business and southeast industrial district was not

the end of lakefront filling, but it serves as a terminus for the first phase
of land modification. By the early twentieth century use of the lakefront
for general urban waste disposal began to wane and federal intervention
forced waste haulers to use other areas for their dumps.[30] Nonetheless, by
1900 land reclamation, using garbage and industrial solids, had created
the foundation of Grant Park, one of the most impressive amenities in the
city; valuable waterfront harbor facilities; and important industrial tracts.

BUILDING A CITY ON WASTES

The second phase of waste disposal overlapped with the first, but continued long after it and brought about environmental change to a much
larger area. It evolved as a result of several fundamental shifts in urban
growth and administration, along with changing concepts of public
health. Expansion of the city's infrastructure to serve a more widely
dispersed residential population occurred in step with encroachment
of urban land uses on extractive activities scattered along the urban
fringe. Public health policy encouraged reconfiguring such obstructions
to urban growth by filling them with refuse.[31] Although they provided
convenient repositories, excavations used as dumps ultimately became
the object of nuisance complaints as population densities increased and
forced disposal activities farther afield. Nevertheless, by capping the objectionable contents, the sites became available for subsequent reuse.

Extensive raising of the land surface also occurred in conjunction
with the laying of the sewerage system authorized in 1855. Due to the
low-lying position of the city, there was inadequate drainage for a
gravity-fed sewer system. To overcome this problem, sanitation officials
laid sewers in the middle of the streets and then raised the grade of the
street over the pipes. Near the Chicago River this called for as much as
ten feet of fill. Much of the material used near the waterway was dredge
spoil derived from a concurrent project to deepen and widen the river.[32]
The river improvement project did not supply enough material to complete the land-raising effort, and the city's Department of Public Works
authorized its garbage carriers to dump their loads in streets and alleys
in the outlying wards.[33] Other fill derived from ash, especially during
cold winters when coal consumption increased.[34] Away from the river
mixed fill generally raised the street level by approximately four feet.
Another major source of fill for the streets and the lakefront was debris
from the fire in 1871.[35] Due to the excess of rubble and ash created by the
fire, the City Council called for raising the grade a full foot and a half
above the previously authorized level for the burned area.[36] Street sweepings, consisting of general refuse and manure, contributed to the filling
of "unimproved streets and lots that were below grade" into the 1890s.[37]

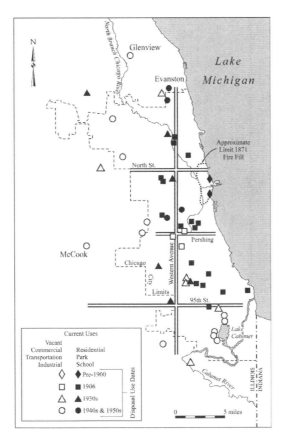

FIG 15.3. Chicago Land Disposal and Post-Fill Land Uses, 1845–1959. *Sources*: Chicago Department of Public Works, *Annual Report* (Chicago, various years); Joseph Patterson, *Report to the Mayor and City Council on the Collection, Removal and Final Disposal of Garbage of the City of Chicago* (Chicago, 1906); Citizens' Association of Chicago, *Garbage Disposal*, Bulletin 28 (Chicago, 1912); Greeley and Hansen Engineers, *Preliminary Report on Refuse collection, and disposal related to matters for the Chicago Department of Streets and Electricity* (Chicago, 1948); Chicago Department of Streets and Electricity, *Annual Report* (Chicago, various years); and *Olcott's Chicago Land Values* (Chicago, 1927 and 1944).

A survey of over three hundred core samples indicates that the use of fill was pervasive within the area bounded by North and Western Avenues and Pershing Road (Figure 15.3). Fill depths vary considerably, but in a transect between Madison and North Avenue, the average depth-of-fill near Michigan Avenue was over twelve feet and fell to slightly more than four feet at the western reaches of the survey. Filling was most extensive along the Chicago River and in the area of the 1871 fire. Accumulations of ash and debris remain in excess of ten feet along the river and an average over five feet through much of the burned district.[38]

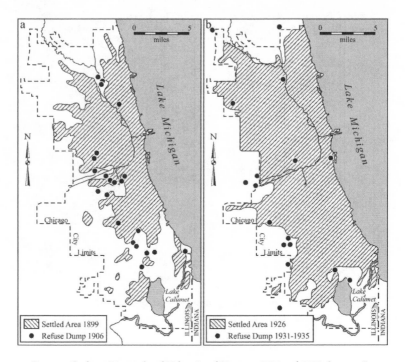

FIG 15.4. Garbage Disposal and Urban Land Uses, ca. 1906 and 1935. *Sources*: For settled areas see Homer Hoyt, *One Hundred Years of Land Values in Chicago* (Chicago: University of Chicago Press, 1933); and for dump sites, see Chicago Department of Public Works, *Annual Reports*, 1931–1935.

Reclaiming pits and quarries within the city was also under way. The city's Department of Public Works reported in 1894 that garbage and ashes "were deposited generally in brickyard clayholes located in sparsely settled districts."[39] By the early twentieth century Chicago was experiencing its first "landfill" crisis, caused in part by the topping off of many of the quarries and the encroachment of residences around dumps. Despite the city's use of disinfectants to minimize the offensive quality of its garbage pits, neighbors filed an increasing number of nuisance complaints. Consequently, the Department of Public Works appealed for funding to secure additional dump sites or for alternate methods of disposal: "The cost of the removal of garbage is largely increased and will continue to increase owing to the fact that the City is compelled to haul a portion of its garbage from eight to ten miles, to wherever a dumping place is available. As the dumping grounds available for the dumping of garbage throughout the City are almost exhausted, arrangements will have to be made for the erection of such plants as will destroy all the garbage wastes, and this the city will be unable to do until more funds are provided."[40]

In 1902 the department reported a modest increase in the total amount of garbage that it handled, and restated its complaint that the department "is greatly hampered for dumping space and it is absolutely necessary that something be done in the way of providing garbage crematories."[41] By 1904 steps were under way to employ a new technology to handle a portion of the city's garbage, although the Department of Public Works declared that "in the extreme outlying wards there is sufficient swampland to take care of the output as it comes out at the present time for several years or until we can get this other scheme [a reduction works] in good, working order."[42]

Several years later the critics declared the city's garbage removal system a failure. Well-to-do residents hired private scavengers who offered more reliable garbage pickup, while other neighborhoods had to rely on "more or less" faithful and "more or less" regular municipal crews. This erratic service transported a mixture of ashes, garbage, and rubbish to some eight dumps, mostly former clay pits, and another twelve swill dumps. Encroaching residential land uses brought larger numbers of people into proximity with the offensive refuse pits. According to a contemporary critic, "the wind being strong, a dump can nauseate readily at over a mile," yet children regularly picked over the garbage heaps for food.[43] The Citizens' Association of Chicago also characterized garbage disposal methods as "primitive and unsanitary" and claimed that the use of clay holes within the city limits had created "an unbearable nuisance.[44]

This situation led to the adoption of a garbage reduction system in 1906.[45] The experiment with a privately operated facility was not wholly successful and dumping continued even as the Chicago Reduction Company became the end point for a portion of the urban refuse. The reduction facility handled only "pure garbage," or biological wastes, while the city continued to heap ash, building debris, and other refuse into dumps.[46] When contractual problems arose with the reduction company, the city resorted to dumping garbage into pits again in 1913. Although the city took over control of the reduction facility and operated it into the 1920s, periodic use of pits for garbage disposal continued.[47]

Despite experiments with new technologies, including incineration during the 1930s, local opposition to offensive conditions continued to accompany existing dumps. In one of its more noble experiments at solving a nuisance problem, the city undertook a project to reclaim Bubbly Creek, a sluggish tributary of the South Branch of the Chicago River. It drained the meatpacking district and served as a drain for domestic sewage. Dredging and clay excavation had enlarged the channel but largely eliminated the gradient necessary to drain the effluent poured into it.

The stagnant channel had become a giant, open-air "septic tank" covered with a thick scum that local lore claimed was capable of supporting the weight of a man. By 1916 conditions were so unbearable the city began work on a conduit to divert industrial and domestic sewage away from Bubbly Creek and began filling the channel with a mixture of refuse and earth.[48] This use of garbage to abate a nuisance highlights the official approval of land dumping within the city limits even in the face of public displeasure and laws forbidding it.

Reclamation of the early twentieth-century dumps took several forms, of which recreational uses were extremely common. Citizen groups in the Packingtown district spearheaded the effort to convert a perennial nuisance into a neighborhood park. They covered the garbage heaped in a former quarry, graded the surface, and created a centerpiece for community recreation.[49] A portion of Harrison Park, situated in a working-class neighborhood on the Near Southwest Side, sparked local opposition as well. A 1938 court order banned the dumping of objectionable material in a former quarry at Nineteenth and Wolcott. Two years later the court modified its order and permitted dumping to resume with the stipulation that operators follow a "scientific process" that included periodic chlorine applications.[50] Within the next decade the municipal parks department gained title to the property and expanded a neighboring park over the former quarry dump.[51] Conversion of dumps to parks transformed nuisances into public amenities in areas where little other land was available for such purposes. In fact, it was extremely common to convert pre-1950 dumps to parks or school grounds.

Almost all of the garbage and refuse dumping done within the city was contrary to municipal law. Municipal ordinances had long prohibited the disposal of offal and other offensive wastes within the city.[52] By 1922 the code had extended the prohibition of dumping to include ashes, refuse, cinders, soot, sand, street sweepings, "or any other substance that may contain disease germs or is subject to be carried by the wind, or any like substance that may decompose or become filthy, noxious or unhealthful" on private or vacant land. The local ordinance, however, gave the public health officer authority to approve dumping which he did not deem a nuisance, and with the landowner's consent.[53] This in effect provided a legal loophole for continued dumping within the city, particularly in wetlands and clay pits—a practice endorsed by local sanitary engineers who touted the value of land reclamation.[54] In addition, this allowed further land reclamation along the lakefront between 1920 and 1922 where huge quantities of the city's waste, along with ash transported from the downtown hotels and office buildings, created the land that supports Soldier Field.[55] The Department of Public Works reported disposing of over 2 million cubic yards of garbage annually during the early 1930s, mostly in dumps within the city. With two exceptions, the 1930s municipal dump sites were farther from the city center than the 1906 sites, lying near the city's extremities.[56]

In the 1940s new terminologies and methods entered the field of land disposal of garbage. The sanitary landfill, adopted by American sanitary engineers during World War II, began to replace the open dump. The landfill differed from the traditional dump in several respects: Proper techniques included daily compaction of the refuse by heavy equipment and the application of an earthen cover. These procedures extended the capacity of a landfill site and most importantly prevented the escape of wind-blown trash and offensive odors while minimizing insect and rodent populations. Early descriptions of operating landfills touted their ability to transform useless land into viable real estate, but the U.S. Public Health Service warned that leachate from landfills could pollute surface and groundwater.[57] Early investigations also warned against constructing houses or commercial structures on top of landfills owing to the potential for methane gas to escape or to the combined impacts of compaction and subsidence.[58]

The sanitary landfill presented an attractive option, particularly for Chicago. It permitted the continued use of land disposal while eliminating troublesome nuisance conditions, but the city was reluctant to adopt new techniques. When neighbors of a dump in northwestern Chicago (Narragansett and Wrightwood Streets) forced its temporary closure in 1940, the courts gave brief hope to the proponents of sanitary landfills. The city shortly thereafter won a temporary stay which allowed it to revamp its operation to conform with sanitary landfill methods, but ultimately it won the case and abandoned any semblance of accepted landfill practices.[59]

By its actions over the next decade, the city continually rejected sanitary landfilling techniques. In 1943 the city disposed of all of its 3.2 million cubic yards of garbage and refuse in five active dumps.[60] Four were in excavations and one in an extensive wetland, which indicates that all were placing garbage in the water table. The attraction of using former pits was its economy. A 1948 report prepared for the Department of Streets and Electricity noted that "land fill is the least expensive method of refuse disposal in Chicago," although it urged the city to improve its practices.[61] Shortly thereafter, a massive U.S. Public Health Service (USPHS) report concurred with the local engineers by citing landfills and dumps as the "most economical method available."[62]

The federal survey, although critical of some elements of Chicago's garbage disposal practices, implicitly endorsed the status quo. The report indicated that the dumps in mined areas were allowing trash to make direct contact with groundwater, an undesirable condition, and daily cover was thin, if provided at all. The large landfill opened in the northern section of Lake Calumet (see Figure 15.2) presented the potential for surface water contamination, so the city constructed a huge dike to separate the wastes from the lake. Use of permeable slag and a breach in the dike in 1945 made this barrier largely ineffective.[63] Furthermore, the federal review found evidence of rat infestation, offensive odors, and numerous open fires at dumps used by the city. Escaping methane gas, a natural by-product of the decomposition process, burned continuously at one dump. Of the city-owned site near Lake Calumet, which had a full list of offenses, the report stated, "No city should permit a dump of this type to exist." While federal investigators characterized only four of the city-supervised disposal sites as being in "fair" condition, they described several private dumps as violating every conceivable public health standard.[64] In sum, the dumps were breeding grounds for insects, veritable banquet halls for rats, sources of offensive and potentially harmful odors and gases, and reservoirs of pollutants. The federal study concluded, however, that disposal in the city could be made satisfactory with several minor modifications, the most important being the use of sanitary landfill techniques.[65]

Abysmal conditions left federal investigators baffled by the lack of legal action against operators. Their review of local ordinances revealed that jurisdiction of dumps fell under the commissioner of buildings, who had the authority to grant permits. The municipal code specified requirements for a permit and clearly stated that dumping without one was a nuisance. Despite a regulatory framework, the investigators discovered that the commissioner of buildings had never issued a permit, nor had his office called for the closure of illegally operated dumps. The USPHS report went on to suggest that even the municipal dumps were illegal, given an ordinance prohibiting refuse disposal within the city limits without approval by the Board of Health, which had no records of reviewing or condoning the existing dumps.[66] Not only were the responsible agencies ineffective, but the division of authority hindered coordination and enforcement.

The fact that Chicago residents consumed water from Lake Michigan led the city and federal surveyors to overlook the environmentally inappropriate selection of landfill sites. Beyond public nuisances, the choice of excavations and wetlands for disposal grounds presented serious threats to shallow groundwater. All of Chicago's thirty-three existing or potential disposal sites in the late 1940s permitted refuse to make direct contact with the water table. Yet the federal report emphasized that future use of these sites required only consideration of such factors as transport costs and available cover.[67] The economics of sanitary engineering was obvious, although good engineering was not. Cost-effective operation was paramount, while geologic considerations were secondary or absent altogether.

By the late 1940s Chicago's waste disposal situation was nearing a critical juncture. Public administration of disposal sites within the city was ineffective, and nuisance conditions abounded. Furthermore, the city-owned dumps were nearly full. In fact, the quarry pit at the House of Corrections reached capacity in 1947, and the city discontinued dumping there.[68] In 1950 only one municipally owned disposal site within the city continued to accept mixed refuse. Municipal crews diverted garbage from western and northern suburbs to disposal sites in adjoining communities (McCook, Glenview, and Evanston), while additional solid wastes served as fill for railroad construction projects.[69] Within only two years, however, public displeasure with the Glenview dump inspired court action that forced its temporary and eventually its permanent closure in 1953.[70] By 1954 the city was left with only one operating dump, the much-criticized Lake Calumet facility.[71]

Chicago at this point had nearly exhausted its internal land disposal options. Use of quarries and wetland sites had provided convenient and inexpensive repositories, despite the fact that city ordinances banned their use and they created undesirable conditions. Furthermore, their use ran counter to fundamental principles of public health. Nevertheless, with its options nearly depleted, the city turned to a new set of excavations, beyond the city limits, during a renewed phase of refuse disposal expansion.

The former dumps soon acquired new land uses. Many of the sites used before 1940 were converted to parks, school grounds, or residential areas, while post-1945 dumps commonly became sites for commercial, industrial, or transportation facilities. The one remaining pocket of active disposal in the city, in and around the relatively secluded wetlands and mined-out clay pits near Lake Calumet, did not undergo any significant change, although it was often eyed as a site for redevelopment schemes such as an airport and even a ski slope (by 2014 much of it had been redeveloped as golf course).

EXPANSION TO EXURBAN SITES

Diminished landfill capacity within the city forced Chicago to export a larger share of its wastes by the 1960s—nearly three million cubic yards a year to a group of far-flung disposal sites in nonurbanized territory.[72] The metropolitan planning commission viewed this situation as an

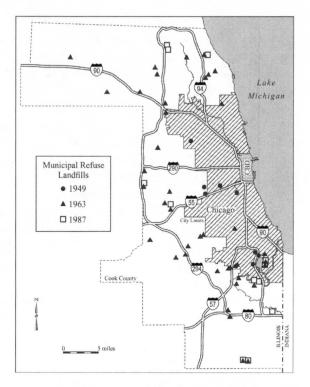

F I G 15.5. Urban Waste Disposal Sites in Cook County, Illinois, 1949, 1963, and 1987.
Source: USPS 1949; Northeast Illinois Metropolitan Planning Commission (here-
after NIPC), *Refuse Disposal Needs and Practices in Northeast Illinois,* Technical
Report 3 (Chicago: NIPC, June 1963), and 1987.

impending garbage crisis and took stock of existing capacity. It found
that fifty-six of the seventy-two operating landfills in the metropolitan
area were open or modified dumps. Only sixteen adhered to sanitary
landfill guidelines. Among the publicly owned landfills, less than one
quarter followed satisfactory practices. The study concluded that the
widespread use of open dumps reflected public indifference to refuse disposal and defied advice to avoid using former excavations.[73]

Within the city of Chicago, three of the active landfills were former
excavations and the others were all marsh sites in the Lake Calumet
area. Throughout the remainder of Cook County, an additional fifteen
excavations served as repositories for urban refuse.[74] The selection of
former quarries and clay pits was particularly ill-suited to the geology
of northeastern Illinois. For more than a decade officials of the Illinois
Department of Public Health had pointed out the potential for pollution
of groundwater when wastes were buried in mined-out areas or too near
exposed rock strata, especially the limestone that is found in northern

Illinois.[75] Nevertheless the 1960s inventory continued to identify low areas, swamps, or pits as suitable sites for landfills.[76]

The period of rapid expansion occurred without any comprehensive oversight or consistent administrative checks. Dispersal of wastes throughout Cook County avoided municipal regulations and controls. It was not until 1966 that the state Department of Public Health issued statewide regulations for sanitary landfills.[77] Although its advice included long-standing public health principles, such as avoiding placement of wastes in the water table, enforcement was limited.

In the absence of internal capacity, Chicago sought technological fixes to its garbage crisis, and incinerators became the choice for reducing landfill space needs. Between 1956 and 1962 the Department of Public Works put three incinerators into operation with a combined capacity of over three thousand tons of garbage a day. Two of the incinerators were in industrial districts near the city limits, and a third was more centrally situated, but also in a manufacturing area. Burning did not completely eliminate the demand for landfills, and the city dumped incinerator ash into an old quarry with the intention of converting it into a park.[78] Furthermore, the trash burning units did not handle all the city's garbage, and peripheral landfills continued to serve as repositories. Several suburban landfills, used during the 1960s, received Chicago wastes, and their leachate has produced previously "unimportant" problems, namely groundwater contamination.[79]

The displacement of refuse disposal to suburban and fringe areas has introduced public health issues not faced by the city itself. The Illinois State Geological Survey reported in 1967 that "the aquifers of northeastern Illinois present in the glacial drift or shallow bedrock are particularly susceptible to contamination from near-surface refuse disposal." This led them to recirculate an old agreement stating that since the suburban communities relied heavily on groundwater for public water supplies, refuse disposal should be limited to sites above the water table or in areas with glacial till soils of low permeability.[80] Yet as late as 1971 most sites proposed for suburban landfills were mined-out quarries or gravel pits.[81]

Again in 1987 the Northeastern Illinois Planning Commission conducted a survey of landfill space, this time with the expressed intent of discovering potential threats to groundwater. It identified several active landfills in the Lake Calumet marshlands, but found all other active sites beyond the city limits of Chicago.[82] Currently operating landfills, although largely adhering to state and federal guidelines, are far larger than their historic counterparts. Furthermore, some began operation before the systematic enforcement of current standards and may have engineering shortcomings such as garbage placed in contact with groundwater, an ineffective lining, or an inadequate leachate collection system. So although the actual number of disposal sites has decreased, both the inactive sites and the remaining landfills with their larger capacity present greater risks in suburban areas.

The current distribution indicates that the process of exporting garbage wastes to the suburban fringe, along with its consequences, is nearly complete. The most recent landfills generally are filled well above the previous land surface; so unlike pre-1960 dumps, they will tend to deflect other urban land uses. Furthermore, they bespeak a dramatically visible element in the topography of suburban areas: in many cases they have become the largest human-made structures. In addition to inhibiting competing land uses, both the perceived and real risks posed by landfills have become a significant factor in defining surrounding land-use patterns. The risk gradient, in addition to more traditional economic land-rent gradients, has become a very real influence in the continuing process of urban growth well beyond the city limits.

In conclusion, a large portion of Chicago is built on what has been called "made land," that is, soils created by human activity. A portion of the made land is dredge spoil or lake-bed sediments, but a large share is also refuse, garbage, or ash. Each zone of active disposal represented a waste frontier. Initially, land creation was concentrated along the lakefront. Later it moved inland along the river system, and it eventually extended outward, keeping pace with the urban infrastructure. Until 1905 mixed refuse was the most significant component of the fill used in raising the street grades. This filling "reclaimed" vast acreages of prairie wetlands. But land reclamation did not stop at the curbs or the lakefront. The rich clay and limestone resources of the Chicago region provided building materials for the commercial towers and brick flats of the city, and the excavations that yielded these resources left gaping cavities in the landscape. Garbage and ash were used to raise or level the topography in accordance with changing land-use needs. Reclamation of this sort reconfigured the land surface in dozens of locations in the vicinity of Chicago. In addition, industrial wastes built up hundreds of acres in southeast Chicago alone. Such modification of the environment is hardly insignificant and served as the essential foundation for the economic and industrial cores of the city.

The use of mining scars for the interment of urban wastes reflects Chicago's chronic struggle to accommodate burgeoning quantities of refuse. In spite of public displeasure with local waste disposal and experiments with technological fixes, land dumping has remained the principal method of waste disposal. Reclamation of pits, quarries, and wetlands opened up pockets of real estate along the rim of urbanization. These tracts have served the municipal park system and permitted the addition of commercial property in the midst of developed areas. Residential housing, which claims a huge share of the total urban land surface, occupies only a few parcels of recycled real estate; fears of methane gas and subsidence largely precluded housing construction, particularly since 1950. Nevertheless, the filling of former excavations was a vital component of urban refuse management as it reconstituted the land surface.

Finally, largely unseen environmental impacts have become most pronounced in recent years as the city exported vast amounts of refuse to fringe areas. Surrounding most Chicago-area landfills are zones of potential groundwater contamination which will ultimately affect future development within these zones. The large number of fringe sites used during the late 1960s created a ring of dispersed zones around the city. In recent years the trend toward consolidation has slowed the dispersal and created fewer zones of risk around mega-landfills. Nonetheless, the 1960s-era landfills, which employed inadequately engineered safety features, continue to disrupt urban growth in areas beyond the reach of direct urban administration.

Chicago's waste frontier reflects a common process of urban growth that can yield insight into city development. Many unanswered questions remain concerning the historical impact of nuisances on land values in distorting or deflecting other land uses and the relationship between waste disposal and the social geography of cities. Perhaps this initial inquiry into the basic elements of waste disposal will illustrate the linkage between environmental modification and the urban fabric, and open the way for additional work on this fundamental aspect of urban geography.

CHAPTER 16

Mrs. Block Beautiful

African American Women and the Birth of the Urban Conservation Movement in Chicago, 1917–1954

SYLVIA HOOD WASHINGTON

A number of histories about the movement of African Americans to Chicago during the Great Migration periods have elucidated the formation and maintenance of strictly enforced racially segregated ghettos during the earlier half of the twentieth century. In *Black Chicago*, Allan Spears observes, "The most striking feature of the Negro housing . . . was not the existence of slum conditions, but the difficulty of escaping the slum."[1] The use of legal and extralegal measures such as restrictive covenants, zoning, bombing, and race riots resulted in the formation of a tightly bound and extremely dense Black Belt which although only three miles in length contained almost 85 percent of the African American population of 110,000 in 1920.[2] Thirty years later, the historian Arnold Hirsch argued that Chicago census figures "revealed not a city undergoing desegregation but one in the process of redefining racial borders after a period of racial stability. Black isolation was, in fact, increasing even as the Black Belt grew."[3]

Little has been written to date about the environmental consequences of racial segregation in Chicago. A direct consequence of this segregation elucidated in my monograph, *Packing Them In*, was high mortalities and morbidities among both adults and children from environmental diseases like typhoid and tuberculosis.[4] Tuberculosis was directly connected to marginalized housing, especially kitchenette apartments. Their lack of air and light was the optimal breeding ground for the tuberculosis

pathogen. These housing conditions, with fire-prone, dilapidated structures overrun with rats, were an environmental nightmare for African
Americans. The death rate for adult African Americans varied between
four to six times that of adult whites between 1900 and 1950, although
they represented only 4–6 percent of the total population.

Mrs. Block Beautiful was the official name given to the married African American women who won pageants in the Chicago Urban League's
annual Block Beautiful competition. Created by the National Urban
League, this competition was the manifestation of the first urban conservation movement in the United States to address the environmental
plight and public health dilemmas faced by racially segregated urban African American communities. This national urban conservation movement was strongly influenced by and modeled after the Chicago Urban
League's efforts. Mrs. Block Beautiful symbolized the critical role that
women played in a movement whose primary objective was to ameliorate or mitigate these morbid living conditions. These women and their
activities typified African American club women and their efforts in the
urban Midwest. According to the historian Darlene Clark Hine, these
club women did not possess the financial resources of Andrew Carnegie,
John D. Rockefeller, or Julius Rosenwald, but "their giving of time and
effort and commitment to racial uplift work . . . and their endless struggle to create living space for the segregated, often illiterate, unskilled and
impoverished black Americans were as valuable as were the two-room
Rosenwald schools . . . the libraries funded by Carnegie and the Rockefeller supported black medical schools."[5]

The environmental conditions that African Americans endured and
that the women of the Block Beautiful movement sought to address
were lucidly described in Richard Wright's *Twelve Million Black Voices*.
Wright, a migrant to Chicago in 1927 at age nineteen, lamented, "We remain to live in the clinging soot just beyond the factory areas, behind the
railroad tracks, near the river banks, under the viaducts, by the steel and
iron mills, on the edges of the coal and lumber yards. We live in crowded,
barn-like rooms, in old rotting buildings where once dwelt rich native
whites of a century ago. Because we are black . . . because the outdoor
boisterousness of the plantation still clings to us . . . white people say that
we are destructive and therefore do not want us in their neighborhoods."[6]

This essay illuminates the vital role that African American women
played in Chicago in launching the first national urban conservation
movement as a means of surviving and optimizing the dismal environmental surroundings with which they contended in the "land of hope."
Their efforts were a direct response to the compromised ecological and oftentimes deadly public health conditions in northern African American

communities that persisted for more than fifty years. These circumstances arose as a direct result of the legal segregation of African Americans from the white social body through racial covenants and racial zoning in the aftermath of the Supreme Court decision of *Plessy v. Ferguson*.

African American women from all disciplines (teachers, laborers, social workers, health care professionals, homemakers, and club women) and social strata were critical to this movement. Their organization and management of the grassroots neighborhood conservation councils made the movement a reality in Chicago. Their hard work also led to the movement becoming national in scope, one of the National Urban League's crown jewels. Redeveloping environmentally salient communities continued until African Americans in Chicago gained access to healthier neighborhoods after the legal dismantling of racial segregation policies that slowly followed the 1954 *Brown v. Board of Education* decision. Their story in many ways captured the African American club women's motto Lift as We Climb, except in this case they realized that they could not "lift" or "climb" themselves completely out of Chicago's environmentally oppressive and racially restricted ghettos. One press release of the period succinctly summarized the reasons behind the movement: "[In] one of Chicago's typical transitional neighborhoods . . . several white families getting tired of garbage heaps in alleys, roaches, alley rats and the like [sell] their homes to Negro families. The new Negro property owners, realizing that the evils of restrictive covenants prevented them from just wearing out a neighborhood and moving into another, decided they would better their neighborhood where they were . . . [and] they began to clean up their own alleys and sweep their own streets."[7]

The Mrs. Block Beautiful contest was initially posed to Miss Alva B. Maxey, director of the Chicago Urban League's Community Organization Department by her supervisor, Frazer T. Lane, in 1954: "A group of judges will be appointed to evaluate the most beautiful who will reign as queen at the final meeting when prizes are awarded to the best block effort [and] the qualifications require that she be a married woman."[8] The idea of a Mrs. Block Beautiful arose as a means of counteracting the waning interest of the African American communities in participating in the Block Beautiful contest. This was the same year that the Supreme Court's *Brown v. Board of Education* decision struck down the ideology of "separate but equal" societies. This decision may have led many African Americans to believe that they would soon be able to escape their racially induced deadly ghettos. The lapse in interest in the Block Beautiful contest also may have been related to the fact that this was the year that the Urban League disbanded because of internal and external turmoil surrounding its controversial executive director, Sidney R. Williams, and director, Earl B. Dickerson, who were perceived as "too radical." Dickerson and Williams had been critical in creating and maintaining the League's Block Beautiful Movement.

According to Lane it was essential that Mrs. Block Beautiful be married. Although the historical record offers no specific reason for this criteria, it was perhaps tied to the fact that the success of the larger Block Movement from its inception was due primarily to the efforts of African American women, especially married women. They became the movement's grassroots activists, organizers, and leaders. Hine also emphasizes that during this period, "the larger society viewed black women as whores and prostitutes."[9] The desire to have a "married" community queen also may have also been influenced by a need to promote a respectable image of African American women to counter such debased images. The effort to promote a more favorable portrait of African American women was consistent with the "initial orientation and objectives of early black women's clubs" that included "creating a positive image of their sexuality."[10] The idea of a Mrs. Block Beautiful was not a random concept but one based on the success of other beauty contests in the city at the same time. According to Lane's memo, the nine-year contest, which was originally cosponsored by the Chicago League and the *Chicago Defender*, needed a "gimmick" to keep it going. The idea of a queen for the contest came from his observations of local businesses in the area that had successfully employed beauty contests to increase sales among their clients.[11]

The annual Block Beautiful contest began in 1945 as part of the Chicago Urban League's Five Year Civic Improvement Plan and was part of its Block Beautiful Movement. The primary objective of the Five Year Plan was to "help Negroes help themselves."[12] For the League this effort was critical for African American communities because of the large influx of in-migrants who were poor and illiterate and for whom "the problems of housing, sanitation, and right living is, for the most part, kept in the background, either through ignorance or the lack of interest."[13] World War II labor demands prompted huge population increases that taxed these communities. With restrictive covenants still in place and serious acts and threats of white mob violence to maintain them, African Americans were densely crowded into the deadly and decaying neighborhoods of Chicago's Black Belt, which became known for its notorious kitchenette apartments immortalized by Lorraine Hansberry's play, *A Raisin in the Sun*. Wright described the impact of kitchenettes:

> The kitchenette, with its filth and foul air, with its one toilet for thirty or more tenants, kills our black babies so fast that in many cities twice as many of them die as white babies. The kitchenette is the seed bed for

scarlet fever, dysentery, typhoid, tuberculosis, gonorrhea, syphilis, pneumonia and malnutrition. . . . [It] scatters death so widely among us that
our death rate exceeds our birth rate, and if it were not for the trains and
autos bringing us daily into the city from the plantations, we black folks
who dwell in northern cities would die out entirely over the course of a
few years.[14]

The Block Movement was initiated by the National Urban League's
affiliate in Pittsburgh's Hill District between 1915 and 1920. The formation of the Block Movement was consistent with the three-pronged objective of the Urban League to help migrating African Americans with
employment, housing, and assimilation.[15] The movement involved the
formation of block clubs "designed to be a bulwark of strength for communities to urbanize people in a better way of life."[16] Lane formed the
first block club in Chicago—the South Side Block Club.[17] The League's
first secretaries were responsible for organizing and supporting the early
neighborhood improvement clubs.[18] Within a year of its formation, the
Chicago Urban League recognized that poor housing was a critical problem for migrating blacks and the rapidly expanding black population. As
early as 1917 it began utilizing African American women to do "block
work," which entailed them entering new arrivals' homes and giving
them advice "about health, cleanliness, deportment in public places, care
of children, overcrowding and efficiency."[19]

Between 1920 and 1924 the Urban League dissolved its civic department, reforming it in 1925–1926 as the Department of Civic Betterment
to address the perpetual issue of unsanitary housing and the concom-
itant environmental and health troubles associated with this problem.
Considered by the League to be its "outstanding achievement during the
period," the department's first director and chairperson were women:
Maude A. Lawrence and Mrs. Frank Brown. An interracial civic committee, which advised the department, contained eighteen members;
all but two were women. They represented the most politically influential social agencies, women's clubs, and civic organizations in Chicago:
the West Side Women's Club, the North Side Community Center, the
Neighborhood Improvement and Protective Association, the Chicago
Federation of Colored Women's Clubs, United Charities, the Social Hy-
giene Council, the Lower North Community Council, the Association of
Commerce, the Community Center Council, the Elizabeth McCormick
Memorial Fund, the Chicago Woman's Club, the Chicago Commons, the
Northwestern University Settlement, and the Chicago Council of Social
Agencies. The first-year efforts of the newly revised department concen-
trated on civic activities, such as thrift, health, recreation, neighborhood

clubs, speakers bureau, and participation in a housing conference that had been called by Mary McDowell.[20]

The League's block work persisted but never really flourished on a sustained basis until the mid-1940s, when it was revived in its most successful form by two women: Mrs. Maude Lawrence on the South Side and Mrs. Rachel Ridley on the West Side under the executive secretaryship of Mr. A. L. Foster.[21] The Block Beautiful Contest became the centerpiece of the new Block Beautiful Movement, and it would thrive for almost ten years under the leadership of the League's executive director, Sidney R. Williams, in cooperation with the League's various Community Organization Department directors: Lillian Proctor, Alva B. Maxey, and Ridley.[22]

SOWING THE SEEDS OF A BEAUTIFUL MOVEMENT

The success of the Block Beautiful Movement of the 1940s and 1950s was undoubtedly tied to the Chicago Urban League's revived efforts during the Depression. The League benefited from "a million dollars in Government funds through the Works Progress Administration, National Youth Administration and AEP" that was poured into African American communities. In the mid-1930s the League began to receive institutional support in the form of workers from these agencies. With this support, the League's Civic Improvement Department was able to file "more organized complaints from residents in regard to zoning violations, vice and crime."[23]

On Chicago's South Side during the Depression, the League joined several government-sponsored community improvement projects led by women on a day-to-day basis: Dr. Ruth Howard, director of the National Youth Administration (NYA) B-153 Health Project, and Mrs. Naola Smith, director of the Better Conduct Program of the NYA B-153 project. The Chicago Urban League's Civic Improvement Department sponsored all NYA projects and its director, A. L. Foster, acted as the general supervisor of the B-153 project. The objectives of the NYA's B-153 project were to supervise "health and better conduct activities for youth whose families were on relief."[24] Although the historical record does not identify the percentages of women supporting or participating, contemporary photographs of these Chicago organizations show only African American women and girls.

Initiated in 1936, the League's Work Progress Administration Project #2526, also referred to as the Community Improvement Project #2526, aimed to develop "new community standards and ideals through Recreation and Education programs." The project had 140 workers at the beginning of 1938 with "at least thirty of them devoting their time to

organizing and instructing adults in the community north of 43rd street."
The program started to decline that same year, following a decrease in
the allotment of Works Progress Administration (WPA) funds.[25] The
importance of WPA funding in advancing community conservation
was reflected in the 1938 *Civic Improvement Department Report*, which
stated, "Through the activities of the Works Progress Administration
workers on project #2526, 40 neighborhood improvement clubs were organized and functioned to some extent. With the loss of these workers,
many of these groups ceased to function." By 1938 the League proudly
claimed that it had "organized the City-Wide Association of Neighborhood Improvement Clubs and through it many creditable things were
accomplished." Unfortunately this success would last only "about six
months . . . due to poor leadership [that] became inactive."[26]

At this time the League first received help for its nascent Block Movement from white or predominantly white women's organizations and
clubs in the city. The Chicago Women's Club's Celia Wooley Committee
supported a small Better Block Contest that selected the block which had
planted the most grass. "In this effort, 100 pounds of grass seed [was]
donated by the Celia Wooley Committee . . . and distributed free to those
who felt they were unable to buy seed." The Committee also "distributed
certificates . . . to those taking interest in improving their premises."[27]
Lacking sufficient paid staff to implement a citywide Block Movement,
the Urban League relied heavily on volunteers, especially female volunteers. Consequently, most of the League's Block Movement activities
reflected the perceived interests of women who constituted the majority
of volunteers. The League sponsored its first Flower Show and Neighborhood Fair during the Depression, demonstrating to attendees "how to
make flowers, gardens and grass and how to solve some of the problems
that are conducive to the creation of slums." This fair also featured a
"Queen of Honor" and had "moving pictures of beautiful spots on the
south side, and beautifully decorated booths set up by commercial flo-
rists, lumber companies, hardware stores, wall paper and rug cleaning
establishments."[28] The League formed two volunteer women's groups:
the Woman's Division and the Junior Woman's Auxiliary of the Chicago
Urban League. The constitution of the latter group, whose members were
between the ages of eighteen and thirty, stated that its purpose was "to
help promote and advance the work of the Chicago Urban League and
the Woman's Division in whatever way possible."[29]

IRENE MCCOY GAINES

The Chicago Urban League enlisted prominent African American
women to promote its community conservation program. In 1938 Earl

Dickerson, the League's vice president, participated in a radio interview, "What the Urban League Means to Chicago," conducted by Irene Mc-Coy Gaines, an influential African American club woman and activist. Gaines introduced herself as "a representative of the general public, and particularly of the women who are so vitally interested in the social, economic and civic status of the citizens," humbly understating her position in Chicago's African American community.[30]

Like many of Chicago's African Americans, Gaines was a southern migrant. Born in Ocala, Florida, in 1892, she was the niece of "George Washington Ellis [who] had served with the U.S. delegation to the West African Nation of Liberia between 1902 and 1910."[31] A Fisk University graduate, she trained as a social worker at the University of Chicago. Her husband, Harris Barrett Gaines, was a noted lawyer and real estate man who served two terms in the Illinois legislature. Irene was a recruiter for the Women's Trade Union League and in 1920 served as the industrial secretary for the first "Negro branch" of the Young Woman's Christian Association in Chicago. By the late 1920s she had become a member of the Woman's City Club of Chicago. From 1924 to 1935 Gaines was the president of the Illinois Federation of Republican Colored Women's Club. One of her crowning achievements was her presidential appointment to the Housing Commission by Herbert Hoover in 1930.[32] By 1938 she was the Director of the Girls Work Division War Camp Community Service, the president of the Northern District Association of Colored Women (NDACW), and an active member of the Chicago Urban League. Gaines wielded influence among African American women who might be productive volunteers for the League. The NDACW's Housing Committee's 1938 report repeated concerns about African Americans' housing conditions that Gaines had raised in her radio interview: "Federated club women have been greatly alarmed over the acute housing condition in Chicago, particularly in the so-called 'black belts' where are found the most blighted and deteriorated areas. We therefore addressed ourselves to the study of these conditions in several of our committees and joined other civic organizations by letters and telegrams in a demand upon the Mayor of this city for the appointment of a Negro to the Housing Board. It was with great disappointment that we read of his recent appointment of white men to the two vacancies on the housing authority."[33] The NDACW's Health and Hygiene Committee's report mirrored the concerns of the League's Civic Improvement Department. Their report urged "that for the welfare of the community, we must keep up our neighborhood in housing. Keep a constant check on garbage removal and general health conditions. Read literature on better homes; listen to radio broadcasts on prevention of disease and methods of disinfecting and disposing of refuse."[34]

Dickerson's radio appearance aimed at encouraging the continued support of Chicago women. He acknowledged that the League was "indebted to the organized women of Chicago because it was largely through their efforts that the Urban League was started." Indeed, the Federated Women's Club had provided two of the League's founders, Jo-anna Snowden and Ruth Standish Baldwin.[35] Obtaining the support of Snowden legitimized the League's efforts, given her social standing in the African American community. Snowden came from one of the city's native African American families known as "Old Settlers." Her parents, Joseph Henry and Anne Elizabeth Lewis Hudlum, lived in the "first house built for colored people and owned by them in 1857."[36] Born in 1864, Joanna attended Chicago's Bryant and Stratton Business College, the School of Civics and Philanthropy, and the University of Chicago after graduating from Englewood High School. She married Joseph Ross Snowden in 1884. Snowden created the Northwestern Federation of Col-ored Women's Clubs (comprised of twenty-four states at that time). She also organized the National Association of Colored Women and served as a deputy recorder for Cook County from 1924 to 1927. An extremely active club woman, Snowden was one of the founders and the secretary and director for the Home for the Aged and Infirm Colored People and treasurer for the Phillis Wheatley Home for Colored Girls.[37] Snowden's endorsement meant that club women would become active in making conservation efforts a reality for the entire community.

NURTURERS OF A BEAUTIFYING MOVEMENT

The League launched its most successful Block Movement between 1945 and 1954. Rachel R. Ridley and Maude Lawrence renewed the move-ment, becoming its bedrock along with Lillian Proctor and Alva B. Maxey. Ridley and her group of female volunteers on Chicago's West Side became key driving forces in the movement. Ridley was the direc-tor of the League's West Side Women's Division and the creative and organizing force for block clubs in this section of Chicago. These West Side League women aggressively and successfully recruited women block club leaders who would eventually represent over half of the block clubs in Chicago for most of the last ten years of the Block Beautiful Movement.

Ridley was one of the most critical persons in the success of the Block Beautiful Movement. When she died in 1986, Ridley was recognized as "a pioneer in women's rights and human rights [who] always worked at getting people to work together."[38] Born in 1911 in Hannibal, Missouri, she had lived in Chicago's West Side since 1918. After graduating from McKinley High School, she attended several colleges before graduating from Roosevelt University in 1946 with a BA in sociology. Both a mother and wife during her activism, she worked as a teacher in a Federal Adult Education Project from 1932 to 1934. She joined the Chicago Urban League in 1937, where she "sponsored NYA programs recruiting, organizing and supervising youth groups in the West Side area." With promotion to a staff position, she sponsored WPA recreational programs for adults and youths on the West Side in 1938 and 1939. When the League's federal funding ended in 1942, Ridley "was appointed Director of the West Side activities by the Urban League, supervising a staff of two professionals and four volunteers." In this position, she was responsible for "initiating block clubs as a vehicle for urbanizing newcomers."[39] She also organized the Midwest Community Council, whose original emphasis was on ensuring law and order in the community. When Ridley left the League in 1952, twenty-seven of her female League colleagues and friends hosted a farewell tea. The program booklet summarized her contributions and their gratitude: "We, of the Chicago Urban League are fortunate to have had the opportunity to fellowship with our good friend 'Rachel' through the years. Through her wisdom and tireless energy, she has enriched our progress and created goodwill for the League."[40] Within two years of Ridley's departure, the Block Beautiful Contest and Movement began to falter.

THE FRUITS OF CIVIC LABOR: THE BLOCK BEAUTIFUL MOVEMENT

In the beginning the League and the *Chicago Defender* cosponsored the Block Beautiful contest, encouraging participation from other agencies and businesses regardless of race.[41] They created the Block Beautiful Movement to motivate African American communities to participate in making the physical and environmental conditions of their neighborhoods cleaner and safer. The movement developed from a "desire for better health conditions, cleaner politics, parks and playgrounds, more schools efficiently equipped with space for all pupils and strict law enforcement."[42] One of the primary environmental concerns of the movement was the threat of fire in the slums. On January 26, 1947, for example, the *Chicago Bee* featured the article, "West Side Slums Fire Claims Four," revealing that community groups conducted investigations and sought prosecutions "over the appalling conditions in over-crowded neighborhoods [which] permitted the death of 11 persons in murderously destructive fires during the past two weeks." The most recent fire included the deaths of four children between the ages of three and fourteen who had been living in "a west side tenement attic apartment which fire inspectors had reported as 'insufferable and dangerous.'"[43]

The threat of fire was not the community's only environmental concern. Rat control topped the list for community activists, citizens, and
the city government. In 1947 "thousands of southsiders joined the citywide war on rats [under] Mayor Kennelly's 'Rat Extermination Week.'"
The campaign to remove the "disease carrying rats" had been motivated
by attacks on fifteen babies who had "been bitten and maimed by rats
during the year."[44] The featured speakers for the mayor's "rat campaign"
included prominent African American women such as Gaines, Loraine
Green, and Mrs. Mildred Casey. These women were chosen along with
prominent male civic leaders such as Frayser Lane and city aldermen
such as Archibald Carey and William Harvey. Loraine Green's husband
was Wendell Green, a prominent African American University of Chicago graduate, attorney, former assistant public defender (1929–1930), appointed Civil Service commissioner (1935–1942), and elected municipal
judge (1942–1950). Loraine and Wendell were southern migrants from
Topeka and Kansas City, Kansas, respectively.[45] Holding a PhB, MA, and
PhD in sociology from the University of Chicago, Loraine received a research assistantship for the University's Institute of Social Research after
graduation and eventually became an Institute member.[46] Highly respected among club women, she was a member of the Board of Directors
of the Urban League, Girl Reserves, YWCA, Woman's City Club, Chicago's Board of Health, and Welfare Council of Metropolitan Chicago.[47]

One objective of the Block Movement was to combat negative racial stereotyping and to demonstrate the worthiness of social equality
for African Americans by creating cleaner and safer communities. The
movement's success would help with "our journey along the road towards full participation and unrestricted participation in the American
way of life."[48] This objective was manifested by League's motto, We Fight
Blight. A 1951 League press release and promotion flier titled *A Sight
Worth Seeing* was specific about its concerns and complaints about African American neighborhoods in Chicago: "In much of Chicago's Negro
community, the casual visitor is appalled at the many evidences of filth
and squalor, the carelessly littered streets and alleys, the buildings unsightly and forlorn from abuse, and neglected repairs, and the grounds
from [which] grass and flowers have since vanished. This we all know
about to our shame."[49]

Five neighborhood councils, totaling nineteen block units, joined the
renewed Block Movement by the Chicago Urban League and the *Defender* in 1940. They were the Snowdenville, DuSable, Abbottsford, Bethune
Progressive, and the West Side Community Improvement Councils.
The League expected all of its fieldworkers to "familiarize [themselves]
with the data description of [their] territory." They were also required to

conduct thorough surveys and to "know [the] neighborhood as well as or better than any resident of it."[50] The League encouraged block clubs and neighborhood councils, once formed, to have at least six standing committees to ensure the neighborhood's physical and social integrity. The majority of the committees were environmental in scope and consisted of the "Youth Activities, Zoning, Health and Sanitation, Foods and Nutrition, Streets and Alleys and Garbage Collection" committees.[51]

Block or neighborhood clubs created ads and fliers which promoted participation in the Block Beautiful contest. A flier by the Langley Avenue Neighborhood Improvement Club boldly announced that "Judgment Day Has Come!" and promised its neighbors that they could win a prize in the Urban League contest if they would all "Cut grass, Wash windows, clean [the] street in front of your premises [and] tidy up your alley." Like others in the movement, this club conducted block parties that offered participants "fun, music, prizes and noted speakers" in an effort to urge participation in their cleanup campaigns. The campaigns not only targeted garbage and debris as environmentally undesirable but also people like "the pitiful drunks who loiter and block the sidewalks at 47th and Langley." Clubs also wanted to make it clear that other institutions and organizations in the city like the "Health Department, Alcoholics Anonymous, the Psychiatric Institute, the Anti-Saloon League and the Police Department" supported their programs.[52] Neighborhood clubs held fund-raising campaigns among community members "to continue our campaigns for a cleaner neighborhood and to purchase the necessary equipment in order to perform this task."[53] Donations purchased lids for garbage containers and grass seed for lawns.

The League encouraged the neighborhoods' councils to join not only in its contest but also in a number of citywide cleanup efforts, like Mayor Kennelly's drive to "Make Chicago Shine in '49." The League sent out letters to its community leaders asking for their participation in the city's Clean Up parade. The League wanted parade participants to "march with brooms, mops, pails and appropriate banners."[54] The League also wanted the block organizers to "Contact land owners as well as tenants and show them the need of improvement techniques" and to impress on "tenants whose land owners are absent . . . that through cooperative effort, conditions may be rectified."[55]

The Block Beautiful Movement frequently conducted public workshops (that included women) in conjunction and in cooperation with other civic organizations to promote conservation efforts. At least nine of the twenty-six speakers and consultants for a community workshop sponsored by the West Side Urban League, West Side Principals' Luncheon Group, and the West Side Ministers Council were women, and over 90 percent of these were married women. The speakers included "Mrs. Evangeline Fahy, President of the Chicago Principals' Club, Mrs. Eleanor Dungan, Education Director, Commission Human Relations . . . Mrs. Marianna Bell, National President, Junior League and Teacher Emerson School . . . [and] Mrs. Tarlease Bell, President of the West Side Community Division, Chicago Urban League."[56] They were all invited to publicly answer the following questions: "What are the basic needs of a community? Who is responsible for developing them?"[57]

Six years into the Block Beautiful Movement, the League encouraged participation by other communities by communicating that its success in improving environmental conditions in the neighborhoods was due to the involvement of women. In 1951 the League bragged, "There is a block which boasts the cleanest alley in the City, because over a period of three years the housewives in each building have assembled regularly twice a week at their back gates, and have swept their alleys. . . . There is a block in which the women residents sold dinners and raised money which was used to fence in vacant lots in their neighborhood, as a means of combating the problem of broken sidewalks which had formerly resulted from the overnight parking of trespassing cars and trucks."[58]

By the end of 1952 there were ninety-seven block clubs, chiefly organized by female volunteers. The League's success was based on their decision to "further implement the process of improving communications [among African Americans] by sponsoring tours of the neighborhoods for groups desiring to see firsthand what the block clubs were doing." The League continued to emphasize the importance of its primarily female fieldworkers. In 1951 the League "decided that the worker should continue to visit in the block. She would busy herself with collecting information about the residents, and search for a person who had status with both the more stable residents and the more mobile group." The League also concluded from experience that married female residents were effective community organizers. One report revealed, "Finally [a] person was found who seemed to possess the qualities needed to act as a cohesive force. She was the daughter of a property owner who had lived in the block for many years, but was herself a tenant, mother and housewife. . . . She offered her home as a meeting place and later she, the worker and a neighbor distributed notices of a planned meeting. The following week the block was organized."[59]

Environmental concerns remained at the heart of the Block Beautiful Movement. The League's Community Organization Department efforts included the "7th Annual Block Beautiful Contest, tours and demonstrations for outer community groups, parked car removal campaigns, participation in the Mayors' Cleaner Chicago Week [and] the Board of Education's Clean Up Campaign."[60] The League's 1954 promotional booklet, *We Fight Blight*, reiterated the Movement's environmental concerns by stating that the "League was the first to organize and to use the block groups as a medium to prevent physical decay of neighborhoods." The pamphlet also reported that the League had 175 organized blocks which took part in its 8th Annual Block Beautiful contest in 1953 and that it had "constructively influenced 53,000 residents . . . [because] 450 meetings by organized block groups were held in 1954 [that] worked on problems common to their particular block."[61]

At least another two decades passed before Chicago's African Americans were able to escape racially segregated communities as a direct consequence of the 1964 Civil Rights Act and the 1968 Fair Housing Act, although the 1954 case of *Brown v. Board of Education* may have been a factor in the lapse of interest in the League's Block Beautiful Movement. The 1954 decision signified for many African Americans across the country that a "new day" was coming with changes in a life of race-based segregation. The Supreme Court decision created this hope because it dismantled the legal justification of a "separate but equal" society which had hardened over the decades. From the turn of the twentieth century to the 1954 decision, African American communities clearly understood that they were literally and figuratively locked into racially segregated housing and communities that were environmentally marginalized because of racism. The Mrs. Block Beautiful contest was part of the Chicago Urban League's Block Movement, and it was specifically geared to help African Americans improve the living conditions of their racially segregated and dense living spaces.

Mrs. Block Beautiful embodied the mature African American women who tried to salvage and transform the brutal environmental conditions in which most African Americans were trapped. She was also a metaphor of the importance of self-help in communities that existed in the "promised land." The "land of hope" provided African American migrants with access to greater economic prosperity than they had enjoyed in the land of Dixie, but the trade-off for this freedom was an unanticipated environmental disenfranchisement which they rarely faced in the South. Rather than capitulate, the African American community relied on women like Mrs. Block Beautiful to educate and guide their families and neighbors on how to create beauty and health in the beastly and oftentimes deadly urban ghettos of the North.

PART V

REENVISIONING THE LAKE AND PRAIRIE

The industrial era wrought fundamental change in the region and in how residents interacted with the natural environment. While many urbanites saw only the briefest glimpses of nature at the height of industrialization, University of Chicago botanist Henry Chandler Cowles, who helped develop the field of ecology, immersed himself in the Indiana Dunes, just a short train ride away along the Lake Michigan shoreline. The landscape architect Jens Jensen, who worked alongside Cowles, noted that "the primitive prairies of Illinois have not been entirely destroyed. Here and there has been left something of the primitive that the plow had not turned under. It seems a pity, rather a stupidity, that some section of this marvelous landscape has not been set aside for future generations to study and to love—a sea of flowers in all colors of the rainbow."[1]

Residents of the Chicago area during the twentieth century joined ecologists and landscape architects in seeking to preserve notable remnants of the natural landscape, as well as to foster its restoration. Citizen activists worked through new and existing public and private organizations, as well as founding new ones. For instance, in 1922 Joy Morton founded the Morton Arboretum to promote the study and conservation of trees. Openlands was established in 1963 to promote urban conservation efforts of the remaining prairie, wetlands, and open spaces. In his contribution, William Barnett shares the story of May Watts, a writer and educator at Morton Arboretum who studied with Cowles and worked with Jensen. Developing a distinctly midwestern strand of environmentalism, she galvanized suburban women in a grassroots movement to protect prairies and other open space from highways and sprawl.

Other Chicago citizen activists joined post–World War II efforts to restore what had been lost or damaged through urban industrialization. Robert Gioielli examines the work of Businessmen for the Public Interest (BPI) to protect Lake Michigan from the environmental risks of a lakefront airport or nuclear energy plants, demonstrating an important shift in environmentalism away from grassroots activism to institutionalization.

Mark Bouman appropriately closes this volume with his discussion of efforts to preserve and protect the Calumet region. Containing some of the dunes which Cowles studied, natural wetlands and reconstructed rivers, and heavy industry and reclaimed brownfields, the Calumet region encapsulates many of the broader themes of landscape change in Chicago. Efforts to celebrate and preserve this landscape of fragile ecosystems and abandoned steel mills offer a path to a deeper appreciation of the complex mixture of natural and human history that exists across twenty-first-century Chicago.

CHAPTER 17

May Theilgaard Watts and the Origins of the Illinois Prairie Path

WILLIAM C. BARNETT

In September 1963 the *Chicago Tribune* published a reader's letter entitled "Future Footpath?" calling for a multiuse nature path to replace an
interurban railroad across the western suburbs which had closed due
to competition from new highways. May Theilgaard Watts, a seventy-
year-old retired woman, wrote the letter, and Chicagoans had little reason to expect her vision would become reality. But Watts was a remarkable and beloved naturalist who was well positioned to serve as a bridge
between earlier conservationists and the emerging 1960s environmentalism. This movement was in its infancy—Rachel Carson's *Silent Spring*
had appeared twelve months earlier, and Earth Day would not be celebrated until 1970. But many new ideas were coming together, and Watts
provided a distinctly midwestern strand of environmentalism. She
sparked a grassroots campaign by suburban women, and in under three
years they had negotiated a lease for the railroad right-of-way. The Illinois Prairie Path's 1966 agreement with DuPage County gave the nonprofit group its first twenty-seven miles of trail, and a 1972 agreement
with Kane County extended two branches of the trail westward to the
Fox River. In less than a decade, this grassroots effort led by suburban
women planned, negotiated, and built the Illinois Prairie Path, a pioneering rails-to-trails project in the United States.[1]

Now half a century old, the Prairie Path is a network of sixty-one
miles of trails for walking, biking, and horseback riding which stretches

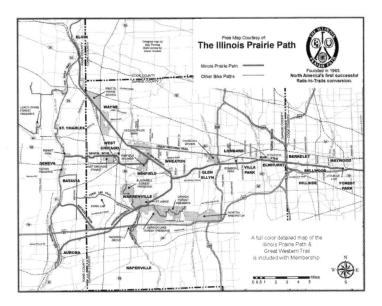

FIG 17.1. *The Illinois Prairie Path*, 2009. Map courtesy of Illinois Prairie Path, a nonprofit corporation.

from Cook County across DuPage County into Kane County. The multiuse trail forms a "Y," with its intersection in Wheaton, Illinois. One branch heads due east toward Chicago, terminating in the Cook County town of Forest Park. The other two branches go west to the Fox River, with the southwest branch ending in Aurora and the northwest branch in Elgin. The path crosses through residential areas, forest preserves, wetlands, farms, and restored prairies, so it protects and links critical green space in Chicago's western suburbs. But how did Watts's proposal become a reality? What was her vision for the Illinois Prairie Path? Why did many suburban women join in? And what do these grassroots activists reveal about ideas about nature in the Chicago area, and about the origins of the national environmental movement?[2]

Much analysis on the origins of environmentalism examines two topics: Rachel Carson's 1962 book *Silent Spring* and the first Earth Day in 1970. Carson's book was a critical turning point, informing readers that toxic chemicals were damaging ecosystems and posing health risks. And Earth Day showed politicians the widespread interest in cleaning up the nation, leading to bipartisan legislation. But broad interest was needed before 1970 to get millions to demonstrate, and Carson did not raise the alarm by herself. Americans reexamining relationships with nature were influenced by many writers. Rachel Carson, Stewart Udall,

Paul Ehrlich, and William Whyte called attention to an impending crisis, while nature writers such as Aldo Leopold, Edwin Way Teale, and May Theilgaard Watts, in her 1957 book *Reading the Landscape: An Adventure in Ecology*, increased awareness of local ecosystems and introduced ecology to the public. Both efforts were needed to shift ideas about nature from conservation to the new environmentalism.[3]

There are four noteworthy aspects of the Prairie Path campaign that helped it succeed and that reveal environmentalism to be a national movement with deep local roots. First, May Watts, who was born in Chicago in 1893 and died in Naperville in 1975, was a bridge between the ideas of early twentieth-century conservationists and the new environmentalists. She was an undergraduate at the University of Chicago with Henry Chandler Cowles, a renowned founder of the field of ecology, and she worked with landscape architect Jens Jensen and his conservation group Friends of Our Native Landscape in the 1930s.[4] Watts looked to the past in her *Tribune* letter: "We need a footpath. Right now there is a chance for Chicago and its suburbs to have a foot path, a long one. . . . The right of way of the Aurora electric railroad lies waiting. If we have the courage and foresight, such as made possible the Long Trail in Vermont, and the Appalachian Trail from Maine to Georgia . . . then we can create from this strip a proud resource."[5] Her 1960s appeal had parallels to these pioneering trails planned in the 1920s because she shared the values of men like Benton MacKaye, whose Appalachian Trail was a response to concerns about cars and urban sprawl.[6]

A second defining trait is that while Watts shared conservationist goals, her efforts were bottom-up, not top-down, with a local community emphasis. Most conservationist leaders were wealthy, powerful men such as Teddy Roosevelt, Gifford Pinchot, and Benton MacKaye, who directed federal agencies to preserve wilderness on a vast scale, often in the western mountains.[7] The women that Watts organized also prized open space, but they were private citizens protecting more ordinary Midwest landscapes. Their campaign was local, not national, and their path was meant for neighbors, not vacationers. Watts highlighted rural Illinois scenery, describing a path through "a forest of sugar maple trees, dipping into a hollow with ferns . . . and a long stretch of prairie, tall grass prairie."[8] Middle-class women protecting local landscapes created the Prairie Path, placing it in the category of grassroots environmentalism.

A third defining quality was that Watts opposed the rapid landscape change in 1950s and 1960s Chicago, and mourned the replacement of farms and prairies with highways and suburbs.[9] The Illinois Prairie Path's name emphasized the state's disappearing prairie ecosystems and called for preserving open spaces. Watts feared another highway could be built along the interurban railway, ending her *Tribune* letter with a call to action: "Right now the right of way lies waiting, and many hands are itching for it. Many bulldozers are drooling."[10] She used the bulldozer as a symbol of the damage done in the name of progress, but she had to work to get others to value midwestern landscapes. The vast stretches of flat farmland in Illinois did not generate as much interest as mountain wilderness, but Watts celebrated the rural Midwest and argued for tallgrass prairie, as did Aldo Leopold of Wisconsin and Edwin Way Teale of Joliet, Illinois.[11] Watts admitted she had to learn to love this ecosystem, writing that "The prairie which seemed so cold and empty on my first exposure, has become warm and full of interest."[12] While she pushed back against Cold War growth, her analysis was deeply local, highlighting prairies and farms threatened by sprawl, and saying less about pollution and nuclear threats.

The campaign's final defining feature was its focus on nature education and ecology. Watts studied ecological principles on undergraduate field trips to Indiana's Dunes and Michigan's Upper Peninsula, and found her calling in these outdoor classes. She became a teacher after graduation and continued in the role of educator for decades, teaching classes for Friends of Our Native Landscape in the 1930s and running Morton Arboretum's education programs throughout the 1940s and 1950s. Her 1957 book *Reading the Landscape* and the hikes she led along the proposed Prairie Path were all structured as scientific field trips. Excursions into nature were her core teaching method, as she brought adults and children to prairies, wetlands, and forests to understand ecology. Her 1963 proposal laid out who would use and maintain the path, naming garden clubs, scouting groups, birdwatchers, and students, and showed that the path was a classroom for citizens to encounter native species through activities such as "identifying and listing plants."[13] The community of nature enthusiasts that Watts built around her Arboretum classes, and affiliated groups, would form the core of an effective grassroots campaign.

THE DEVELOPMENT OF
MAY THEILGAARD WATTS'S IDEAS

May Watts is best known for the work she did in her sixties and seventies, but her passion for nature study spanned her entire life. Going back to her childhood and undergraduate education reveals the roots of her fascination with ecology and the impact of influential mentors who recognized her talents as a naturalist. Her parents implanted a love of nature and of drawing that stayed with her for life. Herman and Claudia Theilgaard and their four daughters lived in Chicago, and Watts recalled, "My botanical background began with my father who was trained as a gardener in Denmark and with whom I gardened, learning scientific names, when I was very young indeed."[14] May's sister Laura described a home filled with plants and artwork, saying, "Papa taught us how to garden. We each had a plot about 3 × 6 feet. . . . Sunday afternoons we usually drew pictures. . . . Sophie and May excelled in this activity."[15] Education was a top priority, and Sophie was a high school principal with a PhD from Northwestern, while Ethel taught in Chicago for nearly forty years.[16]

May Theilgaard followed her sisters into teaching, starting in a one-room schoolhouse after high school. She then taught high school science while taking summer classes at the University of Chicago. She was a top student, graduating Phi Beta Kappa with a major in botany in 1918. The meticulous notebooks she saved are filled with beautifully detailed drawings of trees, pine cones, and flowers, and her papers on Lake Superior and the Indiana Sand Dunes are illustrated with drawings and photographs.[17] Later in life Watts often discussed these life-altering explorations, particularly in her book *Reading the Landscape*.

Watts spoke admiringly of multiple professors, but always described the botanist Henry Chandler Cowles, a pioneer in ecology, as particularly influential. She wrote, "I was fortunate enough to study with Dr. Henry Cowles, and to take many of his field trips including a five-week trip to the Lake Superior region." She saved a tribute to Cowles that praised his ability to inspire student research and teach educators how to use the outdoors as a classroom.[18] Watts summed up his impact, calling Cowles the "great and first American ecologist" and declaring, "He taught me how to read."[19] Much of what she did in the next six decades—as educator, author, and activist—was to re-create her own powerful learning experiences so others could learn about ecology.

May's older sisters never married and devoted their careers to teaching, but May and Laura did marry, with May meeting Raymond Watts on an Indiana Dunes outing. Her 1924 marriage to the engineer and World War I aviator ended her teaching career and brought her to Highland Park's wooded Ravinia neighborhood.[20] Watts pursued gardening and writing with characteristic skill, speaking at garden clubs and publishing poems in the *Chicago Evening Post*. Several poems she wrote as a young mother focus on the beauty in nature and in domesticated female spaces like the kitchen. Her poem "Vision" describes discovering beauty all around her, "Today there have been lovely things / I never saw before; / Sunlight through a jar of marmalade." It continues, "The underside of a white oak leaf / Ruts in the road at sunset; /An egg yolk in a blue bowl." The poem ends with "My lover kissed my eyes last night," suggesting that people both observe and form part of this natural beauty.[21]

Watts wrote a more pointed poem, ironically titled "On Improving the Property," questioning the impact on nature as Highland Park's population doubled in the 1920s. The poem catalogs Ravinia's wild-flowers and laments homebuilders burying these treasures under green lawns. Each stanza praises specific flowers but ends by saying they are gone—all replaced by grass. The poem closes with the bitter statement, "People pass by and say 'Just look at that grass— / Not a weed in it. It's like velvet! / (One could say as much for any other grave.)"[22] Watts led a campaign, writing and illustrating a 1936 booklet titled *Ravinia: Her Charms and Destiny*, published "for the preservation of Ravinia's natural beauty." Newcomers building homes among the area's ravines were the audience, and Watts wrote, "This little book seeks to point out to such new neighbors the things that are probably on their property," with drawings of wildflowers to protect. The book asked newcomers not to clear what they might view as "underbrush," and hoped they would realize "we are tired of the neat smug scenery of Rogers Park. . . . Here is a different beauty—a tangled richer loveliness."[23] These writings reveal her critical view of 1920s suburban growth and the lack of wild nature in Chicago neighborhoods like Rogers Park, and she returned to these issues after World War II.

In Highland Park Watts was a neighbor of Jens Jensen, and the famed landscape architect became another key mentor. Jensen had shaped the design of many of Chicago's major parks as superintendent of the West Park System. In 1920 he went into private practice as a landscape architect and was recognized nationally for designing city parks and private estates for Henry Ford and others. Watts and Jensen both loved local plants, and both admired Henry Cowles. Native plants were central to Jensen's "prairie school" of design, and he wrote, "Every plant has its fitness and must be placed in its proper surroundings so as to bring out its full beauty. Therein lies the art of landscaping."[24] Jensen had much in common with May's father—both were Danish-born immigrants who used European training to work as gardeners in Chicago. They were of the same generation and were influenced by Danish folk schools offering adult education for all—the model for Jensen's Clearing Folk School in Wisconsin. Jensen's home and studio sat on several acres on a ravine's edge, and the Watts house was a mile away, by a wooded stream. Highland Park was a progressive place with writers and artists, including symphony and opera at Ravinia. Jensen was a central community figure, and friends like Watts gathered around his council ring fires for outdoor pageants and solstice celebrations.

May Watts raised four children in the 1930s and found time to speak at garden clubs and to Jensen's Friends of Our Native Landscape, a group seeking to convince Illinois officials to preserve threatened landscapes. She led field trips to help Jensen protect local ecosystems and provide adult education.[25] Jensen ran weeklong nature study programs, and Watts taught at The School in the Dunes for Nature Study in the late 1930s and a similar program at Starved Rock State Park, ninety miles southwest of Chicago, in the 1940s.[26] Leading field trips in natural areas such as Indiana's Dunes, Watts was starting a new cycle, returning as a teacher to sites where she had rich learning experiences.

In the 1930s her talents caught the eye of Jean Morton Cudahy, daughter of Morton Arboretum's founder. Watts helped train teachers at the Arboretum and then began a full-time position creating education programs in 1941.[27] Three mentors, Herman Theilgaard, Henry Cowles, and Jens Jensen, had helped her become an outstanding naturalist and teacher, and their influences are difficult to untangle. May's sister wrote a manuscript on the family, calling May "a rare teacher." After visiting Denmark, Sophie marveled at how May's outdoor classes paralleled that country's folk schools, although May had never seen Denmark. Sophie wrote, "This wooded circle was like an outdoor lecture place in a Danish folk school. Here, too, was a school where no credits were given, and no diplomas awarded, yet groups of women, season after season, worked seriously and eagerly to learn." Herman Theilgaard and Jens Jensen came from that tradition, and Sophie viewed May's programs as "an amazing fulfillment of her father's wish," saying their deceased father "had reached across the years . . . to plant garden after garden."[28]

Watts's Arboretum programs wove together varied strands, with gardening and drawing from her family, ecology from Cowles, and skits from Jensen's adult programs. The Arboretum's Carol Doty wrote, "What a busy place it was, as Mrs. Watts applied her creative ingenuity to getting everybody interested in the outdoors: teachers, scouts and their leaders, homeowners, garden club ladies, Saturday-afternoon hikers, everyone." Watts put her artistic skills to use, drawing maps and plant identification guides that became published books. Doty admired the sense of fun Watts created: "She used her artistic talents to develop worksheets, chalk talks. . . . She invented finger plays, games, songs, skits," and noted the programs' communal spirit, stating "a frequent instruction was, 'Bring your lunch.'" The Arboretum still has many of her teaching materials, from artwork to songs like "Know, Know, Know Your Oaks."[29]

May Watts wrote an article on her educational philosophy's core principles, explaining that "The watchword has been Participation." She stressed community: "It is measurably apparent that there is more learning in a friendly informal group than in one that is coldly formal. . . . Our groups are warmed and knit together by games, songs, puzzles, treasure hunts, discussions."[30] One student described her charisma: "The most outstanding teacher in my experience has been May Theilgaard Watts. . . . Infected with her rare enthusiasm for the life to be found everywhere around us, we fell permanently under the spell of this tall, regal lady whose intellect and humor enabled her to transform the study of nature's intricacies into a delightfully challenging game that she liked to call 'reading the landscape.'"[31]

Watts was educated by men, but built a community of women teaching other women. Garden clubs, a popular activity for wealthy women, were there from the start. Watts added programs for mothers, teachers, families, and scouts, plus a book club.[32] She offered educated suburban housewives in the 1940s and 1950s something special—intellectual stimulation, a like-minded community, and inspiration. One student said Watts supplied hope in the Cold War, "What you do dear Mrs. Watts . . . you fill minds with a strong desire to study and to work in this heavenly direction. What this must mean just now when we are all in such troubles, with shadows darkening daily, no one can measure." Watts identified a simple yet profound goal: "glimpses of what Emily Dickinson called 'Truth's superb surprise.'"[33]

May Watts had devoted female friends who supported her Arboretum work and the Prairie Path. The longest friendship was with Helen Turner, who also studied under Cowles and taught at Morton Arboretum. This sisterhood around Watts extended to her female patrons at the Arboretum. Joy Morton's daughter Jean Morton Cudahy hired Watts and arranged hotels when Watts visited museums and botanic gardens. Suzette Morton Zurcher celebrated Watts's retirement: "Some twenty-two years ago, my Aunt, Jean Morton Cudahy, then Chairman of the Arboretum, believing that the activities should be supplemented by a popular education program had the great good fortune to find the perfect person to implement her idea—Mrs. May T. Watts."[34] Surviving documents contain no discussion of feminism, but both the Arboretum and the Prairie Path were led by strong women. Watts was a generation older than Betty Friedan, whose 1963 book sparked modern feminism; she had more in common with Rachel Carson, whose 1962 book triggered environmentalism, as both writers had a gift for communicating scientific concepts to the public. Watts and Carson appealed to readers as female authors who balanced caring for children with caring for nature. The two nature writers were not as radical as Betty Friedan, but all three authors were role models for women in the 1960s.[35]

THE PRAIRIE AND THE BULLDOZER

In 1941 the Watts family moved to Naperville, some thirty miles west of Chicago, after May began work at Morton Arboretum. Naperville, then a quiet village of five thousand surrounded by farmland, could stand in for the community that Rachel Carson invented in *Silent Spring*: "There was once a town in the heart of America where all life seemed to live in harmony with its surroundings. The town lay in the midst of a checker-board of prosperous farms."[36] Alfred Etter, who succeeded May Watts as Morton Arboretum naturalist, said Watts "came to live on the edge of the prairie in Naperville . . . long before development began to threaten the integrity of the town."[37]

Much of DuPage County was removed from the rapid suburban growth north of Chicago, but the area would change dramatically over Watts's thirty-five years there. As she built a network of female friends, including classmate Helen Turner, who lived next door, Naperville annexed farmland and grew into a sprawling suburb of over thirty thousand with huge subdivisions. By the 1970s Watts was a longtime resident who questioned prodevelopment policies in the Republican town. In 1974 the *Naperville Sun* printed her poem "Lament for Our Central Park," which criticizes the steady loss of open space in the town. It begins, "You were a piece of the prairie once—lonely, / But alive!" and goes on, "Then wise men set you apart, to be / Always, the big green heart / Of the small community— / Naperville's bounty." Her account of this prairie grows angry, as the park shrinks to make way for buildings. The poem's dark ending pronounces the prairie dead: "Soon, the bulldozers drool again, / And the money-changers smile, and then / Erect a high rise over your miserable carcass, / And move, seeking another town to improve."[38]

This rhetoric of bulldozers drooling over the prairie may seem excessive today, but it was a central image for Watts and also characteristic of Earth Day critiques of Cold War growth.[39] Watts wrote about bulldozers in this poem and her Prairie Path proposal, and criticized developers discussing "improvement" back in her 1930s writings. Her use of bull-dozers as villain and prairie as victim must be seen in the context of the whirlwind of development she saw in DuPage County. The county's population grew from one hundred thousand in 1940 to about five hundred thousand when she died in 1975. In a single decade, from 1958 to 1967, all the landmarks that define DuPage County today appeared. The East-West Tollway, now I-88, opened in 1958 and sparked a high-tech corridor. Oakbrook Center, one of the nation's largest shopping malls, opened along I-88 in 1962. O'Hare became an international airport in 1958, expanded to 7,200 acres in Cook and DuPage Counties in 1959, and became the world's busiest airport. The Cold War was central to this boom, as much of Argonne Lab's 1,700-acre facility was built in the 1950s, and Fermilab's 6,800-acre campus was begun by 1967. Watts expressed her feelings about construction around Morton Arboretum in a talk titled "Our Changing Landscape": "To study it at this time, one almost feels that he must put his foot on it and grasp it firmly to keep it from changing before he has had a good look."[40]

Naperville sat between Argonne and Fermilab, and began to boom, with the developer Harold Moser building large subdivisions in the 1950s. Naperville's mayor summed up Moser by saying, "He had a vision for Naperville. He could look out at a field where you and I would see nothing but cows, and he'd see homes." Moser developed twenty-four subdivisions containing ten thousand homes on four thousand acres, a footprint rivalling Argonne and Fermilab.[41] This growth reveals that Watts did inhabit a landscape where bulldozers suddenly arrived, and huge complexes or subdivisions replaced one farm after another. DuPage County had 1,503 farms in the 1945 agricultural census, but when Watts died thirty years later only 266 farms remained, a loss of more than ninety thousand acres of farmland. The Cold War boom of suburbs, highways, and high-tech research brought great prosperity, but Watts saw great loss in the reduction of open space and pastoral landscapes. The village surrounded by farms was gone, and in the words of Rachel Carson, "The people had done it to themselves."[42]

Reading the Landscape, published in 1957, emphasizes prairies more than bulldozers, offering rich explorations of midwestern landscapes, with more muted commentary on suburban sprawl. The book, completed when Watts was sixty-four, looks to the past more than it raises alarms about the future. It provides lessons from her Arboretum classes, showing readers how to understand ecological processes shaping the landscape. Each chapter goes to a specific location, including treasured sites where Watts first learned to read the land with Henry Cowles. Over and over Watts uses the pronoun *we* instead of *I* in describing a group of travelers on a field trip or a hike. Her use of *we* invites readers to join her in interpreting the landscape's stories. The preface shows her modesty: "If these stories are good ones . . . it is because I, the narrator, am bestriding the shoulders of a giant."[43]

Cowles specialized in ecological succession, and the dynamic, windswept Indiana Dunes was his laboratory for analyzing how plants colonize challenging habitats. Watts wrote her chapter titled "Picnic in a Gritty Wind" based on "forty years of intermittent botanizing in the dunes." It recalls her excitement "on the many trips when with eager ears cocked forward, we followed in the quick steps of Doctor Cowles, we were still much too young and vigorous, and too enchanted with the new field of ecology, to care about sun or wind, or sand between our teeth." Every chapter shares this way to study nature: "There is good reading on the land. . . . The records are written in forests, in fence-rows . . . in bogs

. . . in tree rings." She summarizes, "Interpreting this reading matter, in place, on the land, seeing living things in their total environment, is an adventure into the field that is called ecology."[44]

Her book examines people's impact on ecosystems, noting the damage done to fragile dunes. Watts criticizes new parking lots, "I was careful not to catch a glimpse of that ominous stabilization lest it spoil my appetite." She also recalls childhood trips to vanished beachfront dunes replaced by Chicago's Edgewater Hotel, explaining in 1957, "Many tall apartment buildings, acres of cement walks and highways now fill that flattened surface." Watts tried to re-create earlier excursions, often revealing damage from economic development. An essay on Michigan forests highlights how much was lost in her lifetime. On a bittersweet Upper Peninsula expedition, she wrote, "World War II, and the building boom that followed, had incited more cutting of forests. That we might succeed in camping, for even one night, in the forest of my memories, was almost too much to hope for. But I hoped for it, nevertheless, because I wanted to show it to my daughter before it should have disappeared utterly."[45] The final chapter, "Looking Down on Improved Property," criticizes suburban development, as Watts narrates her dismay at her view of the Midwest from an airplane: "From Chicago to Dayton, we looked down on a battle, the battle between man's passion for square corners and Nature's penchant for curves." She views a highway as "cut like a gash" across the land, and summed up suburbia as "a snarl" of streets, calling built-up areas "the fouled nests of humans."[46]

Watts digs deeper into farm landscapes in "Prairie Plowing Match," analyzing the replacement of native prairies by agricultural crops and changes as farming became mechanized. She uses the program from the plowing competition in Wheatland and its timeline of the event's past seventy-five years to explore landscape change since 1877. This voyage back in time, probably inspired by Aldo Leopold's "Good Oak" essay examining tree rings to explore Wisconsin's past, is a search for Illinois prairie. Watts notes innovations like DDT in the 1940s, hybrid seeds and the last horse-drawn plows in the 1930s, cars and steam plows around 1900, and barbed wire in the 1880s. She discusses pioneers breaking the prairie before the Civil War, notes Indians using fire to create that prairie, and ends with Marquette's 1684 travel account. Her goal is "to explain where, and when, the natives had gone." Exploring Wheatland's vegetation, she yells, "Oh! There they *are*!," thrilled to discover remnant prairie, first in a foot-wide gap along a fencerow between a farmer's field and a road, and then behind a cemetery on a hill too steep to mow.[47] Her "Prairie Plowing Match" is a clever environmental history essay combining documentary and landscape analysis to examine farms replacing tallgrass prairie.

Searching for remnant prairies was a common activity for naturalists like Watts, and the abandoned interurban offered a place to find prairie grasses.[48] In *Siftings*, Jens Jensen wrote, "Great prairies appealed to me. Every leisure moment found me tramping through unspoiled bits," and noted, "Along our railroad rights-of-way one meets the last stand of these prairie flowers."[49] Edwin Way Teale, the widely read nature writer who won a Pulitzer Prize in 1966, also observed that railroad corridors offer naturalists a view of the past. Describing a visit to woods from his Joliet childhood, Teale wrote, "I walked back along the edge of this same woods, along the same Michigan Central Railroad that once had provided a kind of highway into wildness," noting, "A railroad right-of-way, to a returning native, is one of the most permanent of landscapes."[50] Watts shared this understanding, recognizing that the abandoned interurban could become a victory for open space in an era of heavy losses since the Forest Preserve District of DuPage County began in 1915 and Morton Arboretum formed in 1922.

Watts's path proposal was a call to action from a beloved community leader, and her constituency went to work. One woman described her network: "Mrs. Watts is well known to thousands of . . . housewives. . . . She is known to teachers, artists, Scouts and Scout leaders, garden club people, club groups, and the many who encountered her as naturalist-lecturer."[51] These women were well educated and some were wealthy, and Alfred Etter noted, "Mrs. Watts' students compose a formidable body of individuals imbued with her ideals and observations."[52] Suburban women ran this coalition, calling on husbands like Raymond Watts for legal and technical expertise. One man who was not a spouse, Gunnar Peterson of the new nonprofit group Openlands, became a crucial ally. Peterson provided an office staff, links to philanthropists, and a male representative to send to meetings with businessmen.[53]

The day-to-day logistics of advancing the Prairie Path from popular proposal to reality is its own complicated story. It took many volunteers, with Helen Turner providing as much leadership as May Watts, including researching legal deeds and writing a regular newsletter.[54] A core group of women created a publicity campaign, giving well over one hundred presentations at libraries, schools, and town halls.[55] Lengthy and complex negotiations with the railroad, utility companies, and local governments followed before the organization achieved its 1966 lease with DuPage County for the first twenty-seven miles of trail, with significant assistance from Gunnar Petersen and Openlands. In 1972 two branches of the trail were extended into Kane County, to end points in Elgin and Aurora. There was also a great deal of physical labor—years before Earth Day—as teams of volunteers removed garbage and planted native species, with more-skilled groups building pedestrian bridges.[56] Illinois Prairie Path volunteers took responsibility for building and maintaining the trail for twenty years, from 1966 to 1986, when rising insurance costs led the nonprofit group to agree to DuPage County taking over trail maintenance.[57] Today's Illinois Prairie Path is the result of countless hours of volunteer work across decades and a fitting testament to dedicated grassroots environmentalists.

The legacy of May Watts and the Prairie Path still shapes the story of green spaces around Chicago. The region's network of trails continues to expand, with the Grand Illinois Trail stretching to the Mississippi River. Since 1963 large additions have been made to a web of open spaces, with national recognition for midwestern landscapes such as the Indiana Dunes National Lakeshore in 1966, the Illinois and Michigan Canal National Historic Area in 1984, Nachusa Grasslands in 1985, and Midewin National Tallgrass Prairie in 1996. Some of these sites are places Watts, Cowles, and Jensen treasured, such as the Indiana Dunes, while others represent the realization of their advocacy for tallgrass prairie restoration. The nonprofit group Openlands, founded in 1963 alongside the Prairie Path, helped achieve several of these victories and continues to do important work.[58]

May Watts can be understood as part of a multigenerational chain of people working to preserve midwestern green spaces. She was not just a bridge to nineteenth-century conservationists. Her 1960s Prairie Path campaign focused on ecology and on grassroots organizing by women, which directly connects with the work of twenty-first century environmentalists. In 2015 and 2016 Illinois residents celebrated the birth of bison calves on tallgrass prairies at Nachusa Grasslands and Midewin that naturalists had been restoring for decades.[59] Expanding herds of reintroduced bison grazing on northern Illinois prairies are signal achievements in ecological restoration and the preservation of open space. The growing network of green spaces around the Chicago metropolitan area reveals that May Watts's work raising awareness of local ecosystems and protecting open space from sprawl continues, and her ideas have become deeply held values.

CHAPTER 18

"Hard-Nosed Professionals"

Gordon Sherman, Businessmen for
the Public Interest, and Environmentalism
in 1970s Chicago

ROBERT GIOIELLI

In the spring of 1972 David Comey, director of environmental research for Businessmen for the Public Interest (BPI), looked down on Chicago's Earth Day protest. In contrast to 1970, when thousands of people showed up for events around the city, just 150 appeared for a small rally at Chicago's Civic Center plaza, for about one hour. A friend of Comey's commented on the sad state of the environmental movement, but, as the environmentalist told *Newsweek* magazine that year, he was much more optimistic. Groups such as BPI, composed primarily of scientists and lawyers who investigated and then litigated environmental cases, were much more important and effective than mass demonstrations. "In 1970 on Earth Day we had a lot of ignoramuses screaming their heads off for a clean environment. In 1972 we have a bunch of hard-nosed professionals plotting how to nail polluters," Comey said.[1]

Although dismissive of protestors, Comey's analysis was largely correct. Environmentalism's moment as a grassroots, national mass movement was relatively short-lived. By 1972 it had quickly become institutionalized, both within a series of well-funded, white-collar nongovernmental organizations, as well as government regulatory agencies. Comey's organization, the Businessmen for the Public Interest (later changed to Business and Professional People for the Public Interest, but still abbreviated BPI), was a Chicago-based public interest law firm. Founded in 1969 and modeled after "Nader's Raiders" and the other public interest firms created during this era, BPI is primarily known today in Chicago for its civil rights work, especially advocacy on behalf of criminal justice reform and affordable and fair housing.

But in its first decade, and especially in its first few years, BPI litigated a number of major environmental cases and helped coordinate some important environmental campaigns in the Chicago region. The story of BPI's early years helps fill in an important gap in the history of environmental activism, both in Chicago and around the country. Although much of our memory of the rapid emergence of environmentalism in the 1960s and early 1970s surrounds national-level events and discussions—the publication of Rachel Carson's *Silent Spring*, Earth Day, the creation of the federal Environmental Protection Agency, among others—a large amount of the real work of trying to solve pollution problems, increase green space, and provide more sustainable transportation and housing options occurred at the local level, especially in America's cities.[2] The story of BPI also provides an important local illustration of the longer-term trends within the national movement, particularly in terms of the funding, staffing, and operation of environmental organizations and their relationship to the general citizen ferment. Although Earth Day and a general rise in environmental concern sparked a wave of grassroots activism, it was difficult to bring about changes with protest alone in the emerging environment-regulatory state. New laws and bureaucracies, such as the Clean Air Act and the EPA, were highly technical, and thus successful activism to defend or encourage enforcement required education and expertise. On the national level these circumstances privileged organizations such as the Environmental Defense Fund and the Natural Resources Defense Council, which were geared toward expertise-laden litigation and bureaucratic negotiation. On the local and regional level grassroots and citizen-based organizations would have more success, but nevertheless the system favored groups like BPI which had full-time staffs of well-trained experts.

This reliance on experts would have a number of consequences for American environmentalism. These organizations, made up of "hard-nosed professionals," were, like Comey, oftentimes dismissive, and sometimes outright hostile, to grassroots activism, encouraging a schizophrenic character within American environmentalism that still exists. The public face of environmentalism is the well-educated, full-time professional activists who spend most of their time lobbying governmental regulatory agencies and politicians. However, they often have little connection to the disparate, tenuous but still vibrant local activism conducted by part-time, volunteer, citizen organizers. This divide has made the American environmental movement appear moribund and sclerotic when viewed from the national level, even though grassroots activism is constant, lively, and widespread.[3] It has also shut out many alternative voices that have a different environmental experience from the white, middle-class, and overwhelmingly male staffers of groups like BPI. Women, working-class people, and racial and ethnic minorities were not only a rare sight in the offices of these organizations, but perhaps more importantly, so were their concerns. During the 1970s environmentalism quickly became a universalized politics that focused on general issues of pollution and harm to the natural world. Within most major environmental organizations, there was very little recognition of inequalities, both in terms of access to amenities (clean air and water, parks and open space) as well as exposure to hazards (toxic dumps, lead paint in homes, etc.), based on race, class, and gender. These divisions have significantly shaped the politics of environmental activism in the United States. The environmental justice movement, which emerged in the 1980s, was in many ways not a series of new issues (environmental inequalities have been widespread in America since the nineteenth century) but an attempt by people of color and other marginalized groups to create a forum for their concerns, which received little traction in environmentalist circles.[4]

This division was oftentimes between national and local groups, but sometimes it was evident solely on the local level, which was the case with BPI. That makes this story not only helpful to understanding the development of environmental activism in Chicago, but also as a case study of larger national trends. BPI's founders modeled it after the national public interest law firms being established at the time, and it thus represents many of these organizations' strengths and weaknesses as agents of social and environmental change. BPI's primary early funder was Gordon Sherman, a fascinating Chicago businessman and philanthropist whose life, interests, and activism represent the widespread nature of, and limits to, the reform impulse of the late 1960s.

AMBULANCE CHASE FOR SOCIAL CHANGE

A long golden muffler stands upright, on its own, against a white background, with a bouquet of summer flowers sticking out of the tailpipe. "The quietest cars on the road have a Midas Muffler" is all the headline says. A few more words at the bottom emphasize how a Midas muffler is clean, installed professionally, guaranteed, and available in "stores coast to coast."[5] The advertisement is smart and clear, and follows all of the rules of modern marketing: show the product, and have as few words as possible. It does not simply hawk the muffler; it says everything the Midas Muffler corporation was about: selling auto parts and service not as a niche or technical product to the grease monkey or gearhead (who was usually male), but as a mass-market item to the massively expanding base of American consumers (many of whom were female).

Midas, which stood for Muffler Installation Dealer's Association, but was really named for the mythical King Midas who could turn anything to gold with a simple touch, was the brainchild of Gordon Sherman, the second-generation scion of a midwestern auto parts dealer. In the early 1950s young Sherman, fresh off a degree from the University of Chicago and a few years in the Navy, joined the family parts distribution business, one of the biggest in the country. Almost immediately, he brought two key ideas to the table. The first was that loud, smelly, dirty auto shops and the services they offered were often intimidating and unwelcoming to many consumers, especially women. What was needed was a shop that was clean and professional, where people could see the work being done "while they waited"—something attractive to America's rapidly expanding middle class. The second was that the best way to expand and manage that business was through a franchise. Commonly associated with fast-food restaurants, franchising allowed Midas to quickly turn itself into a national brand.[6]

A gifted musician and humanities major at the University of Chicago, Sherman did not just keep his love of the great books and classical music as an avocation when he entered the corporate suite—he worked to merge the two together. Stories about Sherman from the late 1960s inevitably mention how he actively recruited psychologists, sociologists, and other intellectual types to work in what was typically considered to be the hyper-masculine, cutthroat world of auto parts. Although such actions made Sherman appear to be a turtleneck-wearing, countercultural ne'er-do-well, his management philosophy represented the larger instinct that he deployed in developing the Midas concept. Running a modern business was as much about handling people as it was about making a particular thing. Reflecting this idea, Midas consciously recruited franchisees with no experience in the auto parts business. "We've found that former auto people are too thing-oriented rather than people-oriented. They take a great interest in pipes and clamps, but they don't want to answer the phone," Sherman told a reporter in 1970.[7]

Sherman's heterodox management style probably also raised more eyebrows because it was accompanied by active and vigorous support for progressive political causes. Like many Americans, his political awakening occurred quite rapidly in the second half of the 1960s, spurred primarily by the Vietnam War. "Until about three years ago, I was kind of a passive hawk," he said in a profile in the *Chicago Tribune*. "Then I got some mailings about the war, and they provoked me into reading everything I could. . . . I began to see the fraudulent arrogance, the downright mendacity with which our government had tried to keep a bad thing going, the incredible way those sanctimonious men in power try to adorn their folly with 'wisdom.'" He soon became involved in the antiwar movement through a national group called Business Executives Move for Viet Nam Peace, eventually becoming chair of the Chicago chapter.[8]

Sherman's activism continued with a focus on Chicago organizations, primarily as a donor and funder, both personally and through the Midas Foundation, which he established and controlled. His first major grants were to Saul Alinsky to help provide permanent funding for the Industrial Areas Foundation (IAF). Although created in 1940, when Alinsky sought to expand his activism and organizing model beyond the Back of the Yards neighborhood in Chicago, IAF was a small organization, primarily supporting the work of Alinsky and his team of associates and acolytes in Chicago and in selected campaigns around the country. Impressed with Alinsky's organizing model, in 1968 Sherman agreed to provide significant funding to transform the IAF into a national training center, where community organizers from across the country could learn Alinsky's techniques. One year later he provided the majority of initial funding for BPI.[9]

Understanding Sherman's background, management approach, and antiwar politics, funding these two major progressive organizations does not seem out of character, but they are also representative of a larger progressive, and even activist, moment in American philanthropy that peaked in the 1960s and early 1970s. American philanthropy entered its modern era around the turn of the twentieth century, when foundations established by Andrew Carnegie, John D. Rockefeller, and Russell Sage sought to not just provide alms for the poor, but permanently improve their condition through improved health care, education, and social research. This progressive spirit spread throughout all of the philanthropists' endeavors, and by World War II the major foundations were an important part of American public life, conducting cutting-edge research, helping to shape public policy, and privately funding projects that were too politically dangerous for state actors to underwrite.[10]

Despite this rise to prominence, the large foundations and philanthropists still considered it politically risky to become directly involved in social justice activism, preferring to work on the margins to alleviate the conditions caused by segregation and income inequality. Think of Julius Rosenwald, who used his fortune from the Sears Corporation to fund black schools in the South, but never directly attacked Jim Crow. This traditional hesitancy changed in the late 1950s, when smaller organizations began actively funding the civil rights movement. The Taconic Foundation, created by heirs to two large banking fortunes, made numerous donations to civil rights groups throughout the 1960s, especially the Student Nonviolent Coordinating Committee's Voter Education Project in Mississippi. This activist spirit then spread to the Ford Foundation, America's largest and most powerful philanthropy. Under the leadership of McGeorge Bundy, former national security advisor and architect of the Vietnam War, Ford began funding activist and minority organizations in the late 1960s. Many efforts focused on supporting civil rights lawyers, and then expanded into major support for a variety of different types of public interest law firms, including the country's most important environmental law firms.[11] This foray into public interest and environmental law reflected the foundation's larger approach to social change, and set patterns for other public interest firms across the country. Although it gained a reputation as the country's most liberal large foundation, the Ford Foundation, especially at the peak of its influence in the 1960s, was always firmly part of the American social and political mainstream. This status meant funding for activist or social justice organizations focused on institutionalization, bringing concerns and grievances off the streets and into the courts, universities, government bureaucracies, and the corporate boardroom.[12]

This general philosophy was reflected in Ford's environmental law program, which led to the creation of the Natural Resources Defense Council, as well as major funding for the Environmental Defense Fund and the Sierra Club Legal Defense Fund. Although these groups liked to portray themselves as crusaders, they were staffed by Ivy League–educated litigators who focused on working within the system, rather than trying to overturn or undermine it.[13] Environmental law specifically, but also public interest law in the 1970s more generally, was built on the belief that the American regulatory system was flawed but could be fixed. Instead of representing the interests of the general public, agencies at both the state and national level had been co-opted by industry. The primary people regulators interacted with, whose views they heard and listened to on a regular basis, were corporate representatives. They were thus no longer looking out for the larger interests of American citizens or consumers. Correcting this situation did not require rethinking the system, but simply inserting another variable into it: public interest law firms. These "public attorneys general" would litigate in the interests of citizens, holding agencies accountable and forcing them to follow their own regulations. They would also sue government agencies, especially

Fig 18.1. *Don't Do It in the Lake!* This flier was part of BPI's successful 1970 campaign to stop construction of the proposed Lake Michigan Airport. Courtesy, BPI.

those engaged in large, poorly conceived infrastructure projects such as dams and nuclear power plants.[14]

In addition to the broader ferment around public interest law, Sherman was also directly influenced by Ralph Nader's activism. After the publication of *Unsafe at Any Speed* (1965), his scathing critique of the auto industry and its cozy relationship with federal regulators, gave him national prominence, Nader went on to attack the Federal Trade

Commission for its lackluster regulatory efforts, the first of a series of salvos that his study group of lawyers and investigators, dubbed "Nader's Raiders," lobbed against a lack of government accountability in air and water pollution, food safety, and interstate commerce. In a late 1960s speech at a Chicago-area legal convention, Sherman praised Nader and challenged those present to follow his model. "How long will Ralph Nader, at his solitary vigil, have to atone for you? He has set a style of activity which, apart from your views on his intent, is a breakthrough in orderly protest. Ralph Nader operates under the amazing notion that there is a certain amount of irresponsibility in government and business. . . . Indeed, ladies and gentlemen, you could go out and (you should pardon the expression) ambulance chase for social change."[15]

Soon after this speech Sherman worked to create Businessmen for the Public Interest. In later interviews he argued that his motivation came from watching police repression of the protests at the 1968 Democratic National Convention in Chicago. However, his vision for the organization, and his model for social change, was much different from the street-level activism he saw during Chicago's "Days of Rage." Although not nearly as deliberate as the Ford Foundation, which oftentimes researched an issue for years before developing a grant program, Sherman nevertheless believed the proper route to reform was to employ full-time, educated activists who worked within the system. We can see evidence of that above, where he expressed admiration for Nader not only for his results, but also because his strategy was a "breakthrough in orderly protest." Later comments about BPI confirmed these intentions. "Since the perpetrators of evil in our society are skilled, experienced professionals, who ply their nefarious trades full time, they should be engaged not by callow youths, fresh and unlearned from college, not by part time volunteers, torn between the demands of their coffers and their consciences, but by equally skilled, equally experienced professionals, serving full time in the public interest," Sherman said in an early interview.[16]

Sherman wanted BPI to be a "Nader's Raiders of the Midwest" that would be funded by the private sector, but would pursue projects and litigate cases primarily in the public interest. He recruited a couple dozen other business owners to help fund this venture, although he was by far the most prominent donor. Most of the others owned small- to medium-size businesses within the Chicago area, since the large Chicago corporations and traditional, big-money philanthropists would not touch the organization. According to Sherman, they were loath to be involved in something that had an activist bent, and that could create conflicts with their existing businesses. For the first couple of years Sherman was listed as "president" of BPI, and the only named "businessman" officially listed on its board. Nevertheless he, like most of the original donors, took a hands-off approach. "Woody Allen's line is 'take the money and run.' Mine is 'give the money and run,'" he told a newspaper reporter.[17]

With an initial yearly budget of around $250,000, the engine behind BPI was Marshal Patner, a well-known Chicago litigator. In its first year BPI was the primary "client" of Patner and his partner, James Karaganis, but by 1970 the group established a more traditional nonprofit structure. Alexander Polikoff became the full-time executive director and Patner the general counsel. BPI's activities in its first few years were split between environmental cases and civil rights work focused on fair housing, especially *Gautreaux v. Chicago Housing Authority*, America's foremost public housing discrimination lawsuit. Filed by Polikoff in 1966 with the ACLU, the lawsuit alleged that the city violated black tenants' civil rights by placing thousands of public housing units solely in African American neighborhoods. Although a federal judge ruled in favor of the original plaintiff in 1969, it took Polikoff and BPI almost twenty years—and additional lawsuits against the city and the federal government—to force the city to adopt a nondiscriminatory, scattered-site housing program.[18]

BPI's environmental work began with assisting the Citizen's Action Program (CAP), which was the first step in a broader plan Saul Alinsky developed in the late 1960s to begin organizing in white working- and lower-middle-class communities. Alinsky's experience in trying to integrate white and black neighborhoods on the South Side of Chicago had shown him that in order to achieve any sort of progressive goals, whites had to be part of broader activist and community coalitions. His plan got its start in Chicago with CAP, which was created along the outer "bungalow belt" on the Southwest Side of the city at the end of 1969. Following the Alinsky model, organizers worked to identify a common issue that united residents in communities like Garfield Ridge and West Lawn. What they found was that many residents were concerned about pollution from the largely unregulated factories and coal-fired utility plants abutting their neighborhoods of single-family homes. CAP quickly became a major activist force in Chicago politics, with a large organization and charismatic community leaders aggressively lobbying the city to enforce and strengthen local pollution control regulations.[19]

Although Alinksy had built CAP based on his model of community organizing, the original goal was for the organization to work in tandem with BPI. Once the two groups had identified an issue, CAP would handle the community organizing, creating a base of support for activism and protest, while BPI would pursue the more technical aspects of litigation. But this partnership quickly fell apart. Dick Harmon and Ed Chambers, CAP's primary organizers and veterans of numerous activist campaigns with Alinsky in Chicago and around the country, could not stand Karaganis, whom they saw as a "legaldonna" with "the organizing sense of a helium balloon." They not only had a more confrontational style, with activists storming city offices demanding immediate enforcement of moribund regulations, but also believed community members should be driving the activism, not the lawyers. BPI believed in working within the system, using the courts and high-level negotiations to hold government and corporations to account. Karaganis, for example, wanted CAP only to show up occasionally at public hearings, and conduct orderly marches outside public buildings, while BPI handled negotiations with city officials.[20]

After Polikoff assumed the leadership of BPI and it became a fully independent firm, the group distanced itself from CAP. The partnership was the first and only time BPI attempted to work closely with a grassroots organization on a coordinated campaign. Thereafter they originated all projects or only partnered with community organizations as part of a coalition, coming in near the later stages to handle litigation and other legal matters. Much of this activity focused on efforts to protect Lake Michigan from large, complex, expensive infrastructure projects that threatened the lake's ecology: a new airport and nuclear power plants.

By 1960 Chicago's population exceeded 3.5 million, and the larger metropolitan area neared 6.8 million people. It also had become America's major airline hub, with thousands of passengers just passing through the city every day. Chicago's airports could barely keep up with the booming local and national air traffic demands. Although only eleven miles from the Loop, Midway Airport was surrounded on all sides by residential neighborhoods, and thus became almost unusable in the early 1960s, when new jet airplanes required longer runways. This shifted almost all traffic to O'Hare Airport, which quickly became the busiest airport in the world. Squeezed for space at both airports, and not wanting to construct a third in the suburbs, in the mid-1960s the city conceived a bold new plan: build a new airport in the lake. Not a lakefront airport, like the old Meigs Field downtown, but an eleven-thousand-acre, round, artificial island, three and a half miles directly east of downtown. Even with a $400 million site cost, more than double the other proposed locations, Mayor Richard Daley, city officials, and business leaders insisted that the "airport in the lake" was the best option. Skepticism emerged quickly about the airport's costs and especially about its environmental impact. Critics worried about noise, air, and especially water pollution. Five regional congressmen, including Sidney Yates who represented the northern part of Cook County, wrote a letter to federal water regulators opposing the project based on concerns about sewage overflow, industrial waste, and storm water runoff into the lake. Public opposition was broad as well, including both white, middle-class north shore Chicagoans and African American South Siders. "How can we possibly consider the double standard of an airport in the lake and abolishing pollution of lake and air," airport opponent Richard McDorman wrote in a letter to the *Chicago Tribune*. A study by the Open Lands Project argued that, although some mitigation was possible, the cumulative ecological impact on the lake would be significant, and hard to foresee. "In creating the proposed airport in the lake, the City of Chicago may forfeit its greatest resource and thereby suffer an incalculable economic and social loss."[21]

With evidence mounting about the environmental risks as well as the astronomical costs of the airport plan, BPI began a successful public information campaign: Don't Do It in the Lake! Coined by Patner, the simple phrase adorned t-shirts, bumper stickers, and clever advertisements that highlighted how the airport would sully the lake and cause a tremendous ancillary and destructive commercial development along Chicago's south shore. Their most interesting and entertaining brochure had a cartoon sketch of what the operating airport would look like, with more than two dozen notations of all of the different environmental and community impacts. Many emphasized water pollution, including possible sewage overflows and oil spills caused by the pipelines that would carry jet fuel to the airport. Another pointed out that the Lake Michigan coho salmon, already endangered, would probably suffer a "death blow" because of airport construction. For that reason the facility in the cartoon is named "Coho Memorial Airport." By early 1970 Illinois and national politicians realized how unpopular the idea was, withdrawing almost all support and forcing Daley to shelve the plan.[22]

Don't Do It in the Lake! was a quick and successful campaign that took advantage of a clever slogan and a particularly half-baked, unworkable scheme by the Daley administration. But at the heart of the controversy were real issues related to the city's growth in the postwar era. When the jet age brought about an air traffic boom in the 1960s, many American airports were poorly equipped to handle the increased passenger traffic and infrastructure demands of larger, heavier aircraft, and Chicago was no exception. But new airports, in addition to costing a tremendous amount of money, create a gigantic environmental footprint in terms of air and noise pollution and land use, so the solution has often just been reconstruction and expansion of existing facilities. BPI's other environmental campaigns would be built around similar challenges: how to provide needed, large-scale infrastructure for a growing metropolis in a manner that lessened the long-term impact on the local and regional environment. After the success of the airport campaign, the next issue they would become involved with was electricity generation. Postwar economic growth was driven by a consumption boom, and as consumption increased, Americans needed more energy to heat and cool their suburban homes and power all of the appliances and gadgets inside them. With this growth in demand regional power utilities went on a building boom, expanding existing power plants and building new ones. Spurred on by the Atomic Energy Commission (AEC), which central- ized the research and development of nuclear power in the United States, the utilities looked to develop hundreds of nuclear reactors across the country to provide energy that eventually would be "too cheap to meter," in the words of AEC Chairman Lewis Strauss in 1954.[23]

By the late 1960s various regional utility companies had planned nine separate nuclear generating stations in the four states that bordered Lake Michigan. With this flurry of construction activity, much of it rushed and ill-conceived, David Comey, BPI's first (and only) environmental di- rector, saw a chance to force the nuclear industry and the federal Atomic Energy Commission to take ecological concerns and public engagement seriously. BPI first targeted the Zion plant, then under construction fifty miles north of Chicago, just south of the Wisconsin border. However, when an organizing effort to try to spark opposition among Chicago- area residents failed, BPI turned its attention to the Palisades plant, which was also under construction across the lake, in southwest Michi- gan. In that case there was more vocal public opposition from a chapter of the Sierra Club and four other local conservation and sport fishing organizations. Comey and BPI argued that the key issue at Palisades, and by extension all of the other plants being constructed on the shores of Lake Michigan, was thermal pollution. Water used to cool the steam that drove the generating turbines was recirculated back into the lake ten to thirty degrees warmer than when it came out, which can be tremendous- ly destructive to fish habitats. Cooling tower technology was available, but construction companies and utilities argued that it was too expen- sive to build and operate. The AEC maintained that under federal law it had the authority only to regulate radiation, not thermal pollution.[24]

Eventually plant operators gave in to demands for cooling technology at Palisades, and Congress authorized the AEC to regulate thermal pol- lution. But BPI's activism against nuclear power did not stop. Although federal regulators and industry scientists constantly questioned Comey's expertise because he had no advanced training in nuclear or environ- mental science (his degrees were in Slavic languages and political sci- ence), he brought an outsider's perspective to what was a textbook case of what public interest lawyers like Nader called "industry capture," when a regulatory body becomes captive to the interests of the industry it is trying to regulate. Since the Manhattan Project, the American nuclear industry was comprised of a small group of politically well-connected companies. The AEC was tasked with simultaneously regulating *and* promoting civilian nuclear power. Moreover, publicly available research on the health and safety of nuclear facilities was virtually nonexistent, and the industry's projections on the operational capabilities of civilian power plants, which were often just scaled-up versions of smaller plants built for weapons production, remained untested.[25]

Within this context BPI and other environmental law firms raised issues that had never been discussed. In the late 1960s the federal government authorized construction permits for the first four plants to line Lake Michigan. Once he examined the plans, Comey identified seemingly obvious problems. For example, he argued that the Donald C. Cook plant, also being built in southwest Michigan, lacked any containment system if the steam pipes that powered the turbines ruptured. "All reactors are dangerous, but Donald C. Cook is something extraordinary. It doesn't have those big domes to contain steam if the pipe ruptures," he explained in a 1975 BPI promotional article. Instead it had little domes lined with ice, because "somebody thought that would be cheaper." Comey's fear, and the fear of many critics of nuclear power, was of a "China incident," whereby the nuclear reactor's cooling mechanisms would malfunction, overheat (going from five hundred to three thousand degrees Fahrenheit in a matter of minutes), and melt through the reactor floor and any subsurface material. "It becomes so intensely hot that it melts . . . through the foundation and then down into the earth. There is no telling how far it will go. This is why the nuclear industry calls it a 'China accident,'" Comey said. The concern was that once the core hit the groundwater it would heat it into giant steam clouds, which would shoot out of the plant, soaking the surrounding countryside and community with radioactive fallout.[26]

The ice-coolant technology designed for the Cook plant was unproven, and not actually less expensive than the other remediation options, containment domes, or cooling towers. BPI filed suit against the granting of the operating permit, eventually forcing the Indiana and Michigan Electrical Company to lower the average power output (to reduce the risk of overheating), and also allow for regular inspections by Comey and other outside observers. These interventions in the Palisades and Donald Cook plants were just two of the seven nuclear power plants against which BPI eventually took legal action. Most of these forced modifications to construction plans and operating permits. The one plant BPI helped to defeat completely was the Bailly Nuclear Generating Station, planned by the Northern Indiana Public Service Commission (NIPSCO) for northern Indiana on the Lake Michigan shoreline.

Proposed to the AEC in 1970, the Bailly plant was plagued with problems from the start. NIPSCO wanted to build it on land adjacent to its existing coal-fired power plant, but this location placed it adjacent to the newly created Indiana Dunes National Lakeshore, as well as a recently completed Bethlehem Steel manufacturing complex. In choosing this site NIPSCO was essentially reigniting fires that were still smoldering over the creation of the national lakeshore a few years before, when advocates for the dunes only assented to a new industrial port and steel mill once Congress authorized the purchase of land for the new park. In addition to the significant industrial and recreational development on both sides of the site, NIPSCO's proposal would have created the closest nuclear power plant to a large city in the United States. Local environmental groups became active once again, particularly the Izaak Walton League and Save the Dunes, and they enlisted support from the U.S. Department of the Interior, looking to protect the national lakeshore, and Bethelehem Steel, which had spent years developing its mill site and did not want the added risk of a nuclear accident right next door.[27]

With environmental groups such as the Izaak Walton League and Save the Dunes as the primary activists, BPI stepped in to serve as legal counsel. The arguments against the Bailly plant were initially aesthetic and ecological—that the risk to the adjoining dunes would be too high, even with cooling towers and a plan to capture and contain radioactive material. BPI protested the construction permit all the way to the U.S. Supreme Court, but the AEC eventually granted approval. Building began in 1976, but the newly reformed Nuclear Regulatory Commission halted construction in 1977, citing concerns about the foundations for the plant's giant cooling towers being placed on sandy soil. This decision led to four more years of BPI and local opponents, as well as the Interior Department, harassing NIPSCO about the environmental impact of the construction process. Citing cost overruns, the utility cancelled the plant in 1981, converting the entire project to a coal-fired facility. This was a major victory not only for BPI, but also for regional environmental advocates. Although situated in Indiana, Indiana Dunes National Lakeshore was an important recreation and natural amenity for Chicago-area residents, especially for those who lived south of the city, a region dominated by steel mills, chemical plants, and some of the densest (and most highly polluting) industrial development in the country.[28]

The defeat of the Bailly plant officially ended BPI's environmental program, although it had not been active for years. Comey had left to become the executive director of a Chicago-area environmental group,

Citizens for a Better Environment, in 1976, but stayed as a BPI board member and continued to be involved in opposition to nuclear power plants until his tragic death in a 1979 car accident. Staff attorney Douglas Cassel took over the Bailly case, but without Comey pushing for involvement in environmental issues as a staff member and then board member, the organization's attention turned to other projects. BPI worked occasionally on environmental justice issues in the 1990s, but became primarily a civil rights and urban affairs firm.[29]

Although the move away from environmental issues was primarily the product of a shift in staff and board interests, funding did play a role. In its early days BPI was a "seat of your pants" organization, according to Polikoff, which reflected the freewheeling spirit of its founder, Gordon Sherman. The staff defined priorities, exercising a flexibility made possible by Sherman's largesse. But the public interest impresario's heterodox methods and politics caught up with him in 1971, when he lost control of Midas, Inc. in a stock proxy fight with his father, Nathan Sherman. Out of a job, he left Chicago and took semiretirement in the San Francisco Bay area, pursuing his interests in music and nature photography. No longer in charge of Midas, Sherman did not have the monies to fund BPI, which had come from the Midas corporate foundation. This forced BPI to have to raise money like a normal nonprofit organization. Although it quickly gained a core of donors, this appears to have made the group more focused on other issues, and environmental ones fell by the wayside.[30]

On balance, BPI's environmental work was tremendously successful. In addition to preventing the construction of the airport in the lake and the Bailly nuclear plant, it also helped increase the safety and decrease the risk of environmental harm among the region's existing nuclear power plants, and raised general awareness about environmental threats to Lake Michigan. These achievements in many ways can be chalked up to the fact that BPI consisted not of students and "long-hairs" protesting in the streets, but "hard-nosed professionals" with a knowledge of the complex intricacies of pollution regulation, lake ecology, and nuclear reactor construction. Comey and other staff members were experts at digging through studies, reports, and memoranda to find flaws, inconsistencies, and glaring mistakes with various plans and programs. Their efforts paid off in real results in the courtroom and with various regulatory agencies.

However, a full accounting of BPI's cases and projects also reveals that despite the preference for "skilled, experienced professionals," these accomplishments also depended on active citizen organization and opposition. The various nuclear power plant cases are the best example. There was no citizen opposition to the Zion plant, and so efforts to mount a challenge to its operation permit quickly died out. In contrast, BPI worked in concert with robust citizen's groups in its campaign against the Bailly, Palisades, and Donald Cook plants. These groups provided energy and interest, fund-raising support that paid legal fees (more than $100,000 for opposing the Bailly plant, for example), but also standing in the courts. Although American courts expanded the definition of who possessed standing to bring an environmental case during this period, successful litigation usually required an active local organization to show what and how harm was being done to the local community and ecosystem.[31]

The issue of standing highlights the broader value and the limitations of the public interest law firms and other forms of "top-down" environmental advocacy. These firms had the expertise and knowledge to properly research cases and litigate them in the courts, which was often the best route for environmentalists in the American regulatory system. Sherman's vision of the public interest firm being the primary organization, rather than working in concert with citizen's group, had flaws, at least at the local level. It privileged expertise, which was necessary when dealing with complex environmental issues, but often proved ineffective unless BPI partnered with a grassroots organization or was part of a broader coalition. These conditions limited the firm's ability to expand or shift the definition of environmental activism, locking in a specific white, middle-class view of environmentalism, and if there were no citizen's groups active on a certain issue, it was actually difficult for BPI to pursue it independently. The history of American environmentalism is too often told at the national level, focusing on general ferment and environmental concern, as well as the actions of national groups. But local groups were responsible for significant and tangible progress in improving local environmental conditions and helping people create healthier and safer communities. Conversely, they also reflected some of the limitations of the national movement, especially when it came to balancing grassroots activism with technical analysis and complex litigation.

BPI also helps us understand Chicago's environmental challenges at the dawn of the environmental decade, and the types of responses those hurdles provoked. Despite popular memory that directly connects environmentalism with the most radical threads of the antiwar movement and the counterculture, in general most American activists and organizations were moderate and reformist in orientation, looking to reconcile the desire for economic security and material prosperity with the need for a healthy environment. Sherman, and BPI, fit securely into this mold. As someone whose personal and family fortunes were built upon abundant American consumerism, Sherman was no radical looking to blow up the franchise that laid the golden muffler. Yet he also understood that the existing system was destructive, and so reform was needed. This would ideally be done through American institutions, such as corporations and the courts, where proper scientific knowledge and technical expertise could be applied to the issue in an orderly and consistent manner. This reformist perspective extended to BPI's environmental campaigns. Don't Do It in the Lake! was never about shutting down airports, or even getting local residents to rethink the environmental sustainability of large-scale jet transit. BPI and other opponents simply said that if Chicago needed a new airport, there were better alternatives, and that Lake Michigan was too important a resource to risk the danger of oil and jet fuel spills and long-term habitat disturbance. The same went for nuclear power. Comey always argued that BPI was never opposed to atomic energy, only to the irresponsible way that regional utilities went about building and operating nuclear generating stations, and the environmental and public health threats brought about by that carelessness.

CHAPTER 19

The Calumet Region

A Line in the Sand

MARK BOUMAN

The Calumet region's sands swirl constantly over the lines drawn between city and nature. As documented by the Calumet Heritage Partnership's (CHP) effort to create a Calumet National Heritage Area (NHA),
the region presents a nationally significant example of what happened
when the forces of mercantile and industrial urbanization met one of the
great swaths of "first nature" in the Midwest. The Feasibility Study from
which this essay is drawn demonstrates that the Calumet area meets the
National Park Service's definition that NHAs are "places where natural,
cultural and historic resources combine to form a cohesive, nationally
important landscape."[1]

The Calumet region suffered shape-changing impacts from a century
of heavy industrial activity and city building that moved rivers, leveled
hills, filled wetlands, and imprinted the actions and lifeways of many
peoples on an area of great biodiversity. But the patchwork industrial
development of the Calumet region did not create wall-to-wall industry. Some land was held by industry for its future use; other areas were
platted for residences but never built on; other land was eyed for future
development but time passed before action could be taken; and some was
protected and restored through tenacious struggle. The effect was that,
amid the scenes of what David Nye has called the American "technological sublime," "nature" persisted. What resulted was the extraordinary
juxtaposition of industry and ecology that characterizes the region today,
a crossroads for nature, industry, and people.[2]

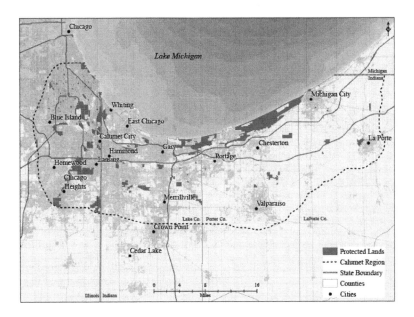

FIG 19.1. *Calumet Regional Map with Protected Lands.* Map by Mark Bouman.

A NATURAL BOTANICAL PRESERVE

At the cul-de-sac of Lake Michigan where the shoreline changes from a north–south to an east–west orientation, the last two thousand years of postglacial lake recession and sand deposition played notes of landscape formation on the lake plain that echo to the present. Sandy lake bottom sediments ferried to the beach were blown into gentle dune ridges, a pattern repeated roughly a hundred times and ultimately yielding the finely textured "ridge and swale" landscape that characterizes the space between Lake Calumet and the high Indiana dunes. This Tolleston strandplain is now home to some of the most significant ecological restorations in the nation, side by side with massive industrial concerns that still call the region home.

The Valparaiso Moraine sits astride the region's midsection, with richly forested and hummocky "knob and kettle" topography. Streams flowing off the moraine's southern slopes carried glacial outwash materials toward the now channelized Kankakee River and ultimately to the Gulf of Mexico. In the wide, flat floodplain of that river, a marsh called the "Everglades of the North" sharply bounded the region until the river was drained in the early twentieth century to make an important farming zone.[3]

The Calumet River rises in hilly moraine country in LaPorte County, Indiana. Once the waters reach the lake plain, both the Grand Calumet and Little Calumet Rivers flow slowly parallel to Lake Michigan, held between intervening beach ridges. The rivers today unite about six miles south of the lake in Illinois, although the main channel through South Chicago was not likely created until the early nineteenth century.[4]

The landscape variations between sand and clay, ridge and marsh, lakeside and landside, set up local variations on grand continental themes and create a place of uncommon—and given the subsequent urban-industrial land uses, unexpected—biological richness. The Calumet region is an *ecotone*—a transition zone where three great bioregions come together. Like clasped fingers held parallel to the lakeshore, one region gradually hands off its characteristics to the other. From the east come the deciduous forests of eastern North America, dominated by oak and hickory in well-drained soils and by beech and maple on wetter ground. The west opens up into the tallgrass prairies of Illinois. Remnants of northern plant types and habitats are reminders of the recent glacial past.[5]

The high dunes on the lakeshore provide a variety of niches for plants and animals. As the pioneering ecologist Henry Chandler Cowles put it in 1916: "Within a stone's throw of almost any spot one may find plants of the desert and plants of rich woodlands, plants of the pine woods, and plants of swamps, plants of oak woods and plants of the prairies. Species of the most diverse natural regions are piled here together in such abundance as to make the region a natural botanical preserve, not only of the plants that are characteristic of northern Indiana, but also of the plants of remote outlying regions."[6] Cowles's studies started a chain of ecological research leading to the present focused on ecological succession and biodiversity at the dunes.

On the lake plain away from the dunes, beach ridges and inter-dunal swales provide a patchwork of diverse habitats. Where water tables are high, marshes, swamps, and wet prairies predominate, with many bird species attracted to food sources and nesting sites. Where sandy beach ridges allow soils to drain, oak woodlands and prairie savannas hold sway. More than seven hundred plant species grace the region, more than eighty-five of which are deemed rare at the state or global level. Among them was a plant seen nowhere else in the world, *Thismia americana*, which was last spotted in 1916 and is now believed to be extinct.[7]

The wetland ecosystems offer excellent sources of food, nesting sites, and resting places for a wide variety of migrating birds traversing the north–south continental flyway. A Field Museum curator wrote in 1909 that the yellow-headed blackbird was "once abundant in the vicinity of

Chicago about Lake Calumet and is still not uncommon in that locality."
Rare today, or completely gone from the region, such species no doubt
graced the daily lives of those who lived there just a century before.[8]

THE WORLD'S LARGEST INDUSTRIAL DISTRICT

Not only does the Calumet region provide textbook geological and biological examples, it also headlines the texts of industrial developers, economic geographers, and urban historians. Most writers on the history of
American urbanization and economy point to the major reorientation
that occurred between the Civil War and World War II. It was based on
the interrelated extraction of coal and iron ore; the rise of integrated production systems especially in the iron and steel industry but also vehicles,
chemicals, machine tools, and electric appliances; the stitching together
of these systems by an integrated national transportation network, especially the railroads; the rise of corporate forms of business organization;
and the recruitment of large labor forces from Europe and the southern
United States. This national reorientation gave rise to the regional dominance of the western Great Lakes, and particularly in the area of steel
production, the dominance of the western anchor of the Manufacturing
Belt—the Calumet region. This region developed into what geographer
Robert Lewis argues was the "world's largest industrial district in the
first half of the twentieth century."[9]

In an influential article written fifty years ago, John Borchert, who
grew up just south of the steelmaking district in Crown Point, traced the
rise of this "Steel Rail" period in American economic history (roughly
1870–1920). The metropolitan areas that "boomed" in this period were
those best positioned within the national railroad network and able to
participate in the production of vast quantities of increasingly inexpensive steel in efficiently laid out mills.[10] The Calumet region had both; it
was central to the rise of the "Steel Rail" period.

The Bessemer converter's introduction in the 1860s vastly increased
the potential to produce large amounts of cheap steel. While it was first
employed by the North Chicago Rolling Mills along the Chicago River's
North Branch, the logic of the new steel-making technology demanded
bigger sites than were available there. The Calumet region awaited, and
by 1881 the company was building its South Works at the mouth of the
Calumet River, joining the Brown Ironworks (1875) which was built up-
stream and later evolved into International Harvester's Wisconsin Steel
Company. After the turn of the century Indiana sites became more important, especially Inland Steel (1901), Gary (1906), and Mark Manufacturing (1914). The evolution of these firms also illustrates the growing
vertical and horizontal integration of the industry characteristic of the

era: the very evolution of the name of South Works into Carnegie-Illinois into U.S. Steel suggests the ever-expanding scope of operations. When it was completed in 1962, Bethlehem Steel's Burns Harbor, Indiana, plant was the last integrated steel facility to be built in the United States where materials moved all the way from raw form to finished product.[11]

In spite of the closures of four integrated mills in Chicago between 1980 and 2001, the region remains critically important to the American steel industry. In 2014 five of the nation's eleven integrated steel mills were located here, including its largest producer, ArcelorMittal's Indiana Harbor. In addition to the large integrated mills, there are several other key producers. For example, A. Finkl and Sons echoed the North Chicago Rolling Mills by moving from a plant by the North Branch of the Chicago River to Chicago's Southeast Side in 2011.

Railroads were central to the ability to assemble raw materials and to distribute finished products to market. Building the lines through the flat Calumet region posed little problem: a map of railroads through the region shows a series of straight-line tangents cutting across the lake plain, and then markedly deviating from the "air line" when encountering the moraine country.

After the turn of the century passengers and freight were increasingly carried through the region on steel-framed railcars. Pullman was an iconic producer and operator of passenger cars, as well as a leading freight car manufacturer. By the 1930s, and through mergers with companies in Hammond and Michigan City, the Pullman-Standard Company was the nation's leading railcar manufacturer. It was joined in the region by other steel railcar producers in Hegewisch, Chicago Heights, East Chicago, Blue Island, and Burnside.

Once established in the region, the steel industry proved to be magnetically attractive to a variety of related businesses. A web of industrial and short line railroads moved steel from mills to fabricators with relative ease. Steel supply companies, refractory manufacturers, and by-products producers burgeoned. Other firms were attracted by the availability of inexpensive steel in the context of location in the Chicago market, by the region's centrality to the national rail network, or by the region's location immediately adjacent to agricultural areas.

Having grown to a critical mass during Borchert's "Steel Rail" period, the coal-steel-rail complex remained crucial during the next phase he names the "Auto-Air-Amenity" period. A close relationship between the steel producers and the petroleum and automotive industries quickly emerged. Standard Oil established a refinery at Whiting in 1889, initially to refine crude from Indiana-Ohio oil fields into kerosene. As the automobile industry grew and the demand for a wider variety of fuels surged, the refinery became the sixth largest in the nation, a major supplier to the midwestern gasoline consumer and the leading supplier of jet fuels to O'Hare and Midway airports. Meanwhile, the automobile industry became a huge customer for the steel industry, and remains so today. Automobiles themselves continue to be produced in the Calumet region at Ford's oldest U.S. operating facility on Torrence Avenue.[12]

REARRANGING THE WATERS
AND REMAKING THE LAND

Making the Calumet region industrially productive required reworking its lands and waters. This meant straightening the Calumet River and connecting it to the Chicago and Des Plaines Rivers system, filling and draining wetlands, and adding land to Lake Michigan. Work began in earnest in 1870, when the Army Corps of Engineers built structures at the mouth of the Calumet River to prevent the relentless sand from obstructing navigation. They next dredged a channel to the junction with the Grand Calumet River. By 1871 the first cargo ships called at Calumet Harbor.[13]

Navigation interests became more organized with the establishment of the Chicago Harbor Commission in 1908. The commission forwarded a number of proposals to improve navigation and create port facilities, but it also oversaw, with the completion of the Cal-Sag Channel in 1922, the reversal of the flow of the Calumet River. Subsequent widenings made the potential reverse flow even greater. By 1965 when the O'Brien Lock and Dam was constructed in Chicago, the drainage system and pattern of flow had been completely altered and boats could transit between Lake Michigan and the Illinois Waterway.[14]

A similar chain of events unfolded in Indiana. In 1901 work began to create Indiana Harbor and to connect it to the Grand Calumet River via the Indiana Harbor Ship Canal. When Gary was developed in 1906, the Grand Calumet River was relocated about a half-mile south of its historic course, and now ran through banks of masonry and slag for several miles. In 1926 the Burns Ditch—now Burns Waterway—connected the upper reaches of the Little Calumet River with Lake Michigan just east of Ogden Dunes. These canals and diversions made port development possible, but also drained the marshes, creating new passages for storm runoff and providing pathways for invasive aquatic species to enter the river system. A check dam at the Indiana Harbor Canal sends most waters east of East Chicago into Lake Michigan via that canal. West of that structure, the Grand Calumet is a part of the Illinois Waterway system. West of Burns Waterway, waters of the Little Calumet River also head toward the Illinois Waterway.

A web of landmark steel bridges soon vaulted across the newly tamed waterways. A variety of designs, including trunion bascule, lift, and trusses carried on high piers ensured that navigation could continue un-impeded while providing striking additions to the built landscape.

Industries soon began to alter their sites. The growing scale and in-tegrated nature of late nineteenth-century industrial operations meant that factories needed docking and rail facilities to receive raw materials and to ship finished products; land to stockpile raw materials and parts; space to expand; and room to dump waste. The Calumet area not only offered lower land costs than congested Chicago, but also sheer size.[15] In Indiana the sandy lakefront could easily be pushed aside to make way for port facilities and urban development. And while Illinois's Public Trust Doctrine required that any lake filling proceed for the public benefit, In-diana allowed private expansion into Lake Michigan.[16]

The railroads were the "first major change agents of the landscape."[17] Railroad rights-of-way had an enormous impact beyond the noise, smoke, and danger of the rolling trains. Their embankments fragmented wetlands and altered drainage patterns. Hot cinders started prairie fires. Maintenance-of-way crews trimmed and pruned back vegetation, in lat-er years applying pesticides to the task.

Railroads hauled "astronomical" amounts of sand from convenient locations in the dunes country. Where clay soils predominated, clay pits and brickyards clustered along the tracks. New wood-frame housing construction was banned within Chicago after the 1871 fire, making the city a huge market for bricks, most of them produced locally. By 1927 Cook County ranked "as the foremost brick-producing county in the United States." Most brickyards have now closed, but the associated pits frequently remain.[18]

Sand was a spectacular resource. Thousands of railroad cars exported sand to help fill Grant Park in postfire Chicago. Railroads required sand for locomotive operations. Much of the material for building Chicago's elevated rail embankments in the first quarter of the twentieth century came from the dunes. By 1927 Fryxell wrote that "whole trainloads are daily hauled away." Glassmaking firms such as Ball Glass of Muncie and Pittsburgh Plate Glass of Kokomo removed huge amounts of sand, much of it in the first two decades of the century from the Hoosier Slide dune that once towered over Michigan City. The result was that huge sections of dune country—as much as a square mile at a time—were stripped of this defining material.[19]

Leveling land on the one hand, industry made land on the other. From approximately Seventy-Fifth Street in Chicago southward, around to Miller Beach in Gary, and hopscotching over Ogden Dunes to Burns

Harbor, the Lake Michigan shoreline is made land. Some of this is civic or recreational space, but the most extensive man-made land was created by industry. From west to east, significant portions of made land and the companies that built them include: U.S. Steel and Youngstown Sheet and Tube in South Chicago, Commonwealth Edison at the State Line, Amoco Oil in Whiting, Youngstown Sheet and Tube and Inland Steel (on its huge peninsula) in East Chicago, and the various facilities of U.S. Steel in Gary. Eleven million cubic yards of sand were moved when the U.S. Steel Gary Works was built, much of it pumped onto the site from Lake Michigan.[20] Dune mining continued into the 1960s at Burns Harbor, where, in addition to removing some of the highest and most spectacular of the sand dunes, Bethlehem Steel built part of its new operations on fill in Lake Michigan. According to historian Kenneth Schoon, between 1900 and the late 1970s, more than 3,775 acres—roughly six square miles—of Lake Michigan in Indiana were filled in.[21]

At times the fill used to create this new land was a by-product of the industrial operations themselves, such as the slag created by blast furnaces. South Works and Inland Steel used slag as lake fill as they extended operations into Lake Michigan. Steel manufacturers who were a few miles from Lake Michigan, like Wisconsin Steel and Republic Steel, simply dumped slag into adjacent wetlands.[22]

Other industries also created striking landscape elements. For example, crude and refined petroleum is stored in extensive tank farms in the Tolleston strandplain. Each tank is set within a bermed, graded, and drained containment area, the entire group constituting a checkerboard-like grid that can stretch for a mile or more. These tank farms cover more than a thousand acres.

In various ways, then, building on the impetus provided by public agencies such as the Army Corps of Engineers, private industries played a geomorphologic role in rearranging the physical landscape of the region: they cut down the heights and filled in the lowlands and thereby flattened an already flat terrain. In the process, soggy land was made dry, shallow waters were made to run more deeply, and the boundary between land and water, formerly subject to great daily, seasonal, and annual fluctuations—if indeed a "boundary" existed at all—was fixed tightly in place.[23]

REMNANT WETLANDS AND NATURAL AREAS

Wetlands had a chance to survive if located at some remove from navigable watercourses and major rail junctions. Even here, however, "survival" might just be a phase in a cycle of land acquisition, subdivision, construction, abandonment, and/or neglect. Indian Ridge Marsh in

Chicago—a haven for marsh birds—was a platted subdivision for most of the twentieth century that was never built out. Street grades, never lined with structures, cut across the ridge and swale topography of the Shirley Heinze Land Trust's Ivanhoe South preserve in Gary, a story repeated at a number of ecological restoration sites. Hegewisch Marsh's one hundred acres were about half covered over with railroad tracks and structures of the nearby steel supply warehouse operation. When they were removed, the marsh slowly recovered. Wolf Lake and Lake Calumet were simply too big to be filled entirely, though they are far smaller now than they were one hundred years ago.[24]

The result of being bypassed by real estate development was to create islands of water or relatively undisturbed vegetation in a sea of dry land and urbanization. "Sportsmen's" clubs sprouted at especially rich locations where hunters espied remaining waterfowl. Others "shot" the birds with spotting scopes or cameras, and a birding avocation took flight through the twentieth century.[25]

Scientists, collectors, and a growing set of Chicago-based natural area advocates were aware of the riches of these places.[26] After the University of Chicago was founded just a few miles to the north in 1892, the area attracted botanizing faculty and graduate students. In 1916 Cowles's colleague Rollin D. Salisbury noted, "The dunes are going and more are to go. I do not think we should stop it altogether, because the head of Lake Michigan is so advantageously situated for industrial development that industries must develop there. . . . [But we can] secure the permanent preservation of a generous and well-selected tract, for the use of ourselves, and of the generations to come."[27] In that year agitation for a Dunes National Park reached its peak, led by the Chicago-based Prairie Club, whose members included Cowles, Carl Sandburg, Jane Addams, and the landscape architect Jens Jensen. The effort enjoyed strong endorsement by club members and the new National Park Service superintendent Stephen Mather, and forty-two people testified strongly in favor of the park at hearings held in Chicago that year. But war intervened, enthusiasm waned, and Congress was not ready to appropriate the funds to purchase the land. However, an alternate strategy of creating an Indiana Dunes State Park did succeed in 1926.[28]

National park advocates finally had their day in 1966 when a characteristically Calumet compromise brokered by Illinois senator Paul Douglas resulted in the creation of the country's first National Lakeshore around the new Port of Indiana and Bethlehem Steel's Burns Harbor plant. Park advocates, led by the Save the Dunes Council, had been pushing for designation since the early 1950s. The new steel mill directly bordered the National Lakeshore's Cowles Bog, a National Natural Landmark.

At Cowles Bog the industrial Calumet region muscled its way into the senses. In 1969 Mayer and Wade wrote, "Fire and smoke charged into the sky as a constant reminder to the world of Chicago's brute industrial strength. Most people were appalled by the dirt, pollution, and ugliness of the scene, but to some there was an elemental beauty to the rough shapes and raw power embodied in this steaming jungle of steel and brick and concrete."[29] The elemental beauty changed again when the fires went out. By the turn of the twenty-first century regional narratives spoke of "rusted dreams," of a "drosscape," and of what anthropologist Christine Walley called "Exit Zero," a place filled with brownfields and industrial relics, and most importantly, dislocated people and communities. How would they respond?[30]

"ON THE PATH TO SUSTAINABILITY"

A textbook landscape with a textbook economic history, including its booms and its busts, the Calumet region now finds itself at the center of efforts to write the new text on the next American city. Celebrating innovation and heritage, new paths to sustainability and land conservation in an urban context marked the region, from brownfield redevelopment to education and stewardship to recreation to creative partnerships.

Rustbelt Challenges

Since 1980 the region's economy changed markedly. Large steel facilities in Chicago such as U.S. Steel's South Works, Wisconsin Steel, Republic/LTV, and Acme Steel closed. Other firms radically downsized their employment as they modernized their facilities. Joblessness, population loss, social dislocation and contaminated brownfields became persistent issues. Suburbanization already in process intersected with issues of race and class to create severe inequalities where those hardest hit with deindustrialization also bore the brunt of environmental disamenities, as in Gary. This fate befell other places in the American Manufacturing Belt, and indeed, what happened to all of them is one of the most significant national stories of the past four decades. The reasons for industrial decline are many. Increased global competition, corporate failure to keep technological pace, difficult choices made by unions, the changing structure of demand, and increased energy costs have all been mentioned prominently in the discussion of the Calumet region's decline.[31]

The Calumet region constituted what Alan Berger called a "drosscape," a "waste landscape" resulting from deindustrialization and suburbanization. It is replete with ongoing industrial releases to air, land, and water; brownfields; superfund sites; and more than its fair share of sanitary landfills. Rising from the drosscape, however, are significant stories of resistance and recovery and new efforts to smudge the line between city and nature.

One key element of the drosscape is continuing pollution-generating economic activity. Numerous industrial and commercial facilities still operate in the Calumet region. By-products of their activity are the wastes emitted to air, water, soil, or groundwater. Health issues of particular importance to residents include lead poisoning, asthma, skin rashes, and pesticide poisoning. Since 1986 companies must report their releases from a list of 755 chemicals to the Toxic Release Inventory. The map of hazardous substance–producing or –using facilities that appear in this annual report neatly marks the industrial core of the Calumet region.[32]

The legacy of polluting industries is registered in the region's large number of brownfield sites and polluted waters. Brownfields occur when unknown levels of contamination from prior activity deter reinvestment and reuse, especially when suburban "greenfields" appear to be less expensive, more extensive, and better connected to the freeway grid than railroad-oriented industrial properties. The most significant brownfields are those Superfund sites where the known contaminants must undergo costly cleanup. Twenty-five of the most seriously contaminated sites are listed on EPA's Comprehensive Environmental Response, Compensation, and Liability Information System (CERCLIS) list.[33] Site-by-site cleanup may still not capture the long-term effects of windblown pollutants. East Chicago's West Calumet neighborhood is now facing housing displacement and disruptive soil cleanup of pollutants emitted by the now closed and cleaned up U.S. Smelting Lead Refining smelter when it was still operating decades ago.[34]

The aquatic equivalent to brownfields are contaminated streams. The Grand Calumet River is nationally significant in being the only one of forty-three "Areas of Concern"—designated by the Great Lakes Water Quality Agreement and required to have a Remedial Action Plan—that fails every one of the criteria of "beneficial use impairments."[35] These include such things as restrictions on fish consumption, fish tumors or other deformities, beach closings, loss of habitat, and tainting of fish and wildlife flavor.

While Chicago area wastes have long ended up in the Calumet region, this trend increased rapidly in the past thirty years with the development of numerous sanitary landfills. Some landfills present serious issues of leaching, slope instability, and odor. The Paxton II landfill grew to a 170-foot height, had unstable and eroding slopes, and needed intervention by the Illinois EPA lest it catastrophically collapse onto nearby roads.[36]

Significantly, and while not minimizing the challenge the region faces to make its lands and waters safe for people and for nature, there is

positive movement to remove each one of these drosscape components.
For example:

- Overall toxic releases are down across the region since 1986. At the site
  of one of the largest emitters, and in response to a consent decree with
  the EPA for Clean Air Act violations, the BP Refinery in Whiting has
  established a fenceline system of air monitors and makes the data collected available to the public.[37]

- Increasing interest in brownfield redevelopment signals a "paradigm
  shift" in urban land use planning: yesterday's liabilities are today's opportunities.[38] Chicago's Brownfields Forum in 1994 prompted new state
  laws that guide "how clean is clean." This "tiered approach to cleanup
  objectives" considers a site's future use: cleanup objectives are less stringent if it is to be industrial than if it is to be residential. New ways to
  restore brownfields have been studied, such as "mining" remnant iron
  from the slag that covers so much of the region or using trees to take
  heavy metals into their vascular systems and then capturing the residue.[39] The world's largest urban solar array covers a former brownfield
  site in Chicago's West Pullman neighborhood.

- Two of the beneficial use impairments have now been delisted from
  the Grand Calumet River. Toxins in the river itself have been either
  capped or removed, and The Nature Conservancy has created sixty-
  five acres of restored open space on adjacent land.

- Gas recovery and recreational development characterize several sanitary landfill sites.

Berger argues that "drosscapes have few stakeholders, caretakers, guardians, or spokespersons."[40] This is not the case in the Calumet region. The
achievements listed above could not have happened without strong action by environmental advocates, such as the Alliance for the Great Lakes,
Chicago Legal Clinic, Grand Calumet Task Force, Hoosier Environmental Council, Sierra Club, and National Resources Defense Council.

Strong locally based organizations are environmental justice pioneers. The struggle for environmental justice in the Calumet area has
national implications, and not only because Barack Obama began his
political career working on environmental issues in the Calumet region.
He worked with Hazel Johnson, who articulated a notion of environmental justice especially relevant to low-income people of color. Johnson
founded People for Community Recovery (PCR), a community-based
environmental organization located at the Chicago Housing Authority's

Altgeld Murray Homes. She referred to Altgeld's regional position, surrounded by landfills, sewage treatment plants, and active industry, as
being in the hole of a "toxic doughnut." PCR and partners like the Hege-
wisch-based Southeast Environmental Task Force are part of a vigorous environmental movement that is alive and well in the region. It can
count some important victories: Waste Management was forced to close
a noncompliant hazardous waste incinerator; large polluting companies
once paying millions of dollars in fines willingly entered Good Neighbor
Dialogues that focus on pollution prevention; and the city was pressured
to sustain its 1989 moratorium on new landfill construction. The Task
Force recently successfully fought the storage of large piles of fine particulate petcoke from the BP Whiting refinery at transfer terminals along
the Calumet River.[41]

Response Rooted in Regional Assets

In 1990 Mayor Richard M. Daley stunningly announced that Chicago
would pursue construction of a Lake Calumet International Airport, intended to match the size and scope of O'Hare.[42] Daley's ambitious plan
entailed lowering landfills, rerouting the Calumet River, and moving
forty thousand people.

The city's proposal linked an effort to stimulate local economic development with a desire to resolve major environmental issues. In opposing
the airport proposal, environmental and economic development advocates found common cause, forming alternate scenarios for development
based in the region's considerable assets. In 1992 the city withdrew its
proposal. By this time the ground had been laid on the city's Southeast
Side for efforts to reenvision Calumet as a region for both people and
nature.

One such vision came from the Lake Calumet Study Committee,
which sought to protect the lake and adjoining marshes and their resil-
ient bird diversity.[43] By 1995 the effort developed into a call for a Calumet
Ecological Park, and the Calumet Ecological Park Association was created to advocate for it. In 1998 the National Park Service considered such
a designation in a Special Resource Study, ultimately pointing out that
its NHA would be a promising avenue to pursue. As a result the Calumet
Heritage Partnership was formed.[44]

Meanwhile, the City of Chicago reframed its post-airport approach
to the region through its Calumet Initiative. Moving in concert with a
State of Illinois Lake Calumet Ecosystem Partnership, created in 1999,
the city began to take stock of the hydrology of the land it owned, to
consider further purchase for conservation, to bring together stakeholders around a cluster of toxic landfills, and to consider future land uses

devoted to recreation, education, and conservation. In 2000 Daley and Governor George Ryan announced a comprehensive rehabilitation plan for the Calumet regional environment, including a Calumet Land Use Plan, an ecosystem management plan, purchase and rehabilitation of two key marshes, and construction of a showcase environmental center in one of them.[45]

A series of regional summits, convened first in 2001 by the City of Chicago to gather and discuss the social, environmental, and economic research of its new regional focus area, gradually evolved in location, scope, and purpose. More than 175 attendees from the bi-state region attended summits in 2013 and 2015, and by then a regional approach had been firmly established. Creating a Calumet National Heritage Area became the group's top goal and provided an overarching framework for the discussion.

"Have you always enjoyed musty, old things?," two leaders of the Calumet Heritage Partnership were asked by a television interviewer.[46] Here lies one popular view, that heritage is ancient and irrelevant. But in the Calumet, both environmental and economic development professionals and grassroots advocates are at work in a landscape that is the living heritage of an epic and continuing encounter between the city and nature. Their effort to blow sand over the line separating the two recalls what William Cronon wrote in recalling his childhood journeys through the "orange cloud" of industrial Calumet to the "green lake" of bucolic Wisconsin: "We can only take them together and, in making the journey between them, find a way of life that does justice to them both."[47]

Notes

INTRODUCTION

1. Published works with insights into the varied communities that make up Chicago include James R. Grossman, Ann Durkin Keating, and Janice L. Reiff, eds., *The Encyclopedia of Chicago* (Chicago: University of Chicago Press, 2004); Ann Durkin Keating, ed., *Chicago Neighborhoods and Suburbs* (Chicago: University of Chicago Press, 2008); Alan Ehrenhalt, *The Lost City: The Forgotten Virtues of Community in America* (New York: Basic Books, 1995); Dominic A. Pacyga and Ellen Skerrett, *Chicago—City of Neighborhoods: Histories and Tours* (Chicago: Loyola University Press, 1986); and Dominic A. Pacyga, *Chicago: A Biography* (Chicago: University of Chicago Press, 2009). Online resources on neighborhoods include the Newberry Library's Chicago Neighborhood Guide, https://www.newberry.org/chicago-neighborhood-guide for historical research; and "The City of Neighborhoods Project," http://www.thecityofneighborhoodsproject.com/the-neighborhoods, to explore today's Chicago.

2. Craig E. Colten, ed., *Transforming New Orleans and Its Environs: Centuries of Change* (Pittsburgh: University of Pittsburgh Press, 2001); Joel A. Tarr, ed., *Devastation and Renewal: An Environmental History of Pittsburgh and Its Region* (Pittsburgh: University of Pittsburgh Press, 2003); Martin V. Melosi and Joseph A. Pratt, eds., *Energy Metropolis: An Environmental History of Houston and the Gulf Coast* (Pittsburgh: University of Pittsburgh Press, 2007); and Char Miller, ed., *On the Border: An Environmental History of San Antonio* (Pittsburgh: University of Pittsburgh Press, 2001).

3. William Deverell and Greg Hise, eds., *Land of Sunshine: An Environmental History of Metropolitan Los Angeles* (Pittsburgh: University of Pittsburgh Press, 2005); Michael F. Logan, *Desert Cities: The Environmental History of Phoenix and Tucson* (Pittsburgh: University of Pittsburgh Press, 2006); Anthony N. Penna and Conrad Edick Wright, eds., *Remaking Boston: An Environmental History of the City and Its Surroundings* (Pittsburgh: University of Pittsburgh Press, 2009); Brian C. Black and Michael J. Chiarappa, eds., *Nature's Entrepot:*

Philadelphia's Urban Sphere and Its Environmental Thresholds (Pittsburgh: University of Pittsburgh Press, 2012); and Christopher J. Castaneda and Lee M. A. Simpson, eds., *River City and Valley Life: An Environmental History of the Sacramento Region* (Pittsburgh: University of Pittsburgh Press, 2013).

4. William Cronon, *Nature's Metropolis: Chicago and the Great West* (New York: W. W. Norton, 1991); and Andrew Hurley, ed., *Common Fields: An Environmental History of St. Louis* (St. Louis: Missouri Historical Society Press, 1997).

5. Cronon, *Nature's Metropolis*, 384.

6. Brian McCammack analyzes the foodways of African American migrants from the rural South in chapter 9; Michael Innis-Jiménez examines housing conditions for Mexican Americans in South Chicago in chapter 10.

7. Mark Bouman examines ecological recovery in the Calumet Heritage Area in chapter 19.

8. Theodore J. Karamanski surveys the changing relationships by Chicagoans and Lake Michigan in chapter 4.

9. Robert Morrissey analyzes early Native Americans in the Illinois tallgrass prairie region in chapter 1.

10. Ann Vileisis, *Discovering the Unknown Landscape: A History of America's Wetlands* (Washington, DC: Island Press, 1997); Craig E. Colten, *An Unnatural Metropolis: Wresting New Orleans from Nature* (Baton Rouge: Louisiana State University Press, 2004); and Philip Garone, *The Fall and Rise of the Wetlands of California's Great Central Valley* (Berkeley: University of California Press, 2011).

11. Wayne Grady, *The Great Lakes: The Natural History of a Changing Region* (New York: Greystone Books, 2007), 5–36, 64–77; Joel Greenberg, *A Natural History of the Chicago Region* (Chicago: University of Chicago Press, 2002), 177–78.

12. "Mississippi River Facts," National Park Service, https://www.nps.gov/miss/riverfacts.htm (accessed Apr. 30, 2017).

13. Greenberg, *A Natural History of the Chicago Region*, 177–80; Louis P. Cain, *Sanitation Strategy for a Lakefront Metropolis: The Case of Chicago* (De Kalb: Northern Illinois University, 1978), 5.

14. Matthew Corpolongo studies the reversal of the Chicago River in chapter 5.

15. William C. Barnett examines May Watts and the Illinois Prairie Path in chapter 17.

16. Conevery Bolton Valencius, *The Health of the Country: How American Settlers Understood Themselves and Their Land* (New York: Basic Books, 2002); Linda Nash, *Inescapable Ecologies: A History of Environment, Disease, and Knowledge* (Berkeley: University of California Press, 2007); Nancy Langston, *Toxic Bodies: Hormone Disruptors and the Legacy of DES* (New Haven, CT: Yale University Press, 2010); and Eula Biss, *On Immunity: An Inoculation* (Minneapolis: Graywolf Press, 2014).

17. Updating booster language from the canal and railroad eras, the nonprofit group World Business Chicago's website describes Chicago as "at the center of the east-west nexus joining Europe and Asia, and the north-south nexus of NAFTA"; http://www.worldbusinesschicago.com/economy/ (accessed Nov. 20, 2019). "Fortune 500 Companies List: 1 out of 3 Are Located in Just 6 Major Cities," *USA Today*, Nov. 1, 2018. Corporations based in the Chicago area include Walgreens, Boeing, Archer Daniels Midland, Caterpillar, Allstate, Abbot Laboratories, and McDonald's, showing agricultural, industrial, service economy, and high-tech sectors.

18. John Broich, *London: Water and the Making of the Modern City* (Pittsburgh: University of Pittsburgh Press, 2013); Stéphane Castonguay and Michèle Dagenais, eds., *Metropolitan Natures: Environmental Histories of Montreal* (Pittsburgh: University of Pittsburgh Press, 2011).

19. Carl Sandburg, "Chicago," *Poetry: A Magazine of Verse*, Mar. 1914, 191–92; Nelson Algren, *Chicago: City on the Make* (New York: Doubleday, 1951).

20. Notable works that interpret Chicago's industrial booms, in addition to Cronon's *Nature's Metropolis*, include Carl Smith, *Urban Disorder and the Shape of Belief: The Great Chicago Fire, the Haymarket Bomb, and the Model Town of Pullman* (Chicago: University of Chicago Press, 1995); and Donald L. Miller, *City of the Century: The Epic of Chicago and the Making of America* (New York: Simon & Schuster, 1996).

21. Peter Nekola and James R. Akerman analyze mapmakers such as Rand McNally and maps of Chicago in chapter 12.

22. Andrew Needham, *Power Lines: Phoenix and the Making of the Modern Southwest* (Princeton, NJ: Princeton University Press, 2016); Christopher F. Jones, *Routes of Power: Energy and Modern America* (Cambridge, MA: Harvard University Press, 2014); and Ted Steinberg, *Down to Earth: Nature's Role in American History*, 4th ed. (New York: Oxford University Press, 2018).

23. On the decades just after World War II in Chicago, from the architecture of Mies van der Rohe to the housing projects of Richard J. Daley, see Thomas Dyja, *The Third Coast: When Chicago Built the American Dream* (New York: Penguin, 2014). On public housing, see Arnold R. Hirsch, *Making the Second Ghetto: Race and Housing in Chicago, 1940–1960* (New York: Cambridge University Press, 1983); and D. Bradford Hunt, *Blueprint for Disaster: The Unraveling of Chicago Public Housing* (Chicago: University of Chicago Press, 2009).

24. Michael Rawson, *Eden on the Charles: The Making of Boston* (Cambridge, MA: Harvard University Press, 2010); Catherine McNeur, *Taming Manhattan: Environmental Battles in the Antebellum City* (Cambridge, MA: Harvard University Press, 2014); Matthew Klingle, *Emerald City: An Environmental History of Seattle.* (New Haven, CT: Yale University Press, 2009); and Christopher W. Wells, *Car Country: An Environmental History* (Seattle: University of Washington Press, 2013).

25. Katherine Macica studies domesticated animals in in urban neighborhoods in chapter 3.

26. Ann Durkin Keating and Kathleen A. Brosnan analyze cholera in early Chicago in chapter 2; Joshua Salzmann examines rising death rates in the industrial era in chapter 13.

27. Thomas G. Andrews, *Killing for Coal: America's Deadliest Labor War* (Cambridge, MA: Harvard University Press, 2008); and Chad Montrie, *Making a Living: Work and Environment in the United States* (Chapel Hill: University of North Carolina Press, 2008). See also Colin Fisher, *Urban Green: Nature, Recreation, and the Working Class in Industrial Chicago* (Chapel Hill: University of North Carolina Press, 2015); and Brian McCammack, *Landscapes of Hope: Nature and the Great Migration in Chicago* (Cambridge, MA: Harvard University Press, 2017).

28. Colin Fisher explores May Day and anarchist views of the natural world in chapter 8.

29. Daniel H. Burnham and Edward H. Bennett, *Plan of Chicago* (Chicago: Commercial Club of Chicago, 1909). See also Carl Smith, *The Plan of Chicago: Daniel Burnham and the Remaking of the American City* (Chicago: University of Chicago Press, 2007).

30. Natalie Bump Vena examines work relief labor in the Cook County Forest Preserves in chapter 11.

31. Steven H. Corey studies air and water pollution in Chicago in chapter 14.

32. Craig E. Colten analyzes refuse disposal and Chicago's wastelands in chapter 15.

33. Robert R. Gioielli examines Businessmen for the Public Interest in chapter 18.

34. Neil M. Maher, *Nature's New Deal: The Civilian Conservation Corps and the Roots of the American Environmental Movement* (New York: Oxford University Press, 2009); Dorceta E. Taylor, *The Rise of the American Conservation Movement: Power, Privilege, and Environmental Protection* (Durham, NC: Duke University Press, 2016); and Robert Gottlieb, *Forcing the Spring: The Transformation of the American Environmental Movement* (Washington, DC: Island Press, 1993).

35. Robert D. Bullard, *Dumping in Dixie: Race, Class, and Environmental Quality* (London: Routledge, 2000); and Sylvia Hood Washington, *Packing Them In: An Archaeology of Environmental Racism in Chicago* (Lanham, MD: Lexington Books, 1994). See also Andrew Hurley, *Environmental Inequalities: Class, Race, and Industrial Pollution in Gary, Indiana, 1945–1980* (Chapel Hill: University of North Carolina Press, 1995); Eric Klinenberg, *Heat Wave: A Social Autopsy of Disaster in Chicago* (Chicago: University of Chicago Press, 2002); David Naguib Pellow, *Garbage Wars: The Struggle for Environmental Justice in Chicago* (Cambridge, MA: MIT Press, 2002); and Dorceta E. Taylor, *Toxic*

Communities: Environmental Racism, Industrial Pollution, and Residential Mobility (New York: NYU Press, 2014).

36. Sylvia Hood Washington examines environmental justice in Chicago in chapter 16.

37. For analysis of urban renewal in twenty-first-century Chicago, see Larry Bennett, *The Third City: Chicago and American Urbanism* (Chicago: University of Chicago Press, 2015). On Millennium Park, see Timothy J. Gilfoyle, *Millennium Park: Creating a Chicago Landmark* (Chicago: University of Chicago Press, 2006).

38. On sustainability ideas, see the city government website https://www .cityofchicago.org/city/en/progs/env.html. On the history of Chicago's parks, see https://www.chicagoparkdistrict.com/about-us/history-chicagos-park.

39. Harold L. Platt analyzes flood risk in the era of climate change in chapter 6.

40. See Jeanne Gang, *Reverse Effect: Renewing Chicago's Waterways* (Chicago: Studio Gang Architects, 2011).

41. Daniel Macfarlane and Lynne Heasley examine today's submerged and hidden infrastructure in chapter 7.

42. Christine Meisner Rosen, *The Limits of Power: Great Fires and the Process of City Growth in America* (New York: Cambridge University Press, 1986).

PART I: WHERE PRAIRIE MEETS LAKE

1. Charles Butler, quoted in Bessie Louise Pierce, *As Others See Chicago: Impressions of Visitors, 1673–1933* (Chicago: University of Chicago Press, 1933), 45.

2. Jacqueline Peterson, "The Founding Fathers: The Absorption of French-Indian Chicago, 1816–1837," in *Native Chicago*, ed. Terry Straus (Chicago: Albatross Press, 2002), 31–66.

3. Kelly Lytle Hernández, *City of Inmates: Conquest, Rebellion, and the Rise of Human Caging in Los Angeles, 1771–1965* (Chapel Hill: University of North Carolina Press, 2017), 7; James Belich, *Replenishing the Earth: The Settler Revolution and the Rise of the Anglo World, 1783–1939* (New York: Oxford University Press, 2009), 159–62; and Walter L. Hixson, *American Settler Colonialism: A History* (New York: Palgrave Macmillan, 2013), 7, 65–66.

CHAPTER 1: NATIVE PEOPLES IN THE TALLGRASS PRAIRIES OF ILLINOIS

1. William Cronon, *Nature's Metropolis: Chicago and the Great West*, 1st ed. (New York: W. W. Norton, 1992).

2. William Cronon, "The Trouble with Wilderness; or, Getting Back to the Wrong Nature," *Environmental History* 1, no. 1 (Jan. 1, 1996): 7–28, https://doi .org/10.2307/3985059; "What Is the Environmental Humanities?" The Environmental Humanities at UCLA, http://environmental.humanities.ucla.edu/?page_ id=52 (accessed Mar. 2, 2017).

3. To be sure, Cronon was clear that the terms were imprecise and ambiguous, precisely on account of the kinds of problems that I aim to outline in this essay. Cronon, *Nature's Metropolis*, xix.

4. As Cronon wrote, "Before Chicago, there was *the land* [emphasis mine]," implying the nonhuman predecessor to the anthropogenic landscape of Chicago. Cronon, *Nature's Metropolis*, 23, 25, 98.

5. Cronon, *Nature's Metropolis*, 26, 98, 266.

6. Cronon, *Nature's Metropolis*, 26, 58.

7. For perhaps the most eloquent statement on the importance of historians' narrative content choices of their histories, see William Cronon, "A Place for Stories: Nature, History, and Narrative," *Journal of American History* 78, no. 4 (1992): 1347–76, https://doi.org/10.2307/2079346. Cronon's idea about "opening scenes" and contrasting "closing scenes" might be especially relevant here. In the narrative of *Nature's Metropolis*, the closing scene is a thoroughly anthropogenic and hybrid landscape—Chicago and the hinterland it shaped. Given this, the "opening" scene is something quite different, a landscape not nearly so anthropogenic. See Cronon, "Place for Stories," 1354–55.

8. Cronon, *Nature's Metropolis*, 26.

9. See Dan L. Flores, "The Great Despoblado and Other Fantasies of Wilderness," in Flores, *Horizontal Yellow: Nature and History in the Near Southwest* (Albuquerque: University of New Mexico Press, 1999), 1–35.

10. My thinking on this subject is shaped particularly by Andrew M. Bauer, "Questioning the Anthropocene and Its Silences: Socioenvironmental History and the Climate Crisis," *Resilience: A Journal of the Environmental Humanities* 3 (2016): 403–26, https://doi.org/10.5250/resilience.3.2016.0403.

11. Cronon, *Nature's Metropolis*, 25; Robert Michael Morrissey, "The Power of the Ecotone: Bison, Slavery, and the Rise and Fall of the Grand Village of the Kaskaskia," *Journal of American History* 102, no. 3 (Dec. 1, 2015): 667–92, https://doi.org/10.1093/jahist/jav514.

12. For a good introduction to prairie ecology, see James Ellis, "Understanding Prairie in the Prairie State," in *Illinois Master Naturalist Curriculum Guide*, ed. Nature Conservancy (U.S.) et al. (Urbana: University of Illinois Extension, 2008); Edgar Nelson Transeau, "The Prairie Peninsula," *Ecology* 16, no. 3 (July 1, 1935): 423–37, https://doi.org/10.2307/1930078; Roger C. Anderson, "The Eastern Prairie-Forest Transition—An Overview," *Proceedings of the Eighth North American Prairie Conference. Western Michigan University, Kalamazoo, MI*, ed. Richard Brewer (Kalamazoo: Department of Biology, Western Michigan University, 1983), 86–92; Roger C. Anderson, "Evolution and Origin of the Central Grassland of North America: Climate, Fire, and Mammalian Grazers," *Journal of the Torrey Botanical Society* 133, no. 4 (Oct. 1, 2006): 626–47, https://doi.org/10.3159/1095-5674(2006)133[626:EAOOTC]2.0.CO;2; John Madson, *Where the Sky Began: Land of the Tallgrass Prairie* (Iowa City: University of Iowa Press,

2004); James Claude Malin, *The Grassland of North America, Prolegomena to Its History* (Lawrence, KS: self-published, 1947).

13. Morrissey, "Power of the Ecotone"; Robert Michael Morrissey, "Bison Algonquians: Cycles of Violence and Exploitation in the Mississippi Valley Borderlands," *Early American Studies: An Interdisciplinary Journal* 13, no. 2 (2015): 309–40.

14. E. C. Pielou, *After the Ice: The Return of Life to Glaciated North America* (Chicago: University of Chicago Press, 1991), 2; Betty Flanders Thomson, *The Shaping of America's Heartland: The Landscape of the Middle West* (Boston: Houghton Mifflin, 1977), 182. The concept of prairies as "wilderness" is still embraced by some writers. See for instance Joel Greenberg's excellent recent collection of nature writing from the Chicago area, *Of Prairie, Woods, and Water: Two Centuries of Chicago Nature Writing* (Chicago: University of Chicago Press, 2008), 119.

15. For a great overview on several disciplines that can serve historians interested in vegetation change over deep time, see Dave Egan and Evelyn A. Howell, eds., *The Historical Ecology Handbook: A Restorationist's Guide to Reference Ecosystems*, Science and Practice of Ecological Restoration (Washington, DC: Island Press, 2005). On the recognition of change as the important constant in the ecological past, see Curt Meine's forward to *Historical Ecology Handbook*. For a primer on palynology, see chap. 9, "Palynology: An Important Tool for Discovering Historic Ecosystems," by Owen K. Davis. A useful introduction to tree ring studies is chap. 8, "Using Dendrochronology to Reconstruct the History of Forest and Woodland Ecosystems," by Kurt Kipfmueller and Thomas Swetnam. A great discussion about how historians might put all of this to use is Samuel E. Munoz et al., "Defining the Spatial Patterns of Historical Land Use Associated with the Indigenous Societies of Eastern North America," *Journal of Biogeography* 41, no. 12 (Dec. 1, 2014): 2195–210, https://doi.org/10.1111/jbi.12386.

16. Anderson, "Eastern Prairie-Forest Transition"; David M. Nelson et al., "The Influence of Aridity and Fire on Holocene Prairie Communities in the Eastern Prairie Peninsula," *Ecology* 87, no. 10 (2006): 2523–36; John W. Williams, Bryan Shuman, and Patrick J. Bartlein, "Rapid Responses of the Prairie-Forest Ecotone to Early Holocene Aridity in Mid-Continental North America," *Global and Planetary Change* 66, no. 3–4 (Apr. 2009): 195–207, https://doi.org/10.1016/j.gloplacha.2008.10.012; Daniel I. Axelrod, "Rise of the Grassland Biome, Central North America," *Botanical Review* 51, no. 2 (1985): 163–201; Joseph J. Williams et al., "Ecosystem Development Following Deglaciation: A New Sedimentary Record from Devils Lake, Wisconsin, USA," *Quaternary Science Reviews* 125 (Oct. 1, 2015): 131–43, https://doi.org/10.1016/j.quascirev.2015.08.009.

17. Anderson, "Evolution and Origin of the Central Grassland," 631.

18. Margaret B. Davis, "Quaternary History of Deciduous Forests of Eastern North America and Europe," *Annals of the Missouri Botanical Garden* 70, no. 3 (1983): 550, https://doi.org/10.2307/2992086.

19. Nelson et al., "The Influence of Aridity and Fire."

20. Axelrod, "Rise of the Grassland Biome," 166.

21. Scott L. Collins and Linda L. Wallace, eds., *Fire in North American Tallgrass Prairies*, 1st ed. (Norman: University of Oklahoma Press, 1990); Roger C. Anderson, "The Historic Role of Fire in the North American Grassland," in Collins and Wallace, *Fire in North American Tallgrass Prairies*, 8–18; Omer C. Stewart, *Forgotten Fires: Native Americans and the Transient Wilderness*, ed. Henry T. Lewis and M. Kat Anderson (Norman: University of Oklahoma Press, 2009).

22. Anderson, "Historic Role of Fire," 18.

23. Much recent research in ecology and environmental history has been dedicated to exploring the profound ways that Native peoples impacted and shaped landscapes in the precontact era. For examples that inspired my own exploration of the tallgrass prairies, see Kat Anderson and Thomas C. Blackburn, eds., *Before the Wilderness: Environmental Management by Native Californians*, Ballena Press Anthropological Papers, no. 40 (Menlo Park, CA: Ballena Press, 1993); Kat Anderson, *Tending the Wild: Native American Knowledge and the Management of California's Natural Resources* (Berkeley: University of California Press, 2005). For a contrary view, see Waldo Rudolph Wedel, "The Central North American Grassland: Man-Made or Natural?," *Studies in Human Ecology* (1957): 39–69.

24. Egan and Howell, *Historical Ecology Handbook*.

25. Louis Vivier, "Letter from Father Vivier to Another Jesuit, June 8, 1750," in *The Jesuit Relations and Allied Documents: Travels and Explorations of the Jesuit Missionaries in New France, 1610–1791*, ed. Reuben Gold Thwaites, vol. 69 (Cleveland: Burrows Bros. Co., 1896), 207.

26. Claude Dablon and Claude Allouez, "Relation of 1670–1671: Events Attending the Publication of the Faith to the Fire Nation, and to One of the Ilinois Nations," in Thwaites, *Jesuit Relations and Allied Documents*, 55.

27. Mary L. Simon, "Reevaluating the Evidence for Middle Woodland Maize from the Holding Site," *American Antiquity* 82, no. 1 (Jan. 2017): 140–50, https://doi.org/10.1017/aaq.2016.2.

28. Charles C. Mann, *1491: New Revelations of the Americas before Columbus* (New York: Knopf, 2005), 251. Of course, it is an oversimplification to say that Native peoples used the prairie only for hunting. For an excellent and fascinating recent look at one Native group's diverse activities and knowledge relating to prairie ecosystems, from the contact era through more modern times, see Michael Paul Gonella, "Myaamia Ethnobotany," Miami University, 2007, https://etd.ohiolink.edu/ap/10?0::NO:10:P10_ACCESSION_NUM:miami1184770633.

29. Nicolas Perrot, "Memoir on the Manners, Customs, and Religion of the Savages of North America," in *The Indian Tribes of the Upper Mississippi Valley and Region of the Great Lakes*, ed. Emma Helen Blair (Lincoln: University of Nebraska Press, 1996), 121–26; Louis Hennepin, *A New Discovery of a Vast Country in America, by Father Louis Hennepin. Reprinted from the Second London Issue of 1698*, ed. Reuben Gold Thwaites, vol. 1 (Chicago: A. C. McClurg, 1903), 149, www .americanjourneys.org/aj-124a/.

30. Alan K. Knapp, ed., *Grassland Dynamics: Long-Term Ecological Research in Tallgrass Prairie*, Long-Term Ecological Research Network Series 1 (New York: Oxford University Press, 1998); Alan K. Knapp et al., "The Keystone Role of Bison in North American Tallgrass Prairie," *BioScience* 49, no. 1 (Jan. 1, 1999): 39–50; N. Thompson Hobbs et al., "Fire and Grazing in the Tallgrass Prairie: Contingent Effects on Nitrogen Budgets," *Ecology* 72, no. 4 (Aug. 1, 1991): 1374–82, https:// doi.org/10.2307/1941109; S. D. Fuhlendorf and D. M. Engle, "Application of the Fire-Grazing Interaction to Restore a Shifting Mosaic on Tallgrass Prairie," *Journal of Applied Ecology* 41, no. 4 (Aug. 1, 2004): 604–14, https://doi.org/10.1111/ j.0021-8901.2004.00937.x; David C. Hartnett, Karen R. Hickman, and Laura E. Fischer Walter, "Effects of Bison Grazing, Fire, and Topography on Floristic Diversity in Tall-Grass Prairie," *Journal of Range Management* 49, no. 5 (1996): 413–20.

31. John White, *A Review of the American Bison in Illinois with an Emphasis on Historical Accounts* (Urbana, IL: Nature Conservancy, 1996).

32. Aldo Leopold, *Game Management* (New York: C. Scribner's Sons, 1933), xviii. My thinking is strongly influenced by Leopold, especially his classic definitions of the "mechanism of game management," chap. 2.

33. For excellent histories that explore human-animal relations in the premodern world, see Harriet Ritvo, *Noble Cows and Hybrid Zebras: Essays on Animals and History* (Charlottesville: University of Virginia Press, 2010); Dorothee Brantz, ed., *Beastly Natures: Animals, Humans, and the Study of History* (Charlottesville: University of Virginia Press, 2010); Virginia DeJohn Anderson, *Creatures of Empire: How Domestic Animals Transformed Early America* (Oxford: Oxford University Press, 2004).

34. For an excellent discussion of this dynamic, see Stewart, *Forgotten Fires*, 120–22. See also Carl O. Sauer, "Grassland Climax, Fire, and Man," *Journal of Range Management* 3, no. 1 (1950): 20, https://doi.org/10.2307/3894702.

35. Stewart, *Forgotten Fires*, 118.

36. Stewart, *Forgotten Fires*, 114.

37. For an excellent and readable discussion of the history of this scientific consensus, see Charles Mann's discussion of "Holmberg's Mistake," in the early chapters of Charles C. Mann, *1493: Uncovering the New World Columbus Created* (New York: Knopf, 2011). Recent work in paleoecology has advanced the reconstruction of Indian environments considerably. See Munoz et al., "Defining the Spatial Patterns of Historical Land."

38. William Cronon, *Changes in the Land: Indians, Colonists, and the Ecology of New England* (New York: Hill and Wang, 1983).

39. The successes of this new transnational early American history begin with Richard White, *The Middle Ground: Indians, Empires, and Republics in the Great Lakes Region, 1650-1815* (New York: Cambridge University Press, 1991). For recent commentary about the promise of continental and transnational early American history, see Allan Greer, "National, Transnational, and Hypernational Historiographies: New France Meets Early American History," *Canadian Historical Review* 91, no. 4 (Dec. 2010): 695-724; Paul W. Mapp, "Atlantic History from Imperial, Continental, and Pacific Perspectives," *William and Mary Quarterly* 63, no. 4 (Oct. 1, 2006): 713-24; Eric Hinderaker and Rebecca Horn, "Territorial Crossings: Histories and Historiographies of the Early Americas," *William and Mary Quarterly* 67, no. 3 (July 1, 2010): 395-432; Joyce E. Chaplin, "Expansion and Exceptionalism in Early American History," *Journal of American History* 89, no. 4 (Mar. 2003): 1431-55; Alan Taylor, "Squaring the Circles: The Reach of Colonial America," in *American History Now*, ed. Eric Foner and Lisa McGirr (Philadelphia: Temple University Press, 2011), 3-23, http://public .eblib.com/EBLPublic/PublicView.do?ptiID=714507. The tendency to locate the "beginnings" of midwestern history in the nineteenth century is a good illustration of Cronon's point about the significance of beginnings and end points in historical narrative. See Cronon, "A Place for Stories," 1356.

40. A good summary of key writings in the development of the Anthropocene concept is Robert S. Emmett and David E. Nye, *The Environmental Humanities: A Critical Introduction* (Cambridge, MA: MIT Press, 2017), 95-104. Although he does not deploy the concept of Anthropocene specifically, John McNeill argues forcefully for a stark break in human history in the twentieth century, characterized most importantly by increased energy intensity and resource exploitation. See John R. McNeill, *Something New under the Sun: An Environmental History of the Twentieth Century World* (New York: Norton, 2000).

41. My thinking here is influenced by James Scott, who advocates for thinking in terms of a "thin anthropocene" beginning four hundred thousand years ago with the human mastery of fire, and a later "thick anthropocene" in more recent times. James C. Scott, *Against the Grain: A Deep History of the Earliest States* (New Haven, CT: Yale University Press, 2017), 2.

42. For an example of restoration ecology's skepticism of the idea of a "baseline" ecosystem prior to human influence, see Egan and Howell, *Historical Ecology Handbook*, xvi. A more subtle approach is to identify differential anthropogenic "rates of change," a practice which does not imagine an untouched pristine state. See Owen K. Davis, "Palynology: An Important Tool for Discovering Historic Ecosystems," in Egan and Howell, *Historical Ecology Handbook*, 250. Recognizing the deep history of human agency in shaping the prairie should also inform discussion not just about baselines but about the very

meaning of restorations and the hybrid status of restored landscapes. For a great recent exploration in comic book form, see Kozik, "Knowing Prairies: An Essay in Comic Form," *Edge Effects* (blog), June 13, 2017, http://edgeeffects.net/uw -arboretum-prairie/.

CHAPTER 2: CHOLERA AND THE EVOLUTION OF EARLY CHICAGO

1. Isaac Rawlings, *The Rise and Fall of Disease in Illinois* (Springfield, IL: State Department of Public Health, 1927), 43. *Pandemic* is a disease outbreak over a wide geographic area, affecting a high proportion of population; *epidemic* refers to a sudden increase in the cases of disease in a given location. See *Merriam-Webster Dictionary*, s.v. "pandemic" and "epidemic," https://www.merriam -webster.com/dictionary.

2. Charles Rosenberg argues, "Cholera, a scourge of the sinful to many Americans in 1832, had, by 1866, become the consequence of remedial faults in sanitation"; Rosenberg, *The Cholera Years: The United States in 1832, 1848, and 1866* (Chicago: University of Chicago Press, 1962), 5. Catherine McNeur countered that even in 1832 "New Yorkers were aware of the environmental causes of disease"; McNeur, *Taming Manhattan: Environmental Battles in the Antebellum City* (Cambridge, MA: Harvard University Press, 2014), 268n31.

3. See Mary A. Penrose's reminiscence in A. T. Andreas, *History of Chicago, from Its Earliest Period to the Present, Volume I Ending with the Year 1857* (Chicago: A. T. Andreas, 1884), 119; and Captain Augustus Walker's reminiscence in the "Early Days on the Lakes," in *Publications of the Buffalo Historical Society*, ed. Frank H. Severance, vol. 5 (Buffalo, NY: Buffalo Historical Society, 1862), 313.

4. Joel Greenberg, *A Natural History of the Chicago Region* (Chicago: University of Chicago Press, 2002), 177–80; William Cronon, *Nature's Metropolis: Chicago and the Great West* (New York: W. W. Norton, 1991), 23–28; and Louis P. Cain, *Sanitation Strategy for a Lakefront Metropolis: The Case of Chicago* (De Kalb: Northern Illinois University Press, 1978), 5.

5. Others were at Mackinac, Green Bay, and Prairie du Chien.

6. Colbee C. Benton, *A Visitor to Chicago in Indian Days: Journal to the Far-Off West* (Chicago: Caxton Club of Chicago, 1957), 79–80.

7. Richard White, *The Middle Ground: Indians, Empires, and Republics in the Great Lakes Region, 1650–1815* (Cambridge: Cambridge University Press, 1991), chap. 3.

8. John Haeger, "The American Fur Company and Chicago of 1812–1835," *Journal of the Illinois State Historical Society* 61 (June 1868): 132–33; Robert Stuart to John Kinzie, Sept. 11, 1825, American Fur Company Papers, Chicago History Museum (hereafter CHM); and Ann Durkin Keating, *Rising Up from Indian Country: The Battle of Fort Dearborn and the Birth of Chicago* (Chicago: University of Chicago Press, 2012), 219–21.

9. Kerry A. Trask, *Black Hawk: The Battle for the Heart of America* (New York: Holt, 2006), 9.

10. See Cronon, *Nature's Metropolis*, 76–78.

11. Cronon, *Nature's Metropolis*, 102; and Paul Wallace Gates, *History of Public Land Law Development* (Washington, DC: Government Printing Office, 1968), 63–65.

12. Trask, *Black Hawk*; and R. David Edmunds, *The Potawatomis: Keepers of the Fire* (Norman: University of Oklahoma Press, 1978), 250–52.

13. John W. Hall, *Uncommon Defense: Indian Allies in the Black Hawk War* (Cambridge: Cambridge University Press, 2009), 8; and Keating, *Rising Up from Indian Country*, 223–34.

14. William Hinkling, *Caldwell and Shabonee* (Chicago, 1877), in Draper 9YY, Microfilm Reel 119, Wisconsin Historical Society.

15. See Milo M. Quaife, *Chicago and the Old Northwest* (Urbana: University of Illinois Press, 2001), 327, for an account of Rev. Stephen R. Beggs, who had settled at Plainfield and fled to Fort Dearborn.

16. Lemuel Bryant Travel Journal, 1832, n.p., CHM.

17. Rosenberg, *Cholera Years*, 3; William K. Beatty, "When Cholera Scourged Chicago," *Chicago History* 11 (Spring 1982): 2–13; and Nancy Tomes, *The Gospel of Germs: Men, Women and the Microbe in American Life* (Cambridge, MA: Harvard University Press, 1998), 26.

18. Guillame Constantin de Magny and Rita R. Colwell, "Cholera and Climate: A Demonstrated Relationship," *Transactions of the American Clinical and Climatological Association* 120 (2009): 119–28, https://www.ncbi.nlm.nih.gov/pmc/articles/PMC2744514/.

19. Walker, "Early Days on the Lakes," 310–11. Also see "Chicago in 1832," *Chicago Tribune*, Feb. 8, 1861, 2.

20. Samuel Manning Welch, *Home History: Recollections of Buffalo during the Decade from 1830 to 1840, or Fifty Years Since* (Buffalo, NY: Peter Paul & Bros., 1890), 266.

21. Augustus Walker, quoted in Andreas, *History of Chicago*, 121.

22. General Winfield Scott as related to John Wentworth, in Quaife, *Chicago and the Old Northwest*, 119–20.

23. Bryant Travel Journal.

24. Andreas, *History of Chicago*, 119; and Rawlings, *Rise and Fall of Disease*, 43–44.

25. Quaife, *Chicago and the Old Northwest*, 335–37.

26. Patrick Wolfe, "Settler Colonialism and the Elimination of the Native," *Journal of Genocide Studies* 8 (Dec. 2006): 387; and Michael Adas, "Settler Colony to Global Hegemon: Integrating the Exceptionalist Narrative of the American Experience into World History," *American Historical Review* 106, no. 4 (Dec. 2001): 1692–720.

27. Francis Paul Prucha, *American Indian Treaties: The History of a Political Anomaly* (Berkeley: University of California Press, 1994), 188–89. Some tribal members, most notably the Pokagon Band, negotiated amendments to the Treaty of Chicago that allowed them to remain on lands in southern Michigan and thus within the realm of Chicago's growing influence. See John M. Low, *Imprints: The Pokagon Band of Potawatomi Indians and the City of Chicago* (East Lansing: Michigan State University Press, 2016), 22–34.

28. *Report of the Board of Health for the City of Chicago for 1867, 1868 and 1869 and a Sanitary History of Chicago from 1833 to 1870* (Chicago: Lakeside Press, 1871), 11–12 (hereafter *Board of Health 1867*); and Bessie Louise Pierce, *As Others See Chicago* (Chicago: University of Chicago Press, 2004), 11.

29. Robin Einhorn, *Property Rules: Political Economy in Chicago, 1833–1872* (Chicago: University of Chicago Press, 1991), 51–52; Cain, *Sanitation Strategy*, 27–30; and Rawlings, *Rise and Fall of Disease*, 101.

30. Rawlings, *Rise and Fall of Disease*, 44.

31. Einhorn, *Property Rules*, 54.

32. Einhorn, *Property Rules*, 99; and John B. Hamilton, "The Epidemics of Chicago," *Bulletin of the Society of Medical History of Chicago* 1 (1911–1916): 73–86. Chicago's approach to municipal services reflected urban patterns across the United States. See Jon Teaford, *The Unheralded Triumph: City Government in America, 1870–1900* (Baltimore: Johns Hopkins University Press, 1984), 217–50.

33. Ann Durkin Keating, *Chicagoland: City and Suburbs in the Railroad Age* (Chicago: University of Chicago Press, 2005), 12–14; and Cronon, *Nature's Metropolis*, 63–85.

34. Clark E. Carr, *The Illini, a Story of the Prairies* (Chicago: McClurg, 1904), 40.

35. *Board of Health 1867*, 18.

36. Thomas Neville Bonner, *Medicine in Chicago, 1850–1950: A Chapter in the Social and Scientific Development of a City*, 2nd ed. (Urbana: University of Illinois Press, 1991), 7–8.

37. John Moses and Joseph Kirkland, *The History of Chicago, Illinois*, vol. 1 (Chicago: Munsell, 1895), 108.

38. *Board of Health 1867*, 22.

39. Sam Bass Warner Jr., *The Private City: Philadelphia in Three Periods of Growth* (Philadelphia: University of Pennsylvania Press, 1968), 3–4. Also see Einhorn, *Property Rules*, 16–17. More recently, Philip J. Ethington, rejects privatism in favor of "republican liberalism" that envisioned a more unitary notion of public good; Ethington, *The Public City: The Political Construction of Urban Life in San Francisco, 1850–1900* (New York: Cambridge University Press, 1994), 8.

40. By 1852 the city physician received reimbursements for services. *Board of Health 1867*, 29.

41. *Board of Health 1867*, 20–23.

42. Constance Bell Webb, *A History of Contagious Disease Care in Chicago before the Great Fire* (Chicago: University of Chicago Press, 1940), 50–53.

43. A smattering of reports survived in Chicago City Council Proceedings Files, 1849–1853, Illinois State Archives, Regional Depository, Northeastern Illinois University Libraries. They are primarily under entries labeled "City Physician's Report on Cholera Cases" or "Reports of Physicians on Cholera Cases," and are mostly for 1849.

44. Thomas Spencer, *An Essay on the Nature of the Epidemic Called Asiatic Cholera* (Albany, NY: Webster and Skinners, 1833), 10–11.

45. Beatty, "When Cholera Scourged Chicago," 8–10.

46. "City Physician's Report on Cholera Cases" or "Reports of Physicians on Cholera Cases," July 10, June 8, August 13, July 13, July 14, June 17, and June 25, 1849, Chicago City Council Proceedings.

47. "City Physician's Report on Cholera Cases," June 29 and June 7, 1849.

48. By 1849, Rosenberg argues, Americans saw a mixture of environmental and personal factors in play, as in the above reports; Rosenberg, *Cholera Years*, 5–7.

49. *Board of Health 1867*, 22, 43. Data on estimated cholera deaths in table 2.1 are drawn from *Board of Health 1867*, 27, 30, 33, 37, 41; and Rawlings, *Rise and Fall of Disease*, 103.

50. Stephanie Coontz, *The Social Origins of Private Life: A History of American Families, 1600–1900* (New York: Verso, 1988), 176.

51. Suellen Hoy, *Good Hearts: Catholic Sisters in Chicago's Past* (Urbana: University of Illinois Press, 2006), 37–40. Also see Act to Incorporate the Mercy Hospital and Mercy Orphan Asylum of Chicago, Feb. 21, 1851, Illinois General Assembly; and Mother Agatha O'Brien, Chicago, to Sister Mary Scholastica, Galena, Feb. 7, 1851, Archives of Sisters of Mercy-Chicago (hereafter ASMC).

52. Mother Agatha O'Brien, Chicago, to Sister Mary Gertrude McGuire, Galena, Nov. 12, 1850, ASMC; [Sister Mary Gabriel O'Brien], *Reminiscences of Seventy Years (1846–1916): Sisters of Mercy–St. Xavier's Chicago* (Chicago: F. J. Ringley, 1916), 51–52; and Kathleen A. Brosnan, "Public Presence, Public Silence: Nuns, Bishops, and the Gendered Space of Early Chicago," *Catholic Historical Review* 90 (July 2004): 485–88.

53. Clare L. McCausland, *Children of Circumstance: A History of the First 125 years (1849–1974) of Chicago Child Care Society* (Chicago: Child Care Society, 1976), 6–10, 12; and Mrs. Charles Gilbert Wheeler, *Annals of the Chicago Orphan Asylum from 1849 to 1892* (Chicago: Board of the Chicago Orphan Asylum, 1892), 16.

54. McCausland, *Children of Circumstance*, 3.

55. Wheeler, *Annals of the Chicago Orphan Asylum*, 21. See Mamie Ruth Davis, "A History of Policies and Methods of Social Work in the Chicago Orphan Asylum" (MA thesis, University of Chicago, 1927); and Kenneth Cmiel, *A Home*

of Another Kind: One Chicago Orphanage and the Tangle of Child Welfare (Chicago: University of Chicago Press, 1992), esp. chap. 1.

56. Ellis S. Chesbrough, *Chicago Sewerage: Report of the Results of Examinations Made in Relation to Sewerage in Several European Cities, in the Winter of 1865* (Chicago: Board of Sewerage Commissioners, 1858), 81.

57. *Board of Health 1867*, 24.

58. Ann Durkin Keating, *Building Chicago: Suburban Developers and the Creation of a Divided Metropolis* (Columbus: Ohio State University Press, 1988), 39–41, 190; and Einhorn, *Property Rules*, 137.

59. Cain, *Sanitary Solution*, 23–30; and Einhorn, *Property Rules*, 138–40.

60. Einhorn, *Property Rules*, 139.

61. Cain, *Sanitary Strategy*, 27–30; Einhorn, *Property Rules*, 139; and Martin V. Melosi, *The Sanitary City: Urban Infrastructure from Colonial Times to the Present* (Baltimore: Johns Hopkins University Press, 2000), 97–98.

62. *Board of Health 1867*, 54.

63. Nathan Davis, quoted in *Board of Health 1867*, 60–61.

64. Bonner, *Medicine in Chicago*, 8.

65. *Board of Health 1867*, 92, 104–6, 111; and N. S. Davis, "Report on the Means of Improving the Sanitary Condition of Chicago," *Chicago Medical Examiner* 6 (Dec. 1865): 705–12, cited in Bonner, *Medicine in Chicago*, 180–81; and Tomes, *Gospel of Germs*, 54.

66. "The Cholera," *Chicago Tribune*, Oct. 15, 1866.

67. *Board of Health 1867*, 108–10, 165 (disinfectants), and 176 (groundwater). See Owen Whooley, *Knowledge in the Time of Cholera: The Struggle over American Medicine in the Nineteenth Century* (Chicago: University of Chicago Press, 2013), 118, 149–52; and Rosenberg, *Cholera Years*, 193–200.

68. "The Health of Chicago," *Chicago Tribune*, Nov. 14, 1866.

69. *Board of Health 1867*, 121–65.

70. Theodore J. Karamanski, *Rally Round the Flag: Chicago and the Civil War* ((Lanham, MD: Rowman & Littlefield, 2006), 93–96. See also Peggy Tuck Sinko, "Jane Hoge," in *Women Building Chicago: A Biographical Dictionary*, ed. Rima Lunin Shultz and Adele Hast (Bloomington: Indiana University Press, 2001), 398–99; and Mary Livermore, *My Story of the War* (Hartford, CT: A. D. Worthington, 1889).

71. Harold M. Hyman, *A More Perfect Union: The Impact of the Civil War and Reconstruction on the Constitution* (New York: Alfred Knopf, 1973), 320.

72. John B. Jentz and Richard Schneirov, *Chicago in the Age of Capital: Class, Politics, and Democracy during the Civil War and Reconstruction* (Urbana: University of Illinois Press, 2012), 13.

73. Einhorn, *Property Rules*, 16–17; and Warner, *Private City*, 3–4. Neither Einhorn nor Warner grapple with the gender implications to any great extent.

74. Walter Nugent, "Demography," *The Encyclopedia of Chicago*, ed. James R. Grossman, Ann Keating, and Janice L. Reiff (Chicago: University of Chicago Press, 2004), 234.

75. Prudential Insurance Company, "Crude Death Rate, 1844–1920," in Grossman et al., *Encyclopedia of Chicago*, 235.

CHAPTER 3: ANIMALS AT WORK IN INDUSTRIALIZING CHICAGO

1. Karen Sawislak, *Smoldering City: Chicagoans and the Great Fire, 1871–1874* (Chicago: University of Chicago Press, 1995), 1–2; Richard F. Bales, *The Great Chicago Fire and the Myth of Mrs. O'Leary's Cow* (Jefferson, NC: McFarland, 2002), 3–6.

2. Sawislak, *Smoldering City*, 43–44; Bales, *Great Chicago Fire*, 60.

3. Many species of wild animals also lived in Chicago and its environs, although the relationship between wildlife and humans was not nearly as meaningful as that with domestic animals during this period. For more on the wildlife native to the Chicagoland area, see Joel Greenberg, *A Natural History of the Chicago Region* (Chicago: University of Chicago Press, 2002).

4. U.S. Bureau of the Census, "Population of Cities Having 25,000 Inhabitants or More in 1900," Census Bulletin no. 11 (Washington, DC: Government Printing Office, 1900), 4; U.S. Bureau of the Census, *Domestic Animals Not on Farms or Ranges*, Census Bulletin no. 17 (Washington, DC: Government Printing Office, 1900), 8. In 1900 there were approximately 176,000 horses, cattle, mules, hogs, sheep, and goats in Chicago and nearly 1.7 million people, compared to New York, with a population of 3.4 million people and approximately 151,000 animals, equaling around 4 percent of the human population.

5. Ted Steinberg, *Down to Earth: Nature's Role in American History*, 3rd ed. (New York: Oxford University Press, 2013), 158–59; City of Chicago, *The Laws and Ordinances of the City of Chicago Passed in Common Council* (Chicago: The Chicago Democrat, 1837), 3 (hereafter *1837 Ordinances*); City of Chicago, *Municipal Code of Chicago* (Cincinnati: American Legal Publishing Corporation, 2017), http://library.amlegal.com/nxt/gateway.dll/Illinois/chicago_il/municipalco deofchicago?f=templates$fn=default.htm$3.0$vid=amlegal:chicago_il.

6. In his book, *Shock Cities*, Harold Platt borrows the term "shock city" from historian Asa Briggs to describe a city experiencing rapid growth and change, along with attendant social and environmental problems. Harold L. Platt, *Shock Cities: The Environmental Transformation and Reform of Manchester and Chicago* (Chicago: University of Chicago Press, 2005), 15.

7. A growing body of literature that examines working animals in nineteenth- and twentieth-century urban America has emerged over the last decade. For example, Clay McShane and Joel Tarr, *The Horse in the City: Living Machines in the Nineteenth Century* (Baltimore: Johns Hopkins University Press, 2007);

and Ann Norton Greene, *Horses at Work: Harnessing Power in Industrial America* (Cambridge, MA: Harvard University Press, 2008), offer overviews of the uses of horses in cities from a national perspective. More recently, historians have focused their attention on the role of animals in particular cities. For example, see Catherine McNeur, *Taming Manhattan: Environmental Battles in the Antebellum City* (Cambridge, MA: Harvard University Press, 2014); and Frederick Brown, *The City Is More Than Human: An Animal History of Seattle* (Seattle: University of Washington Press, 2016). Historians of animals in Chicago have primarily confined their studies to animals in the city's stockyard complex. For example, see Louise Carroll Wade, *Chicago's Pride: The Stockyards, Packingtown, and Environs in the Nineteenth Century* (Urbana: University of Illinois Press, 1987); William Cronon, *Nature's Metropolis: Chicago and the Great West* (New York: W. W. Norton, 1991); and Dominic Pacyga, *Slaughterhouse: Chicago's Union Stock Yard and the World It Made* (Chicago: University of Chicago Press, 2015).

8. U.S. Bureau of the Census, "Number of Specified Domestic Animals in Cities over 25,000," in *Census Reports*, vol. 5, *Agriculture*, part 1, *Farms, Live Stock, and Animal Products* (Washington, DC: Government Printing Office, 1902), 580, table 41.

9. City of Chicago, *The Charter and Ordinances of the City of Chicago (to Sept. 15, 1856, inclusive) Together with Acts of the General Assembly Relating to City, and Other Miscellaneous Acts, with an Appendix* (Chicago: D. B. Cooke, 1856), 327–29 (hereafter *1856 Ordinances*); City of Chicago, *Ordinances of the City of Chicago* (Chicago: City of Chicago, 1881), 380, 343 (hereafter *1881 Ordinances*).

10. U.S. Bureau of the Census, *Domestic Animals Not on Farms or Ranges*, 8; Department of Health, City of Chicago, *Report and Handbook of the Department of Health of the City of Chicago for the Years 1911 to 1918 Inclusive* (Chicago: City of Chicago, 1919), 937.

11. Henry E. Alvord and Raymond A. Person, *The Milk Supply of Two Hundred Cities and Towns* (Washington, DC: Government Printing Office, 1903), 26–27; Department of Health, City of Chicago, *Report of the Department of Health of the City of Chicago for the Years 1907, 1908, 1909, 1910* (Chicago: City of Chicago, 1911), 86.

12. Perry Duis, *Challenging Chicago: Coping with Everyday Life, 1837–1920* (Urbana: University of Illinois Press, 1998), 136; "To Prevent Cholera," *Chicago Daily Tribune*, Apr. 11, 1888.

13. For more on Progressive Era efforts to clean up urban milk supplies, see Kendra Smith-Howard, *Pure and Modern Milk: An Environmental History since 1900* (New York: Oxford University Press, 2014).

14. Department of Health, *Report of the Department of Health . . . 1907, 1908, 1909, 1910*, 85–87; "Wants City Cows for City Babies," *Chicago Daily Tribune*, Feb. 29, 1908.

NOTES TO PAGES 41–43

15. "The City's Milk Supply," *Chicago Daily Tribune*, Mar. 2, 1908; Alan Czaplicki, "'Pure Milk Is Better Than Purified Milk': Pasteurization and Milk Purity in Chicago, 1908–1916," *Social Science History* 31 (Fall 2007): 418.

16. Pacyga, *Slaughterhouse*, 4; Cronon, *Nature's Metropolis*, 223; Chris Philo, "Animals, Geography, and the City: Notes on Inclusions and Exclusions," in *Animal Geographies: Place, Politics, and Identity in the Nature-Culture Borderlands*, ed. Jennifer R. Wolch and Jody Emel (New York: Verso, 1998), 65.

17. Pacyga, *Slaughterhouse*, 37, 93; Chicago Union Stock Yard, *Chicago Union Stock Yard: An Illustrated Description of the World's Greatest Livestock Market* (Chicago: Union Stock Yard and Transit Company, 1900).

18. *Chicago Union Stock Yard.*

19. *Chicago Union Stock Yard*; Pacyga, *Slaughterhouse*, 9–10, 16.

20. *Chicago Union Stock Yard.*

21. Joel Tarr, *The Search for the Ultimate Sink: Urban Pollution in Historical Perspective* (Akron, OH: University of Akron Press, 1996), 324; McShane and Tarr, *Horse in the City*, 188. As early as 1839 Chicago imposed a speed limit of no "faster gait or pace than a common walk"; City of Chicago, *The Revised Charter and Ordinances of the City of Chicago* (Chicago: The Daily Democrat, 1851), 82 (hereafter *1851 Ordinances*).

22. W. K. Everson, "A Week without Horses," *Chicago Daily Tribune*, Nov. 10, 1872.

23. *Chicago Union Stock Yard.*

24. *Chicago Union Stock Yard*; McShane and Tarr, *Horse in the City*, 11.

25. Edwin Griswold Nourse, *The Chicago Produce Market* (New York: Houghton Mifflin, 1918), 23; "Chicago as the Greatest Horse Market in the World," *Chicago Daily Tribune*, June 17, 1900.

26. Tarr, *Search for the Ultimate Sink*, 324.

27. Chicago Transit Authority, *Historical Information, 1859–1965* (Chicago: Chicago Transit Authority, 1966), v.

28. Robert M. Buck, "Motor Sounds Doom of the Fire Horse," *Chicago Daily Tribune*, May 21, 1911.

29. McShane and Tarr, *Horse in the City*, 111.

30. "Mandels Sells Last Horse," *Chicago Daily Tribune*, Jan. 11, 1916; "Autos Replace Last of Tribune's Horses," *Chicago Daily Tribune*, June 13, 1928; Frank Ridgeway, "Farm and Garden: City Work Needs Both Motor Trucks and Draft Teams," *Chicago Daily Tribune*, July 24, 1927.

31. Robert Hunter, *Tenement Conditions in Chicago* (Chicago: City Homes Association, 1901), 128.

32. Nourse, *Chicago Produce Market*, 23–24.

33. McShane and Tarr, *Horse in the City*, 54; Anti-Cruelty Society, *The Fourteenth Annual Report of the Anti-Cruelty Society of Chicago* (Chicago: Anti-Cruelty Society, 1913), 17–18, 27. Numerous examples of accidents caused by

spooked horses are detailed in the *Chicago Daily Tribune* at the turn of the century; for example, "Police Busy Chasing Runaway Horses," *Chicago Daily Tribune*, June 8, 1891; "Horse Stirs a Panic: Furious Runaway Causes a Series of Stampedes," *Chicago Daily Tribune*, Jan. 28, 1898; and "Runaway in Crowded Street: Horse Causes Terror in State Street," *Chicago Daily Tribune*, May 17, 1903.

34. *1837 Ordinances*, 3.

35. *1881 Ordinances*, 357–58.

36. *1856 Ordinances*, 169, 300.

37. Andria Pooley-Ebert, "Species Agency: A Comparative Study of Horse-Human Relationships in Chicago and Rural Illinois," in *The Historical Animal*, ed. Susan Nance (Syracuse, NY: Syracuse University Press, 2015), 159–60.

38. Pooley-Ebert, "Species Agency," 155, 157–58.

39. McShane and Tarr, *Horse in the City*, 46–47.

40. Pooley-Ebert, "Species Agency," 156.

41. Walter Nugent, "Demography: Chicago as a Modern World City," in *The Electronic Encyclopedia of Chicago* (Chicago: Chicago Historical Society, 2005), www.encyclopedia.chicagohistory.org/pages/962.html; Homer Hoyt, *One Hundred Years of Land Values in Chicago* (New York: Arno Press, 1933), 483.

42. Gottfried Koehler, *Annals of Health and Sanitation in Chicago* (Chicago: Department of Health, 1919), 1499; Martin Melosi, *Garbage in the Cities: Refuse Reform and the Environment*, rev. ed. (Pittsburgh: University of Pittsburgh Press, 2005), 20.

43. Tarr, *Search for the Ultimate Sink*, 11.

44. Department of Health, City of Chicago, *Annual Report of the Department of Health for the City of Chicago for the Year Ended December 31, 1893* (Chicago: Cameron, Amberg, 1894), 80.

45. Platt, *Shock Cities*, 345; Dawn Day Biehler, "Flies, Manure, and Window Screens: Medical Entomology and Environmental Reform in Early-Twentieth-Century US Cities," *Journal of Historical Geography* 36 (2010): 72–73; Melanie A. Kiechle, *Smell Detectives: An Olfactory History of Nineteenth-Century Urban America* (Seattle: University of Washington Press, 2017), 235–36.

46. Hunter, *Tenement Conditions in Chicago*, 11–17.

47. *1881 Ordinances*, 333.

48. Sophonisba P. Breckinridge and Edith Abbott, "Chicago Housing Conditions, IV: The West Side Revisited," *American Journal of Sociology* 17, no. 1 (July 1911): 11–12; Grace Peloubet Norton, "Chicago Housing Conditions, VII: Two Italian Districts," *American Journal of Sociology* 18, no. 4 (Jan. 1913): 515.

49. Sophonisba P. Breckinridge and Edith Abbott, "Chicago Housing Conditions, V: South Chicago at the Gates of the Steel Mills," *American Journal of Sociology* 17, no. 2 (Sept. 1911): 173–74; Natalie Walker, "Chicago Housing

Conditions, X, Greeks and Italians in the Neighborhood of Hull House," *American Journal of Sociology* 21, no. 2 (Nov. 1915): 313–14.

50. Grace Abbott, "A Study of the Greeks in Chicago," *American Journal of Sociology* 15, no. 3 (Nov. 1909): 389–90; Milton B. Hunt, "The Housing of Non-family Groups of Men in Chicago," *American Journal of Sociology* 16, no. 2 (Sept. 1910): 156, 160.

51. Breckinridge and Abbott, "Chicago Housing Conditions, V: South Chicago," 173; Howard Wright Marshall, "The Pelster Housebarn: Endurance of Germanic Architecture on the Midwestern Frontier," *Material Culture* 18 (Summer 1986): 67.

52. *1881 Ordinances*, 343.

53. *1849 Ordinances*, 92.

54. Hunter, *Tenement Conditions in Chicago*, 131–32.

55. Hunter, *Tenement Conditions in Chicago*, 132.

56. "Council Passes Tenement Rule," *Chicago Daily Tribune*, Dec. 18, 1902; City of Chicago, *The Revised Municipal Code of Chicago of 1905* (Rochester, NY: The Lawyers' Co-Operative, 1905), 348–52 (hereafter *1905 Ordinances*).

57. Department of Health, City of Chicago, *Report of the Department of Health of the City of Chicago for the Years 1919, 1920 and 1921* (Chicago: City of Chicago, 1923), 251; Biehler, "Flies, Manure, and Window Screens," 76–78.

58. Edith Abbott and Sophonisba P. Breckinridge, *The Tenements of Chicago, 1908–1935* (Chicago: University of Chicago Press, 1936), 477.

59. Department of Health, *Report of the Department of Health . . . 1907, 1908, 1909, 1910*, 86; U.S. Bureau of the Census, "Live Stock on Farms and Elsewhere," in *Fourteenth Census*, vol. 5, *Agriculture* (Washington, DC: Government Printing Office, 1922), 622, table 80; Hoyt, 485.

60. Abbott and Breckinridge, *Tenements of Chicago*, 109.

61. *1905 Ordinances*, 365–66, 482.

62. Susan D. Jones, *Valuing Animals: Veterinarians and Their Patients in Modern America* (Baltimore: Johns Hopkins University Press, 2003), 21.

63. This law remained in the city code until 1975. At that time, the ordinance relating to animals was repealed and rewritten, and included neither a minimum space requirement for cattle and pigs, nor a ban on owning any type of livestock, presumably because such animals had followed the farmers out of the city as the population grew and property values increased. City of Chicago, *Municipal Code of Chicago Containing the General Ordinances of the City of Chicago* (Chicago: Index Publishing, 1984), 1058.

PART II: A FRESHWATER CITY

1. Margaret Fuller, quoted in Bessie Louise Pierce, *As Others See Chicago: Impressions of Visitors, 1673–1933* (Chicago: University of Chicago Press, 1933), 96–97.

2. Willa Cather, *The Professor's House* (New York: Vintage, 1990), 20.

3. William Barnett's Introduction to this volume also speaks to the lake as a natural phenomenon.

4. Andrew Isenberg, "Introduction: A New Environmental History," in Isenberg, *The Oxford Handbook of Environmental History* (New York: Oxford University Press, 2014), 7.

5. Paul Sutter, "The World with Us: The State of American Environmental History," *Journal of American History* 100, no. 1 (June 2013): 96.

CHAPTER 4: AN INLAND SEA?

1. William H. Keating, *Narrative of an Expedition to the Source of St. Peter's River* (Philadelphia: H. C. Carey, 1824), 166–67; "The City Seal," Chicago Facts, Chicago Public Library, http://www.chipublib.org/chicago-facts/ (accessed June 2016).

2. For the importance of canals and environmental planning in Chicago history, see James William Putnam, *The Illinois and Michigan Canal: An Economic History* (Chicago: University of Chicago Press, 1918); Libby Hill, *The Chicago River: A Natural and Unnatural History* (Chicago: Lake Claremont Press, 2000); Louis P. Cain, *Sanitation Strategy for a Lakefront Metropolis: The Case of Chicago* (De Kalb: Northern Illinois University Press, 1978); Richard Lanyon, *Building the Canal to Save Chicago* (Chicago: Lake Claremont Press, 2012); William Cronon, *Nature's Metropolis: Chicago and the Great West* (New York: Norton, 1991). For more on environmental history of Chicago and Lake Michigan, see Harold Platt, *Shock Cities: The Environmental Transformation and Reform of Manchester and Chicago* (Chicago: University of Chicago Press, 2005); Margaret Beattie Bogue, *Fishing the Great Lakes: An Environmental History, 1783–1933* (Madison: University of Wisconsin Press, 2000); Terrence Kehoe, *Cleaning Up the Great Lakes: From Cooperation to Confrontation* (De Kalb: Northern Illinois University Press, 1997).

3. Lake Michigan waters respond to the moon and sun in tidal action; however, they are usually limited to a few centimeters.

4. W. Jeffrey Bolster, "Opportunities in Marine Environmental History," *Environmental History Review* 11, no. 3 (July 2006): 567–97; Louis Hennepin, *A Description of Louisiana*, trans. John G. Shea (New York: John G. Shea, 1880), 95–96; James Fennimore Cooper, *The Pathfinder, or, The Inland Sea* (New York: Sully and Kleinteich, 1876), 120–21; Francis, Count de Castelnau, *Vues et souvenirs de l'Amerique du Nord* (1842), quoted in Ken Wardius and Barb Wardius, *Wisconsin Lighthouses: A Photographic and Historical Guide* (Madison: Wisconsin Historical Society Press, 2013), 71; Herman Melville, *Moby Dick*, quoted in William Ratigan, *Great Lakes Shipwrecks & Survivals* (Grand Rapids, MI: William B. Eerdmans, 1960), 13; Rudyard Kipling, *Letters of Travel, 1893–1913* (New York: Doubleday, 1920), 159.

NOTES TO PAGES 54-57

5. John W. Larson, *Those Army Engineers: A History of the Chicago District of the U.S. Army Corps of Engineers* (Chicago: Chicago District, U.S. Army Corps of Engineers, 1979), 13-35; Diary of Captain Morris Sleight, June 30, 1834, typescript "Excerpts from Capt. Sleights Letters and Diaries," Manuscript Department, Chicago History Museum; Ralph Gordon Plumb, "History of Navigation on the Great Lakes," Hearings of the Commerce of the Great Lakes, Committee on Railways and Canals (Washington, DC: Government Printing Office, 1911), 49.

6. "Democratic Party Platform, 1840," May 6, 1840, American Presidency Project, University of California, Santa Barbara, http://www.presidency.ucsb.edu/ws/index.php?pid=29572; *Buffalo Commercial Advertiser*, May 13, 1840; "Gale on Lake Michigan," *Cleveland Daily Herald*, May 13, 1840; J. B. Mansfield, *History of the Great Lakes*, vol. 1 (Chicago: J. H. Beers, 1899), 634; Larson, *Those Army Engineers*, 54; Isaac Stephenson, *Recollections of a Long Life, 1829-1915* (Chicago: privately printed, 1915), 93; Ralph Gordon Plumb, "History of Navigation on the Great Lakes," *Hearings of the Commerce of the Great Lakes, Committee on Railways and Canals* (Washington, DC: Government Printing Office, 1911), 36.

7. Stephenson, *Recollections of a Long Life*, 93; Theodore J. Karamanski, *Schooner Passage: Sailing Ships and the Lake Michigan Frontier* (Detroit: Wayne State University Press, 2000), 29, 53; Gilbert J. Garraghan, "Early Catholicity in Chicago, 1673-1843, Part II," *Illinois Catholic Historical Review* 1, no. 2 (1918): 160.

8. Joanne Grossman and Theodore J. Karamanski, *Historic Lighthouses and Navigational Aids of the Illinois Shore of Lake Michigan* (Chicago: Chicago Maritime Society and Loyola University Public History Program, 1989), 32; Dennis L. Noble, *Lighthouses & Keepers: The U.S. Lighthouse Service and Its Legacy* (Annapolis, MD: Naval Institute Press, 1997), 8-9; Harriet Martineau, *Society in America*, vol. 2 (London: Saunders and Otley, 1837), 6; W. H. Harding, Address to the Houghton County Historical Society, n.d., Hearding Journal and Papers, Bentley Historical Library, University of Michigan, Ann Arbor. For more on the U.S. Lake Survey, see Arthur M. Woodford, *Charting the Inland Seas: A History of the U.S. Lake Survey* (Detroit: Wayne State University Press, 1991).

9. Gustav Pearson, "Captain George G. Meade and the United States Lake Survey," *Engineer*, Sept./Dec. 2010, 44; Mansfield, *History of the Great Lakes*, 201; Robert W. Merry, *A Country of Vast Design: James K. Polk, the Mexican War, and the Conquest of the American Continent* (New York: Simon & Schuster, 2009), 282-83.

10. *Mississippian* (Jackson), July 23, 1847; *Journal of the Proceedings of the Southwestern Convention: Began and Held at the City of Memphis on the 12th of November 1845* (Memphis: The Southwest Convention, 1845), 17; James L. Barton, *Lake Commerce: Letter to the Hon. Robert M'Clelland, Chairman of the Committee of Commerce, U.S. House of Representatives in Relation to the Value and Importance of the Commerce of the Great Western Lakes* (Buffalo, NY: Jewett,

Thomas, 1846), 17; Stephenson, *Recollections of a Long Life*, 92–94. For more on the role of economic forces in antebellum sectional tensions, see Marc Egnal, *Clash of Extremes: The Economic Origins of the Civil War* (New York: Hill & Wang, 2010).

11. Report of the Milwaukee Harbor Commission, 1842, quoted in William George Bruce, *History of Milwaukee, City and County* (Chicago: S. J. Clarke, 1922), 277–88; Milwaukee Board of Trade, *Annual Report of Commerce, Manufactures, and Improvements for the Year 1855* (Milwaukee: Daily Wisconsin Press, 1855), 7; Millard Fillmore, Second Annual Message to Congress, Dec. 2, 1851, American Presidency Project.

12. William Ratigan, *Great Lakes Shipwrecks & Survivals* (Grand Rapids, MI: Wm. B. Eerdmans, 1977), 39–47.

13. Larson, *Those Army Engineers*, 80–94; Karamanski, *Schooner Passage*, 58.

14. Gustav J. Person, "Captain George G. Meade and the Lake Survey," *Engineer*, Sept.–Dec. 2010, 43–49; "Sketch of Increase A. Lapham, LLD," *Popular Science Monthly*, Apr. 1883, 835–40; Rebecca Robbins Raines, *Getting the Message Through: A Branch History of the U.S. Army Signal Corps* (Washington, DC: U.S. Army Center for Military History, 1996), 46; Paul Taylor, *Orlando M. Poe: Civil War General and Great Lakes Engineer* (Kent, OH: Kent State University Press, 2009), xvii–xviii, 221–31; John W. Larson, *Essayons: A History of the Detroit District, Army Corps of Engineers* (Detroit: U.S. Army Corps of Engineers, 1981), 74–77; Larson, *Those Army Engineers*, 133–51; Karamanski, *Schooner Passage*, 127.

15. Christer Westerdahl, "The Maritime Cultural Landscape," *International Journal of Nautical Archaeology* 21, no. 1 (1992): 5–14; Bradley W. Barr, "Observations on Developing a Maritime Cultural Landscape Approach to Managing US National Marine Sanctuaries," Museum of Underwater Archaeology, http://www.themua.org/collections/files/original/bba1a5ca034192b762b3f3f28f9b90f3.pdf (accessed Oct. 2016); Karamanski, *Schooner Passage*, 127–44.

16. Louis Sullivan, *The Autobiography of an Idea* (New York: Press of the American Institute of Architects, 1924), 196–201, 241, 290.

17. In 1870 the Army Corps of Engineers began work on piers to open the mouth of the Calumet River to commerce, creating a second area harbor south of the city, which later became the main harbor.

18. Willa Cather, *The Professor's House* (New York: Vintage Books, 1973), 30; Carl Sandburg, *Chicago Poems* (New York: Henry Holt, 1916), 8.

19. George W. Hilton, *Lake Michigan Passenger Steamers* (Ford, CA: Ford University Press, 2002), 103–27; Sandburg, *Chicago Poems*, 19.

20. The court reaffirmed that the state of Illinois held permanent title to submerged lands within its borders. *Illinois Central R. Co. v. Illinois*, 146 U.S. 387 (1892).

21. Daniel Bluestone, *Constructing Chicago* (New Haven, CT: Yale University Press, 1993), 186–87; Elliott Flower, "Chicago's Great River-Harbor," *Century*

Magazine, Feb. 1902, 483–92; James P. Barry, *Ships of the Great Lakes* (Holt, MI: Thunder Bay Press, 1996), 172–75.

22. George G. Tunnell, George C. Sikes, Paul J. Goode, and John M. Ewen, *Report to the Mayor and Aldermen of the City of Chicago by the Chicago Harbor Commission* (Chicago: Press of A. G. Adair, 1909), 161, 186, 240.

23. The last major shipwrecks were the SS *Carl Bradley*, lost with most of its crew in 1958, and the *Francisco Morazan*, wrecked off South Manitous Island in 1960. Roughly three hundred recreational swimmers or boaters drown in Lake Michigan every year, similar to the number of passengers and crew who perished in the lake in the antebellum era.

24. Ben Hecht, *A Thousand and One Afternoons in Chicago* (Chicago: Chicago Daily News, 1922), 231–34.

CHAPTER 5: CLEANSING CHICAGO

1. Upton Sinclair, *The Jungle* (New York: Doubleday, 1906), 328.

2. Sinclair, *Jungle*, 328–29.

3. John Ericson, *The Water Supply of Chicago* (Chicago: City of Chicago Bureau of Engineering, 1924), 11.

4. Ericson, *Water Supply of Chicago*, 4. Also see Libby Hill, *The Chicago River: A Natural and Unnatural History* (Chicago: Lake Claremont Press, 2000). Hill offers an analysis of the river's environmental change, both natural and engineered, giving the book a wide thematic and chronological scope. See also Louis P. Cain, *Sanitation Strategy for a Lakefront Metropolis* (De Kalb: University of Northern Illinois Press, 1978). Cain examines advances made in sanitation technology as well as in engineering to trace the changes and continuities in how Chicago's leaders addressed water pollution.

5. Kay J. Carr and Michael P. Conzen, *The Illinois and Michigan Canal National Heritage Corridor: A Guide to Its History and Sources* (De Kalb: University of Northern Illinois Press, 1988), 3–5. The historical geographer Michael Conzen documents the development of the Illinois and Michigan Canal and the environmental conditions that facilitated its success. This source provides primary resources concerning the development of both the I&M and the city of Chicago. The work organizes various maps, charts, newspapers, and other materials relevant to the subject.

6. Harold Platt, *Shock Cities: The Environmental Transformation and Reform of Manchester and Chicago* (Chicago: University of Chicago Press, 2004), 98; Walter Nugent, "Demographics: Chicago as a Modern World City," in *The Encyclopedia of Chicago*, ed. James R. Grossman, Ann Keating, and Janice L. Reiff (Chicago: Chicago Historical Society, 2004), 234.

7. A. T. Andreas, *History of Chicago: From the Earliest Period to the Present Time*, vol. 1 (Chicago: A. T. Andreas, Publisher, 1884), 595–97.

8. The Sanitary District of Chicago, *History of the Chicago Sanitary and Ship Canal* (Chicago: City of Chicago Publishers, 1924), 28. This published report, submitted to the Chicago City Council, the Illinois General Assembly, and the Illinois governor's office, detailed the Sanitary District of Chicago's (SDC) organizational strategy for the construction of the Sanitary and Ship Canal. Although the report compiles a wealth of engineering and technological information, it remains biased toward the SDC and its efforts to improve Chicago's water quality. The data contained in this report are useful, despite a narrative favorable to the SDC and its engineers.

9. Platt, *Shock Cities*, 99. Chicago's elevation above sea level is 673 feet.

10. Andreas, *History of Chicago*, 198–200. Readers interested in Andreas's work can find it in its entirety online: https://www.google.com/books/edition/History_of_Chicago/wP0TAAAAYAAJ?hl=en&gbpv=1&printsec=frontcover.

11. Andreas, *History of Chicago*, 30.

12. Andreas, *History of Chicago*, 29.

13. Louis P. Cain, "Sanitation in Chicago: A Strategy for a Lakefront Metropolis," in *Encyclopedia of Chicago*, ed. Grossman et al. (Chicago: Chicago Historical Society, 2004), 301. Cain details these events in his monograph, *Sanitation Strategy for a Lakefront Metropolis*.

14. "Ellis Chesbrough," *Biographical Sketches of the Leading Men of Chicago* (Chicago: Wilson and St. Claire, 1868), 194.

15. "Ellis Chesbrough," 195.

16. "Ellis Chesbrough," 195.

17. Daniel Burnham, quoted in "Stirred by Burnham, Democracy Champion," *Chicago Record-Harold*, Oct. 15, 1910.

18. *Biographical Sketches of the Leading Men of Chicago*, 32; "Anniversary of the Drainage Canal," *Chicago Daily Tribune*, Aug. 3, 1895, 3.

19. Citizens' Association of Chicago, *The Annual Report of the Citizens' Association of Chicago* (Chicago: Hazlett and Reed, 1877), 12.

20. Citizens' Association of Chicago, *Annual Report*, 13.

21. Citizens' Association of Chicago, *Annual Report*, 13.

22. Citizens' Association of Chicago, "Ship Canal," in *Annual Report*, 21.

23. Citizens' Association of Chicago, "Ship Canal," 22.

24. "And a Flood Came," *Chicago Daily Tribune*, Aug. 3, 1885.

25. "Fortunate Chicago," *Daily News*, Aug. 17, 1885.

26. "Fortunate Chicago," iv.

27. "Fortunate Chicago," iv.

28. Platt, *Shock Cities*, 139.

29. John H. Long, "The Illinois State Board of Health: Report of the Sanitary Investigations of the Illinois River and Its Tributaries," *The Ninth Annual Report of the Illinois State Board of Health* (Springfield, IL: Phillips Bros. State

Publishers, 1889). This report is accessible through the Harold Washington Library Center's Municipal Records Office.

30. Platt, *Shock Cities*, 20.

31. Platt, *Shock Cities*, 139.

32. Daniel Schneider, *Hybrid Nature: Sewage Treatment and the Contradictions of the Industrial Ecosystem* (Cambridge, MA: MIT Press, 2011), 30. See also Antoine Magnin, *Bacteria* (New York: W. Wood, 1884), 65. Magnin explains the classification of many of the various bacteria found in Chicago waterways during the nineteenth century. It can be accessed via Google Books.

33. W. M. Harman, "Proceedings," *Proceedings of the Board of Trustees of the Sanitary District of Chicago* (Chicago: City of Chicago Publishers, Dec. 8, 1890), 295. The Harold Washington Library holds these transcripts in their Municipal Records collection of the Government Information division.

34. Chas. Baly, "First Regular Meeting," *Proceedings of the Board of Trustees of the Sanitary District of Chicago* (Chicago: City of Chicago Publishers, Feb. 2, 1891), 8. The City of Chicago published records of the meetings of the SDC that are held in Government Information at the Harold Washington Library in Chicago, Illinois. This volume details the organization of the SDC and the formative measures taken to begin work on the drainage canal.

35. Baly, "First Regular Meeting," 98; and Richard Prendergast, "Drainage Board," *Proceedings of the Board of Trustees of the Sanitary District of Chicago* (Chicago: City of Chicago Publishers, Nov. 12, 1891), 263. Reporting to Richard Prendergast, the chairman of the SDC's Board of Trustees, Cooley documented all elements of the construction process. Composed of sanitarians, businessmen, and local politicians, the board needed as much information as possible so that it could allocate resources accordingly, while the Board of Sewerage Trustees, the Canal Trustees Board, and the Law Enforcement Board completed the project's legal organization.

36. Sanitary District, *History of the Chicago Sanitary and Ship Canal*, 92–94, 95, 96.

37. Sanitary District, *History of the Chicago Sanitary and Ship Canal*, 405.

38. Richard Prendergast, "Drainage Board," 279.

39. *History of the Chicago Sanitary and Ship Canal*, 408–11.

40. Sanitary District, *History of the Chicago Sanitary and Ship Canal*, 423.

41. Sanitary District, *History of the Chicago Sanitary and Ship Canal*, 425.

42. "For a Great Waterway: Work Begun on the Big Ditch from the Lakes," *New York Times*, Sept. 3, 1892, 6; Sanitary District, *History of the Chicago Sanitary and Ship Canal*, 425; "To Begin the Ditch: Drainage Board to Hold 'Shovel Day' Ceremonies," *Chicago Daily Tribune*, Sept. 2, 1892, 8.

43. For a complete engineering history of the Sanitary Canal construction, see Richard Lanyon, *Draining Chicago: The Early City and the North Area* (Chicago: Lake Claremont Press, 2016). This work is exhaustive in its coverage of the

diversion plan and the technology employed in the project. It is less extensive in its analysis of the political and economic pressures that informed the construction and the social tension that remained after the reversal.

44. Sanitary District, *History of the Chicago Sanitary and Ship Canal*, 424–27.

45. "For a Great Waterway," 6.

46. "For a Great Waterway," 6.

47. Sanitary District, *History of the Chicago Sanitary and Ship Canal*, 106.

48. "To Begin the Ditch," 8.

49. "To Begin the Ditch," 18.

50. "To Begin the Ditch," 10–12; and "Advance on the Drainage Canal," *Chicago Daily Tribune*, Jan. 1, 1895, 12.

51. Sanitary District of Chicago, *A Concise Report on Its Organizations, Resources, Constructive Work, Methods, and Progress* (Chicago: City of Chicago Publishers, 1895), 8–12. These documents are accessible in their original printed form at the Harold Washington Library Center in the Municipal Records collection in the Government Information division.

52. "Advance on the Drainage Canal," 12.

53. Sanitary District, *Concise Report*, 18.

54. William E. Worthen, "Drainage Board," *Proceedings of the Board of Trustees of the Sanitary District of Chicago* (Chicago: City of Chicago Publishers, Dec. 13, 1894), 301.

55. "Advance on the Drainage Canal," 12.

56. Sanitary District, *History of the Chicago Sanitary and Ship Canal*, 106.

57. "Big Drainage Canal Open," *Washington Post*, Jan. 2, 1900, 3; "Chicago Drainage Canal: Water of River Turned into Main Channel," *New York Times*, Jan. 2, 1900, 5.

58. "Big Drainage Canal Open," 5.

59. "Big Drainage Canal Open," 3; "Chicago Drainage Canal," 5.

60. "Big Drainage Canal Open," 3.

61. "Report on Drainage Canal Is Delayed Several Months," *Chicago Daily Tribune*, June 14, 1900, 6.

62. "Gives Facts on Drainage Canal," *Chicago Daily Tribune*, Oct. 30, 1900, 16. Oral histories and personal accounts from South Side residents remain scarce. The author has endeavored to document working-class, immigrant experiences through coverage in newspapers and activist accounts.

63. Jane Addams, *Twenty Years at Hull House* (New York: MacMillan, 1910), 184.

64. Sinclair, *Jungle*. The official name of Bubbly Creek is the South Fork of the South Branch of the Chicago River.

65. The neighborhood became known as "Back of the Yards" in 1913.

66. Mary McDowell, quoted in Addams, *Twenty Years at Hull House*, 185.

67. Mary McDowell, quoted in Addams, *Twenty Years at Hull House*, 185.

68. For the limitations of reform in urban environmental history, see Catherine McNeur, *Taming Manhattan: Environmental Battles in the Antebellum City* (Cambridge, MA: Harvard University Press, 2017), 5.

69. Edward J. Kelley, "Concerning the Drainage and Sewerage Conditions in Chicago and the Diversion of 10,000 C.F.S. from Lake Michigan at Chicago," *Memorandum: For the Chief Engineers of the United States Army* (Chicago: City of Chicago Publishers, 1923).

CHAPTER 6: TOO MUCH WATER

1. Andreas F. Prein, Roy M. Rasmussen, Kyoko Ikeda, Changhai Liu, Martyn P. Clark, and Greg J. Holland, "The Future Intensification of Hourly Precipitation Extremes," *Nature Climate Change*, Dec. 5, 2016, http://www.nature.com.flagship.luc.edu/nclimate/index.html?cookies=accepted, for the quotation; and U.S. National Oceanic and Atmospheric Administration, National Climatic Data Center, "Record of Climatological Observations," http://www.ncdc.noaa.gov., for long-term rainfall data. Also see Elaine Tyler May, *Homeward Bound: American Families in the Cold War Era* (New York: Basic Books, 1988); Wendy Wall, *Inventing the "American Way": The Politics of Consensus from the New Deal to the Civil Rights Movement* (Oxford: Oxford University Press, 2008); and Joel Greenberg, *A Natural History of the Chicago Region* (Chicago: University of Chicago Press, 2002).

2. Z. Eastman, "Ancient Chicago," *Chicago Daily Tribune*, Feb. 22, 1874, (hereafter *CT*).

3. Lizabeth Cohen, *A Consumers' Republic: The Politics of Mass Consumption in Postwar America* (New York: Knopf, 2003), 73, 115–65; Margaret Garb, *City of American Dreams: A History of Home Ownership and Housing Reform in Chicago, 1871–1919* (Chicago: University of Chicago Press, 2005); and Carl Condit, *Chicago, 1930–1970: Building, Planning, and Urban Technology* (Chicago: University of Chicago Press, 1974), 286–87, for housing statistics.

4. Sarah Potter, "Family Ideals: The Diverse Meanings of Residential Space in Chicago during the Post–World War II Baby Boom," *Journal of Urban History* 39 (Nov. 2012): 59–78; U.S. Department of Commerce, Bureau of the Census, *Census Reports* (Washington, DC: Government Printing Office, 1940, 1950, 1960, 1970); Ann Durkin Keating, *Chicagoland—City and Suburbs in the Railroad Age* (Chicago: Chicago University Press, 2005); and Robert Bruegmann, *Sprawl: A Compact History* (Chicago: University of Chicago Press, 2005). Also see Brian James McCammack, "Recovering Green in Bronzeville: An Environmental and Cultural History of the African American Great Migration to Chicago, 1915–1940" (PhD diss., Harvard University, 2012); and Michael D. Innis-Jiménez, "Organizing for Fun: Recreation and Community Formation in the Mexican Community of South Chicago in the 1920s and 1930s," *Journal of the Illinois State Historical Society* 98 (Autumn 2005): 144–61.

5. Ralph E. Tarbett, "Sewerage in Chicago," in *The Chicago-Cook County Health Survey*, ed. U.S. Public Health Service (New York: Columbia University Press, 1949), 98–109, for CSD statistics. Also see Rutherford H. Platt, *Open Land in Urban Illinois: Roles of the Citizen Advocate* (De Kalb: Northern Illinois University Press, 1971).

6. Christopher C. Sellers, *Crabgrass Crucible: Suburban Nature and the Rise of Environmentalism in Twentieth-Century America* (Chapel Hill: University of North Carolina Press, 2012), 12.

7. City of Chicago, Chicago Recreation Commission, and Northwestern University, *Chicago Recreation Survey 1937*, ed. Arthur J. Todd et al. (Chicago: Works Progress Administration, National Youth Administration, and Illinois Emergency Relief Commission, 1937), 1:98; *CT*, Apr. 18, 1948. Also see Jane Addams, *The Spirit of Youth and the City Streets* (New York: Macmillan, 1909); and Robin Faith Bachin, *Building the South Side: Urban Space and Civic Culture in Chicago, 1890–1919* (Chicago: University of Chicago Press, 2004), 127–68.

8. William Bayliss, "Sports on the Production Line," *CT*, Sept. 9, 1945, A3.

9. "Expect Record Crowds, Travel over July 4th," *CT*, July 3, 1946, 11 (quotation). Also see Rita Fitzpatrick, "Unique Dunes Sand Park: Its Sands Sing Own Aria," *CT*, Apr. 13, 1947, 1; and "Ten Year Park Plan," *Park Ways* (Chicago: Chicago Park District, Nov. 1945).

10. City of Chicago, *Chicago Recreation Survey*, 1:164; *CT*, July 1, 1952.

11. "Outings Draw Thousands to 7 Park Picnics," *CT*, July 21, 1947, 22; Galen Cranz, *The Politics of Park Design: A History of Urban Parks in America* (Cambridge, MA: MIT Press, 1982), 110–23; and Sellers, *Crabgrass Crucible*.

12. Donald L. Hey, "The Des Plaines River Wetlands Demonstration Project: Restoring an Urban Wetland," in *The Ecological City: Preserving and Restoring Urban Biodiversity*, ed. Rutherford H. Platt et al. (Amherst: University of Massachusetts Press, 1994), 83–92.

13. Harold L. Platt, *Shock Cities: The Environmental Transformation and Reform of Manchester and Chicago* (Chicago: University of Chicago Press, 2005); Craig E. Colten, "Illinois River Pollution Control, 1900–1970," in *The American Environment: Interpretations of Past Geographies*, ed. Lary M. Dilsaver and Craig E. Colten (Lanham, MD: Rowman and Littlefield, 1992), 193–214; Joshua A. T. Salzmann, "The Creative Destruction of the Chicago River Harbor: Spatial and Environmental Dimensions of Industrial Capitalism, 1881–1909," *Enterprise and Society* 13 (June 2012): 235–75; and Carl S. Smith, *City Water, City Life: Water and the Infrastructure of Ideas in Urbanizing Philadelphia, Boston, and Chicago* (Chicago: University of Chicago Press, 2013).

14. Harold L. Platt, "Chicago, the Great Lakes, and the Origins of Federal Urban Environmental Policy," *Journal of the Gilded Age and Progressive Era* 1 (2002): 122–53; and Horace P. Ramey, "Floods in the Chicago Area" (unpublished transcript, Feb. 25, 1958).

15. Ramey, "Floods in the Chicago Area"; Raymond Leland, "Municipal Sewer Systems in Cook County," in U.S. Public Health Survey, *The Chicago-Cook County Health Survey*, 135–46; and *CT*, Aug. 18, Sept. 29, 1946.

16. John Thomson, "Speed Sewer Job to Abate S. Side Floods," *CT*, July 1, 1945, S1; Ward Walker, "Turns Aims to Major Shake-Up," *CT*, Apr. 23, 1947, 5; "Bids on Sewer Works to Be Opened Soon," *CT*, Feb. 12, 1950; "Sanitary Board Grants Leave to Trinkaus," *CT*, May 8, 1954, A8; Tarbett, "Sewerage in Chicago"; and Ramey, "Floods in the Chicago Area."

17. Oscar Hewitt, as quoted in Frank Cipriani, "Chicagoland's Need—A Super Water Authority," *CT*, July 6, 1947, G7.

18. Hewitt, quoted in Cipriani, "Chicagoland's Need," G12.

19. "Some Light on Sanitary District Robbery," *CT*, Apr. 10, 1947, 18.

20. Arnold R. Hirsch, "Martin H. Kennelly—The Mugwump and the Machine," in *The Mayors—The Chicago Political Tradition*, ed. Melvin G. Holli and Paul M. Green (Carbondale: Southern Illinois University Press, 1987), 126–43.

21. "Chicago Region Flood Loss Set at 25 Millions," *CT*, Nov. 1, 1954, 24; Ramey, "Floods in the Chicago Area," 14; and U.S. Department of the Interior, Geological Survey, Warren S. Daniels, and Malcolm D. Hale, "Floods of October 1954 in the Chicago Area, Illinois, and Indiana" (Washington, DC: Government Printing Office, 1955).

22. Geological Survey et al., "Floods of October 1954," 113–26.

23. Geological Survey et al., "Floods of October 1954," 118–21.

24. "Propose Vast Water Service Authority Unit," *CT*, Feb. 6, 1955, 30; "Flood Dangers," *CT*, May 28, 1957, 12; "Flood Damage in Millions," *CT*, July 14, 1957; "Chicago Flood Threat Eases as Rain Abates," *CT*, June 14, 1958, 5; and "City, Suburbs Mop Up after Record Rains," *CT*, Sept. 15, 1961, A3.

25. Roger Biles, *Richard J. Daley: Politics, Race and the Governing of Chicago* (De Kalb: Northern Illinois University Press, 1995); Barbara Ferman, *Challenging the Growth Machine: Neighborhood Politics in Chicago and Pittsburgh* (Lawrence: University Press of Kansas, 1996); and Center of Neighborhood Technology, *The Prevalence and Cost of Urban Flooding: A Case Study of Cook County, Ill.* (Chicago: self-published, May 2013).

26. See David Alexander Spatz, "Roads to Postwar Urbanism: Expressway Building and the Transformation of Metropolitan Chicago, 1930–1975" (PhD diss., University of Chicago, 2010), for a parallel case.

27. "'Foul': Joliet Cries in Flood Control Plan," *CT*, Apr. 30, 1959, S10; "Flood Ordered for DesPlaines River," *CT*, June 8, 1960, B11; "4 Mile Tunnel to Lake Urged to Ease Floods," *CT*, Oct. 4, 1960, 5.

28. "Nature Preserves," *CT*, May 12, 1961, 12; "Lake County Seeks Land for Parks," *CT*, Aug. 20, 1961, NW1; see "Classes Give TV a Hard Time," *CT*, Sept. 23, 1962, 42; "New City Park Is Urged for River Shore," *CT*, July 28, 1963, B8,

for CPD statistics; "Urges More State Parks near Chicago," *CT*, Oct. 4, 1964, 12; "Press for More Docking Facilities," *CT*, Oct. 17, 1965, E1; and Rachel Carson, *Silent Spring* (Boston: Houghton Mifflin, 1962).

29. "Storing Flood Water," *CT*, June 21, 1960, 12; "Begin Flood Control Project near Joliet," *CT*, Sept. 21, 1961, S2; "Millions Spent, but Flood Peril Lives On," *CT*, Oct. 1, 1961, 44. Also see D. Bradford Hunt and Jon B. DeVries, *Planning Chicago* (Chicago: Planner's Press, 2013), 25–40.

30. "Bacon to Sue on Sanitary Overchanges," *CT*, Mar. 19, 1963, B10; "Bacon Blasts Sanitary Unit Trustee Bloc," *CT*, Dec. 5, 1964, 17; "Bacon Blasts 'Suppression' of Civil Service Survey," *CT*, May 28, 1966, A6; "House Votes to Investigate Sanitary District," *CT*, Jan. 10, 1967, A2; "Award for Bacon," *CT*, June 17, 1968, 14; and "Bacon Gets U. of Wis. Faculty Job," *CT*, Mar. 7, 1970, W1.

31. "Sanitary Board Will Reveal Huge Flood Control Proposal," *CT*, Jan. 25, 1965, B11; "Bacon Unveils Billion Dollar Water Project," *CT*, Apr. 12, 1967, B12; Hazar Engineering Company and Metropolitan Sanitary District of Greater Chicago, *Flood and Pollution Control—A Deep Tunnel Plan for the Chicagoland Area* (Chicago: Metropolitan Sanitary District of Greater Chicago, May 1966); Frank E. Dalton, Victor Koelzer, and William J. Bauer, "The Chicago Area Deep Tunnel Project: A Use of the Underground Storage Resource," *Journal of the Water Pollution Control Federation* 41 (Apr. 1969): 515–34.

32. Frank E. Dalton and Raymond R. Rimkus, "The Chicago Area's Tunnel and Reservoir Plan," *Journal of the Water Pollution Control Federation* 57 (Dec. 1985): 1114–21. Also see Judith A. Martin and Sam Bass Warner Jr., "Local Initiative and Metropolitan Repetition: Chicago, 1972–1990," in *The American Planning Tradition: Culture and Policy*, ed. Robert Fishman (Washington, DC: Woodrow Wilson Center Press and Johns Hopkins University Press, 2000), 263–96.

33. "The Garden City," *CT*, June 6, 1965, 20.

CHAPTER 7: WATER, OIL, AND FISH

1. Richard White, *The Organic Machine: The Remaking of the Columbia River* (New York: Hill and Wang, 1995), 64.

2. Daniel Macfarlane, "Nature Empowered: Hydraulic Models and the Engineering of Niagara Falls," *Technology and Culture* 61, no. 1 (2020): 109–43; and Daniel Macfarlane, "'A Completely Artificial and Man-Made Cataract': The Transnational Manipulation of Niagara Falls," *Environmental History* 18, no. 4 (Oct. 2013): 759–84. On waterways as infrastructure, see also Ashley Carse, *Beyond the Big Ditch: Politics, Ecology, and Infrastructure at the Panama Canal* (Cambridge, MA: MIT Press, 2014); Ashley Carse and Joshua A. Lewis, "Toward a Political Ecology of Infrastructure Standards; or, How to Think about Ships, Waterways, Sediment and Communities Together," *Environment and Planning A* 49, no. 1 (2017): 9–28; Martin Reuss, "Rivers as Technological

Systems," in *The Oxford Encyclopedia of the History of American Science, Medicine, & Technology*, vol. 2, ed. Hugh Richard Slotten (Oxford: Oxford University Press, 2014).

3. On the history of the Chicago River, Chicago Diversion, and Chicago's general water history, see John W. Larson, *Those Army Engineers: A History of the Chicago District U.S. Army Corps of Engineers* (Chicago: U.S. Army Corps of Engineers, Chicago District, 1979); David M. Solzman, *The Chicago River: An Illustrated History and Guide to the Rivers and Its Waterways* (Chicago: University of Chicago Press, 2006); Richard Lanyon, *Building the Canal to Save Chicago* (Bloomington, IN: Xlibris, 2012); Harold Platt, *Sinking Chicago: Climate Change and the Remaking of a Flood-Prone Environment* (Philadelphia: Temple University Press, 2018); Joshua Salzmann, *Liquid Capital: Making the Chicago Waterfront* (Philadelphia: University of Pennsylvania Press, 2017); Libby Hill, *The Chicago River: A Natural and Unnatural History*, rev. ed. (Carbondale: Southern Illinois University Press, 2019); Richard Cahan and Michael Williams, *The Lost Panoramas: When Chicago Changed Its River and the Land Beyond* (Chicago: CityFiles Press, 2011); Matt Edgeworth and Jeff Benjamin, "What Is a River? The Chicago River as Hyperobject," in *Rivers of the Anthropocene*, ed. Jason M. Kelly, Philip Scarpino, Helen Berry, James Syvitski, and Michael Meybeck (Oakland: University of California Press, 2018). On energy use in Chicago, see Harold Platt, *The Electric City: Energy and the Growth of Chicago* (Chicago: University of Chicago Press, 1991).

4. These withdrawals concern North America's two largest rivers, the Mississippi and the St. Lawrence.

5. "Top Ten Public Works Projects of the Century," American Public Works Association, http://www2.apwa.net/about/awards/toptencentury/default.htm (accessed June 12, 2018).

6. The federal government initially set the Lake Michigan diversion at 4,167 cfs, but the allowable volume fluctuated.

7. Daniel Macfarlane, *Negotiating a River: Canada, the US, and the Creation of the St. Lawrence Seaway* (Vancouver: UBC Press, 2014), 35–39.

8. Hydroelectric production downstream was a secondary objective.

9. With a flow of 5,000 cfs, these are the largest engineered diversions into the Great Lakes. On Lake Superior, see Nancy Langston, *Sustaining Lake Superior: An Extraordinary Lake in a Changing World* (New Haven, CT: Yale University Press, 2017).

10. Macfarlane, *Negotiating a River*.

11. Averaged out over a forty-year period.

12. Peter Annin, *The Great Lakes Water Wars* (Washington, DC: Island Press, 2018).

13. Arthur M. Woodward, *Charting the Inland Seas: A History of the U.S. Lake Survey* (Detroit: U.S. Army Corps of Engineers, Detroit District, 1991).

14. The International Joint Commission (IJC) contends that the Chicago Diversion reduced the mean levels of Lakes Michigan and Huron by approximately 2.5 inches, Lake Erie by 1.7 inches, Lake Ontario by 1.2 inches, and Lake Superior by less than an inch. If the diversion reached its 10,000 cfs capacity, about three times its current rate, the impact on the lakes would be proportional. Thus, the early twentieth-century estimate of a half-foot drop is reasonable. International Joint Commission, *Report on Further Regulation of the Great Lakes* (Washington, DC, and Ottawa: International Joint Commission, 1976) (hereafter IJC 1976); International Joint Commission, *Report on Great Lakes Diversions and Consumptive Uses* (Washington, DC, and Ottawa: International Joint Commission, 1985) (hereafter IJC 1985).

15. See the special issue of *Canadian Geographer* 60, no. 4 (Winter 2016), on Great Lakes–St. Lawrence as a system.

16. IJC 1976; IJC 1985.

17. Daniel Macfarlane, *Fixing Niagara Falls: Environment, Energy, and Engineers at the World's Most Famous Waterfall* (Vancouver: UBC Press, 2020).

18. Murray Clamen and Daniel Macfarlane, "The International Joint Commission, Water Levels, and Transboundary Governance in the Great Lakes," *Review of Policy Research* 32, no. 1 (Jan. 2015): 40–59; Daniel Macfarlane and Murray Clamen, *The First Century of the International Joint Commission* (Calgary: University of Calgary Press, 2020).

19. Christopher Jones, *Routes of Power: Energy and Modern America* (Cambridge, MA: Harvard University Press, 2014).

20. Michelle Murphy, "Chemical Infrastructures of the St. Clair River," in *Toxicants, Health and Regulation since 1945*, ed. Soraya Boudia and Nathalie Jas (London: Pickering & Chatto, 2013), 104.

21. Even the permit for this was buried within "a Federal Register notice [that] flew under everyone's radar until the public feedback window expired"; Garret Ellison, "Permit for 98-Year-Old Pipelines under St. Clair River Sparks Alarm," MLive, Mar. 16, 2016, https://www.mlive.com/news/2016/03/st_clair_river_pipelines.html.

22. Presumably named for its intricate navigation of Great Lakes states and waterways.

23. Joel Hood, "4 Inspections of Romeoville Pipeline Found No Potential for Leaks, Company Says: Federal EPA Will Check to See if Any Risks Remain," *Chicago Tribune*, Sept. 16, 2010.

24. The price increased by 16 cents in the Chicago area, by 10 cents farther out.

25. Joel Hood, "Hiccup in Romeoville, Indigestion at Pump: Gas Prices Climb in Wake of Pipeline Shutdown," *Chicago Tribune*, Sept. 15, 2010.

26. The BP Deepwater Horizon catastrophe occurred earlier in the year, and dwarfed even the Kalamazoo spill in terms of damage. For the Kalamazoo River

spill, see National Transportation Safety Board, *Pipeline Accident Report: Enbridge Incorporated Hazardous Liquid Pipeline Rupture and Release Marshall, Michigan July 25, 2010,* Pipeline Accident Report NTSB/PAR-12/01 (Washington, DC: National Transportation Safety Board, 2012). See also Elizabeth McGowan and Lisa Song, "The Dilbit Disaster: Inside the Biggest Oil Spill You've Never Heard Of, Part 1," *Inside Climate News,* June 26, 2012.

27. This includes fines and cleanup costs.

28. Eric Anderson, NOAA-Great Lakes Environmental Research Laboratory, "Predicting Currents in the Straits of Mackinac," https://www.glerl.noaa.gov/pubs/brochures/straits.pdf (Ann Arbor, MI: NOAA-Great Lakes Environmental Research Laboratory) (accessed June 12, 2018); David J. Schwab, *Statistical Analysis of Straits of Mackinac Line 5: Worst Case Spill Scenarios* (Ann Arbor: University of Michigan Water Center, Mar. 2016).

29. In the 1960s Italsider, Italy's largest steel manufacturer at that time (and one of the largest in Europe), made the coated steel pipe for the entire Lakehead System.

30. David Schwab, quoted in "Straits of Mackinac 'Worst Possible Place' for a Great Lakes Oil Spill, U-M Researcher Concludes," *Michigan News* (Ann Arbor), July 10, 2014.

31. Jonathan Oosting and Leonard N. Fleming, "Schuette: Close Line 5 Pipelines on 'Definite Timetable,'" *Detroit News,* June 29, 2017; Keith Matheny, "Straits Pipeline Report Raises Fears of Disaster; Enbridge Says Oil, Gas Lines Supported," *Detroit Free Press,* June 1, 2017; Dynamic Risk Assessment Systems, Inc., *Alternatives Analysis for the Straits Pipeline* (Calgary, Alberta: final report prepared for the State of Michigan), project no. SOM-2017–01, document no. SOM-2017–01-RPT-001, Oct. 26, 2017.

32. Jeff Alexander and Beth Wallace, *Sunken Hazard: Aging Oil Pipelines beneath the Straits of Mackinac an Ever-Present Threat to the Great Lakes* (Ann Arbor, MI: National Wildlife Federation, 2012).

33. Jones, *Routes of Power,* 143–44.

34. In the Kalamazoo River, Enbridge didn't act for about seventeen hours because it thought there was an air blockage instead of a spill.

35. Testimony of David Lodge in *Asian Carp and the Great Lakes: Hearing before the Subcommittee on Water Resources and the Environment of the Committee On Transportation and Infrastructure,* 111th Cong., 1 (2010), 10. On Asian carp, see also Andrew Reeves, *Overrun: Dispatches from the Asian Carp Crisis* (Toronto: ECW Press, 2019).

36. P. J. Perea, "Asian Carp Invasion: Fish Farm Escapees Threaten Native River Fish Communities and Boaters as Well," *Outdoor Illinois* 10, no. 5 (May 2002): 8.

37. James L. Oberstar, quoted in *Asian Carp and the Great Lakes,* 13, 32.

38. National Invasive Species Act, H.R. 4283 (104th)/P.L. 104–332 (Oct. 26,

1996). This was a reauthorization and update of the Indigenous Aquatic Nuisance Prevention and Control Act of 1990.

39. Dameon Pesanti, "Smith-Root Finds Innovation in Science, Business," *Columbian*, Feb. 7, 2016.

40. David V. Smith, Fish Repelling Apparatus Using a Plurality of Series Connected Pulse Generators to Produce an Optimized Electric Field, U.S. Patent 4750451, June 14, 1988. Smith Root, Inc. holds twenty-four patents, starting with Identification Tag Implanting Machine, U.S. Patent 3369525, in Feb. 1968, up to Systems and Methods for Aquatic Electrical Barrier Desynchronization, U.S. Patent 9468198, in Oct. 2016.

41. Smith-Root, Inc. [inventors David V. Smith and Lee Roy Carstensen], Electric Fish Barrier for Water Intakes at Various Depths, U.S. Patent US6978734 B1, Dec. 27, 2005.

42. U.S. Army Corps of Engineers, Chicago District, *Dispersal Barrier Efficacy Study: Efficacy Study Interim Report IIA, Chicago Sanitary and Ship Canal Dispersal Barriers—Optimal Operating Parameters Laboratory Research and Safety Tests* (Chicago: Government Printing Office, 2011).

43. Joel Hood, "Carp Creeps into Lake Calumet: Discovery of Invasive Fish Triggers New Calls to Legal Action," *Chicago Tribune*, June 23, 2010; Michael Hawthorne, "Asian Carp Discovered Close to Lake Michigan as Trump Pushes Budget Cuts," *Chicago Tribune*, June 23, 2017.

44. William Cronon, *Nature's Metropolis* (New York: W. W. Norton, 1992).

45. Paul Sutter identified hybridity as a "defining tendency of recent scholarship in American environmental history"; Paul S. Sutter, "The World with Us: The State of American Environmental History," *Journal of American History* 100, no. 1 (2013): 94–119. Unfortunately we can't do justice to the robust envirotech historiography and recent work that intersect with this study. Mark Fiege, Dolly Jorgensen, Martin Melosi, Joy Parr, Sara Pritchard, Jeffrey Stine, Joel Tarr, and Richard White have been in the vanguard of scholars elucidating hybrid natures. For recent scholarship, see Lynne Heasley and Daniel Macfarlane, eds., *Border Flows: A Century of the Canadian-American Water Relationship* (Calgary: University of Calgary Press, 2016); Martin Reuss and Stephen H. Cutcliffe, eds., *The Illusory Boundary: Environment and Technology in History* (Charlottesville: University of Virginia Press, 2010).

46. Macfarlane, "A Completely Artificial and Man-Made Cataract."

47. Daniel Macfarlane, "Fluid Meanings: Hydro Tourism and the St. Lawrence and Niagara Megaprojects," *Histoire Sociale/Social History* 49, no. 99 (June 2016): 327–46.

PART III: THE NATURE OF WORKING-CLASS CHICAGOANS

1. Hilda Satt Polachek, *I Came a Stranger: The Story of a Hull-House Girl* (Urbana: University of Illinois Press, 1991), 29.

2. James R. Grossman, Ann Durkin Keating, and Janice L. Reiff, eds., *The Encyclopedia of Chicago* (Chicago: University of Chicago Press, 2004).

3. Rosalyn R. Lapier and David R. M. Beck, *City Indian: Native American Activism in Chicago, 1893–1934* (Lincoln: University of Nebraska Press, 2004), chap. 6, 175–76. Regarding efforts to memorialize Potawatomi history in Chicago, including at the Columbian Exposition, see John M. Low, *Imprints: The Pokagon Band of Potawatomi Indians and the City of Chicago* (East Lansing: Michigan State University Press, 2016), 165–80.

4. James B. LaGrand, *Indian Metropolis: Native Americans in Chicago, 1945–75* (Urbana: University of Illinois Press, 2002), 48–61, chap. 6; and Orlando Garcia, "Urbanization of Rural Population: An American Indian Perspective," in *Native Chicago*, ed. Terry Straus, 193–203 (Chicago: Albatross Press, 2002).

CHAPTER 8: MAY DAY

1. Charles Rufus Skinner, ed., *Arbor Day Manual: An Aid in Preparing Programs for Arbor Day Exercises, Containing Choice Selections on Trees, Forests, Flowers, and Kindred Subjects; Arbor Day Music, Specimen Programs, Etc.* (Albany, NY: Weed, Parsons, 1890), 329–30; "Arbor Day," *Chicagoer Arbeiter Zeitung*, Apr. 12, 1889, file, German, Chicago Foreign Language Press Survey, Special Collections, Joseph Regenstein Library, University of Chicago (hereafter CFLPS).

2. "Arbor Day." Note: Foreign Press Survey translators originally used "gallons" instead of "gallows."

3. Geoffrey Blodgett, "Frederick Law Olmsted: Landscape Architecture as Conservative Reform," *Journal of American History* 62, no. 4 (1976): 869–89; Roy Rosenzweig and Elizabeth Blackmar, *The Park and the People: A History of Central Park* (New York: Henry Holt, 1992), 1–262.

4. William Cronon, *Nature's Metropolis: Chicago and the Great West* (New York: W. W. Norton, 1991).

5. Richard Schneirov, *Labor and Urban Politics: Class Conflict and the Origins of Modern Liberalism in Chicago, 1864–1897*, The Working Class in American History (Urbana: University of Illinois Press, 1998), 17–24, 32–33; John B. Jentz and Richard Schneirov, *Chicago in the Age of Capital: Class, Politics, and Democracy during the Civil War and Reconstruction*, The Working Class in American History (Urbana: University of Illinois Press, 2012), 13–51; Roger Daniels, *The Work Ethic in Industrial America, 1850–1920* (Chicago: University of Chicago Press, 1974), 106.

6. Schneirov, *Labor and Urban Politics*, 33–35; David Montgomery, *Beyond Equality: Labor and the Radical Republicans, 1862–1872* (New York: Knopf, 1967), 306–10.

7. Schneirov, *Labor and Urban Politics*, 69–76, 81–94; Jentz and Schneirov, *Chicago in the Age of Capital*, 157–64; Bruce Nelson, *Beyond the Martyrs: A*

Social History of Chicago's Anarchists, 1870–1900 (New Brunswick, NJ: Rutgers University Press, 1988), 9–76; Paul Avrich, *The Haymarket Tragedy* (Princeton, NJ: Princeton University Press, 1984), 26–52.

8. Avrich, *The Haymarket Tragedy*, 55–177; Nelson, *Beyond the Martyrs*, 9–51; Schneirov, *Labor and Urban Politics*, 173–79. On labor and the language of slavery, see David R. Roediger, *Wages of Whiteness: Race and the Making of the White Working Class* (New York: Verso, 1991), 63–94.

9. Gerald Gems, *The Windy City Wars: Labor, Leisure, and Sport in the Making of Chicago*, American Sports History Series, no. 8 (Lanham, MD: The Scarecrow Press, 1997), 77–83; Robin Faith Bachin, *Building the South Side: Urban Space and Civic Culture in Chicago, 1890–1919*, Historical Studies of Urban America (Chicago: University of Chicago Press, 2004), 127–204; Colin Fisher, *Urban Green: Nature, Recreation, and the Working Class in Industrial Chicago* (Chapel Hill: University of North Carolina Press, 2015), 38–63.

10. Gems, *Windy City Wars*, 10; Fisher, *Urban Green*, 43–46. On Boone, see Richard Wilson Renner, "In a Perfect Ferment: Chicago, the Know-Nothings, and the Riot for Lager Beer," *Chicago History* 5 (Fall 1976): 161–70.

11. Renner, "In a Perfect Ferment"; "The Origins of Picnics," *Chicagoer Arbeiter Zeitung*, June 27, 1883, quoted in Keil and Jentz, *German Workers in Chicago: A Documentary History of Working-Class Culture from 1850 to World War I* (Urbana: University of Illinois Press, 1988), 205, 206.

12. On memory, see Fisher, *Urban Green*, 38–63; "The Desplaines Hall Workers Club Picnic in Ogden's Grove," *Der Western*, July 22, 1869, quoted in Keil and Jentz, *German Workers in Chicago*, 207 (quote). On ethnicity, see Werner Sollors, "Introduction: The Invention of Ethnicity," in *The Invention of Ethnicity*, ed. Werner Sollors (New York: Oxford University Press, 1989), ix–xx; Kathleen Neils Conzen, David A. Gerber, Eva Morawska, George E. Pozzetta, and Rudolph J. Vecoli, "The Invention of Ethnicity: A Perspective from the U.S.A," *Journal of American Ethnic History* 12 (Fall 1992): 3–4.

13. John Joseph Flinn and John Elbert Wilkie, *History of the Chicago Police: From the Settlement of the Community to the Present Time, under Authority of the Mayor and Superintendent of the Force* (Chicago: Police Book Fund, 1887), 249.

14. Fisher, *Urban Green*, 114–43.

15. Nelson, *Beyond the Martyrs*, 127–52; Fisher, *Urban Green*, 119–20; "The Central Labor Union," *The Alarm*, June 18, 1885 (quote).

16. Albert Richard Parsons, "The Philosophy of Anarchism," in *Anarchism: Its Philosophy and Scientific Basis as Defined by Some of Its Apostles*, ed. A. R. Parsons (Chicago: Mrs. A. R. Parsons, 1887), 172; "The Indians," *The Alarm*, Nov. 8, 1884; Michael Schwab, "Autobiography of Michael Schwab," in *The Autobiographies of the Haymarket Martyrs*, ed. Philip S. Foner (New York: Humanities Press, 1969), 99–100, 126 (quotes).

17. Schwab, "Autobiography," 126; Albert Richard Parsons, "Albert R. Parsons on Anarchy," in Parsons, *Anarchism*, 102; August Spies, "Philosophical Thoughts," *The Alarm*, Feb. 21, 1885; Dyer D. Lum, "Dyer D. Lum on Anarchy," in Parsons, *Anarchism*, 155 (quote).

18. Schwab, "Autobiography," 109; Johann Most, *The Beast of Property: Total Annihilation Proposed as the Only Infallible Remedy: The Curse of the World Which Defeats the People's Emancipation*, pamphlet (New Haven, CT: International Workingmen's Association, [1883?]). See also L.M.S. [Lizzie M. Swank?], "Factory Girls," *The Alarm*, Nov. 14, 1885.

19. "Death Traps," *The Alarm*, Nov. 1, 1884; L.M.S., "Factory Girls"; Samuel Fielden, "Samuel Fielden on Socialism and Anarchism," in Parsons, *Anarchism*, 90–91; Lum, "Dyer D. Lum on Anarchy," 150; Most, *Beast of Property*; Michael Schwab, "Address of Michael Schwab," in Parsons, *Anarchism*, 67–68 (quote).

20. Lucy Parsons, "To Tramps," *The Alarm*, Oct. 4, 1884; Jacqueline Jones, *Goddess of Anarchy: The Life and Times of Lucy Parsons, American Radical* (New York: Basic Books, 2017).

21. On Thoreau, see George Woodcock, *Anarchism: A History of Libertarian Ideas and Movements* (Cleveland, OH: The World Publishing Company, 1962), 453–55; Peter Marshall, *Demanding the Impossible: A History of Anarchism* (London: HarperCollins, 1991), 184–88. John Zerzan, *Future Primitive: And Other Essays* (Brooklyn, NY: Autonomedia, 1994).

22. August Spies, "August Spies on Anarchy," in Parsons, *Anarchism*, 64 (quote); August Spies, "Autobiography of August Spies," in Foner, *Autobiographies of the Haymarket Martyrs*, 87 (quote); Spies, "August Spies on Anarchy," 64; Schwab, "Address," 68; Most, *Beast of Property*.

23. Avrich, *Haymarket Tragedy*, 138–40; Nelson, *Beyond the Martyrs*, 142–46. William Holmes, "Speech on the Paris Commune," *The Alarm*, Apr. 4, 1885, quoted in Avrich, *Haymarket Tragedy*, 139.

24. Avrich, *Haymarket Tragedy*, 73; Albert Parsons, "The International," *The Alarm*, Apr. 4, 1885, quoted in Avrich, *Haymarket Tragedy*, 73.

25. John P. Clark and Camille Martin, "An Introduction to Reclus' Social Thought," in *Anarchy, Geography, Modernity: The Radical Social Thought of Élisée Reclus*, ed. John P. Clark and Camille Martin (Lanham, MD: Lexington Press, 2004), 3–118.

26. Avrich, *Haymarket Tragedy*, 135. For reprint examples, see Élisée Reclus, "Property," in *The Alarm*, Apr. 3, 1886; and Élisée Reclus, "An Anarchist on Anarchy," in Parsons, *Anarchism*, 136–49; "Poverty," *The Alarm*, Apr. 4, 1885 (quote).

27. Most, *Beast of Property*.

28. Lum, "Dyer D. Lum on Anarchy," 156. On Bookchin, see Janet Biehl, *Ecology or Catastrophe: The Life of Murray Bookchin* (New York: Oxford University Press, 2015).

29. Avrich, *Haymarket Tragedy*, 181–96; Alan Calmer, *Labor Agitator: The Story of Albert R. Parsons* (New York: International, 1937), 79 (quote).

30. On Gori, see Donna T. Haverty-Stacke, *America's Forgotten Holiday: May Day and Nationalism, 1867–1960*, American History and Culture (New York: New York University Press, 2008), 47–50; Philip Foner, *May Day: A Short History of the International Workers' Holiday, 1886–1986* (New York: International Publishers, 1986), 17. On May Day, see Howard Hayes Scullard, *Festivals and Ceremonies of the Roman Republic* (Ithaca, NY: Cornell University Press, 1981), 110–11; Peter Burke, *Popular Culture in Early Modern Europe* (New York: Harper & Row, 1978), 274. On anarchist Maifest, see Nelson, *Beyond the Martyrs*, 135; James R. Green, *Death in the Haymarket: The First Labor Movement and the Bombing That Divided America* (New York: Anchor Books, 2007), 132; John D. Lawson, *American State Trials*, vol. 12 (St. Louis: F. H. Thomas Law Book, 1926), 106, quoted in Green, *Death in the Haymarket*, 158 (quote).

31. Philip Sheldon Foner, *American Labor Songs of the Nineteenth Century*, Music in American Life (Urbana: University of Illinois Press, 1975), 581–82; Fisher, *Urban Green*, 124.

32. Avrich, *Haymarket Tragedy*, 197–214; Green, *Death in the Haymarket*, 174–91.

33. Parsons, Spies, Fischer, and Engels were hung. Lingg committed suicide. The others later received pardons. Avrich, *Haymarket Tragedy*, 215–93; Green, *Death in the Haymarket*, 192–230.

34. "The First of May: The Anniversary of the Labor Movement," *Der Vorbote*, May 4, 1887, reprinted in Keil and Jentz, *German Workers in Chicago*, 225–27; "The Anarchists," *Chicago Tribune*, May 21, 1887.

35. "Labor-Day Parade," *Chicago Tribune*, May 2, 1890; Haverty-Stacke, *America's Forgotten Holiday*, 37–40; Foner, *May Day*, 40–55, 84–85; and Chris Wrigley, "Red May Days: Fears and Hopes in Europe in the 1890s," in *Crowd Actions in Britain and France from the Middle Ages to the Modern World*, ed. Michael T. Davis (New York: Palgrave Macmillan, 2015), 208–23.

36. "Ready for the Parade," *Chicago Tribune*, May 1, 1891; "Favor La Marseilleise," *Chicago Tribune*, May 2, 1891; "Preparing for the May Day Parade," *Chicago Tribune*, Apr. 29, 1892; "Postponed Labor Day Parade Held," *Chicago Tribune*, May 3, 1893.

37. "Police in Readiness," *Chicago Tribune*, Apr. 30, 1892; "Red Flags Seized," *Chicago Tribune*, May 2, 1892; "Kuhn's Park Picnic Rather Quiet," *Chicago Tribune*, May 2, 1892 (quote).

38. Foner, *May Day*, 96; Haverty-Stacke, *America's Forgotten Holiday*, 105–73; Randi Storch, *Red Chicago: American Communism at Its Grassroots, 1928–35* (Urbana: University of Illinois Press, 2007), 73–75; William F. Kruse, "The Flag of May Day," *Young Socialists' Magazine*, May 1918, n.p.

39. Adam Rome, *The Genius of Earth Day: How a 1970 Teach-in Unexpectedly Made the First Green Generation* (New York: Farrar, Straus and Giroux, 2014); Justin Moyer and Nick Kirkpatrick, "Would Earth Day's Creator Have Celebrated This Earth Day?," *Washington Post*, Apr. 22, 2015; Brian Tokar, *Earth for Sale: Reclaiming Ecology in the Age of Corporate Greenwash* (Boston: South End Press, 1997). On the Capitalocene, see Jason W. Moore, *Capitalism in the Web of Life: Ecology and the Accumulation of Capital* (London: Verso, 2015). On solidarity for May Day 2017, see Climate Workers and Movement Generation Justice and Ecology Project, "An Open Letter from Environmental & Climate Justice Organizations on May Day," Apr. 25, 2017, http://www.climateworkers.org/?p=214.

CHAPTER 9: BLACK MIGRANT FOODWAYS IN THE "HOG BUTCHER FOR THE WORLD"

1. Isabel Wilkerson, *The Warmth of Other Suns: The Epic Story of America's Great Migration* (New York: Random House, 2010), 295–96. See also Frederick Douglass Opie, *Hog & Hominy: Soul Food from Africa to America* (New York: Columbia University Press, 2008), 56–57; and William M. Tuttle, *Race Riot: Chicago in the Red Summer of 1919* (Urbana: University of Illinois Press, 1996), 95. On food and travel, see Psyche Williams-Forson, *Building Houses Out of Chicken Legs: Black Women, Food, and Power* (Chapel Hill: University of North Carolina Press, 2006), chap. 4.

2. St. Clair Drake and Horace Cayton, *Black Metropolis: A Study of Negro Life in a Northern City* (Chicago: University of Chicago Press, [1945] 1993), 8–9.

3. Although restaurants represented important sites of cooking and eating on Chicago's South Side, this essay focuses on home cooking and eating because working-class black migrant families ate at restaurants relatively rarely simply to save money. See, for instance, Charles Davis and Clarice Durham, interview with the author, Jan. 20, 2010. More broadly, see Katherine Turner, *How the Other Half Ate: A History of Working Class Meals at the Turn of the Century* (Berkeley: University of California Press, 2014), 59–85. On black Chicagoans and restaurants, see James R. Grossman, *Land of Hope: Chicago, Black Southerners, and the Great Migration* (Chicago: University of Chicago Press, 1989), 150–55; Tracy Nicole Poe, "Food, Culture, and Entrepreneurship among African Americans, Italians, and Swedes in Chicago" (PhD diss., Harvard University, 1999), 33–73; Turner, *How the Other Half Ate*, 73–74; and Chicago Commission on Race Relations (CCRR), *The Negro in Chicago: A Study of Race Relations and a Race Riot* (Chicago: University of Chicago Press, 1922), 309–16.

4. Timuel Black, interview with the author, July 27, 2010.

5. Timuel Black, interview with the author, Jan. 21, 2010.

6. Jimmy Ellis, interview with the author, July 26, 2010.

7. This essay adds to the growing body of food studies scholarship that uses "food as a way to articulate spaces and relationships of power" and answers

Nicolaas Mink's call to "begin in the belly" by "pursu[ing] the role of consump-
tion [rather than production] in environmental history." Robert N. Chester
III and Nicolaas Mink, "Having Our Cake and Eating It Too: Food's Place in
Environmental History, a Forum," *Environmental History* 14 (Apr. 2009): 311;
Nicolaas Mink, "It Begins in the Belly," *Environmental History* 14 (Apr. 2009):
314, 318.

8. Wilkerson, *Warmth of Other Suns*, 339. See also Tracy Nicole Poe, "Food,
Culture, and Entrepreneurship," 48.

9. See Turner, *How the Other Half Ate*.

10. CCRR, *Negro in Chicago*, 162–65, 172–76. See also Leila Houghteling,
The Income and Standard of Living of Unskilled Laborers in Chicago (Chicago:
University of Chicago Press, 1927), 20–21, 94; and Drake and Cayton, *Black Me-
tropolis*, 437.

11. These figures appear to have been fairly consistent nationwide. See Paul K.
Edwards, *The Southern Urban Negro as a Consumer* (New York: Prentice-Hall,
1932), 42; and Richard Sterner, *The Negro's Share: A Study of Income, Consumption, Housing and Public Assistance* (New York: Harper & Brothers, 1943), 95.

12. Emmett J. Scott, "More Letters of Negro Migrants of 1916–1918," *Journal
of Negro History* 4, no. 4 (Oct. 1919): 419, 423.

13. CCRR, *Negro in Chicago*, 172.

14. Eva Boggs, "Nutrition of Fifty Colored Families in Chicago" (MA thesis,
University of Chicago, 1929), 28.

15. See Houghteling, *Income and Standard of Living*, 25; CCRR, *Negro in
Chicago*, chap. 8; and Drake and Cayton, *Black Metropolis*, chap. 9.

16. See CCRR, *Negro in Chicago*, 154–65; and Edith Abbott, *The Tenements
of Chicago, 1908–1935* (Chicago: University of Chicago Press, 1936), 125.

17. CCRR, *Negro in Chicago*, 170–71.

18. Richard Wright, *Black Boy* (New York: HarperCollins Perennial Classics,
[1945] 1998), 278. See also Wilkerson, *Warmth of Other Suns*, 277; Opie, *Hog
& Hominy*, 90; and Davarian Baldwin, *Chicago's New Negroes: Modernity, the
Great Migration, & Black Urban Life* (Chapel Hill: University of North Carolina
Press, 2007), 164–65.

19. Less than twenty percent of white working-class women did the same,
in part because their husbands' wages were higher. Houghteling, *Income and
Standard of Living*, 56–57. On black women's labor, see also CCRR, *Negro in Chi-
cago*, 378–85; Boggs, "Nutrition of Fifty Colored Families," 9; Drake and Cayton,
Black Metropolis, 214–32; and Alma Herbst, *The Negro in the Slaughtering and
Meat-Packing Industry in Chicago* (New York: Houghton Mifflin, 1932), 75–80.
On black women and foodways more generally, see Williams-Forson, *Build-
ing Houses Out of Chicken Legs*; Jessica B. Harris, *High on the Hog: A Culinary
Journey from Africa to America* (New York: Bloomsbury, 2011); Anne L. Bower,
ed., *African American Foodways: Explorations of History and Culture* (Urbana:

University of Illinois Press, 2007); and Jennifer Jensen Wallach, ed., *Dethroning the Deceitful Pork Chop: Rethinking African American Foodways from Slavery to Obama* (Fayetteville: University of Arkansas Press, 2015). More broadly on class, gender, and cooking in the early twentieth century, see Turner, *How the Other Half Ate*, chap. 5.

20. CCRR, *Negro in Chicago*, 96, see also 170–71.

21. CCRR, *Negro in Chicago*, 264. See also Turner, *How the Other Half Ate*, 83–84.

22. The regular column initially appeared in 1922 under the title How to Make; it ran until 1929 before being replaced by Bud's Little Girl Cooks, which ran until the mid-1940s. See, for example, How to Make, *Chicago Defender*, Nov. 4, 1922, 14; and Bud's Little Girl Cooks, *Chicago Defender*, Oct. 12, 1929, A3. These and all subsequent *Defender* citations refer to the national edition.

23. "Girls' Work," *Chicago Defender*, June 10, 1922, A2.

24. "The Defender Cook Book," *Chicago Defender*, Apr. 9, 1921, 5.

25. "Kiddies Make Plans for Bud's Big Picnic," *Chicago Defender*, July 5, 1930, A3. Bud Billiken was a character the *Defender* created for its youth club.

26. Goldie M. Walden, "Children at Picnic Return Home Safely," *Chicago Defender*, Aug. 24, 1935, 5.

27. Dr. A. Wilberforce Williams, "Keep Healthy," *Chicago Defender*, July 26, 1913, 4.

28. Dr. A. Wilberforce Williams, "Dr. A. Wilberforce Williams Talks on Preventive Measures, First Aid Remedies, Hygienics and Sanitation," *Chicago Defender*, Aug. 20, 1927, A2. Elsewhere, some Urban League chapters instructed women on how to purchase food and prepare cheaper cuts of meat. Peter Gottlieb, *Making Their Own Way: Southern Blacks' Migration to Pittsburgh, 1916–30* (Urbana: University of Illinois Press, 1987), 191.

29. See Williams-Forson, *Building Houses Out of Chicken Legs*, esp. chap. 3. While the nature of working-class life drove continuity in foodways between South and North, behavioral expectations could be radically different. See Carl Sandburg, *The Chicago Race Riots, July, 1919* (New York: Harcourt, Brace and Howe, 1919), 13; and Timuel Black, *Bridges of Memory: Chicago's First Wave of Black Migration* (Evanston, IL: Northwestern University Press, 2003), 538–39. More broadly on respectability politics and uplift, see Victoria W. Wolcott, *Remaking Respectability: African American Women in Interwar Detroit* (Chapel Hill: University of North Carolina Press, 2001).

30. Dr. A. Wilberforce Williams, "Dr. A. Wilberforce Williams Talks on Preventive Measures," First Aid Remedies, Hygienics and Sanitation, *Chicago Defender*, Mar. 31, 1917, 10. Studies made similar observations about the caloric demands of African American labor in the South. See Edwards, *Southern Urban Negro*, 56.

31. On food as pleasure versus sustenance, see Nancy Shoemaker, "Food and the Intimate Environment," *Environmental History* 14 (Apr. 2009): 341.

32. Houghteling, *Income and Standard of Living*, 25, 186–217.

33. See Boggs, "Nutrition of Fifty Colored Families," 4–5; and Houghteling, *Income and Standard of Living*, 89, 137, 142. The same was true outside Chicago, North and South. See also Edwards, *Southern Urban Negro*, 54; and Kelsey B. Gardner and Lawrence A. Adams, *Consumer Habits and Preferences in the Purchase and Consumption of Meat*, United States Department of Agriculture Bulletin no. 1443 (Washington, DC: U.S. Dept. of Agriculture, Nov. 1926), 2.

34. Boggs, "Nutrition of Fifty Colored Families," 32, 52.

35. Carter Godwin Woodson, *The Rural Negro* (Washington, DC: The Association for the Study of Negro Life and History, 1930), 4.

36. Woodson, *Rural Negro*, 264. See also Boggs, "Nutrition of Fifty Colored Families," 23, 28–29; Sterner, *Negro's Share*, 115. Salt pork was a cheap cut of meat widely consumed by the working classes, preserved so that it did not require refrigeration. On southern diets, see Charles Johnson, *Shadow of the Plantation* (Chicago: University of Chicago Press, 1934), 100–101; Dorothy Dickins, "A Nutritional Investigation of Tenants in the Yazoo-Mississippi Delta," *Mississippi Agricultural Experiment Station Bulletin* 254 (Aug. 1928): 17, 39; Edwards, *Southern Urban Negro*, 56; Opie, *Hog & Hominy*, 50; David L. Cohn, *God Shakes Creation* (New York: Harper & Brothers, 1935), 188.

37. CCRR, *Negro in Chicago*, 171.

38. Carl Sandburg, *Chicago Poems* (Alexandria, VA: Chadwyck-Healey, [1916] 1998), 4. On the extensive reach of Chicago's meatpacking industry, see William Cronon, *Nature's Metropolis: Chicago and the Great West* (New York: W. W. Norton, 1992), 207–62; and Ann Vileisis, *Kitchen Literacy: How We Lost Knowledge of Where Food Comes From and Why We Need to Get It Back* (Washington, DC: Island Press, 2008), 67–73. On pork consumption in Chicago, see Drake and Cayton, *Black Metropolis*, 608; and Davis and Durham interview, Jan. 20, 2010.

39. Booker T. Washington, *Working with the Hands* (New York: Doubleday, 1904), 14. See also Opie, *Hog & Hominy*, 43–44; Edward L. Ayers, *The Promise of the New South: Life after Reconstruction* (New York: Oxford University Press, 1992), 188.

40. CCRR, *Negro in Chicago*, 87. See also Grossman, *Land of Hope*, 4. Thousands of black Chicagoans worked in the stockyards during the Great Migration years, but that connection to meat production did little to change the ways those migrants purchased and consumed meat. For more on black Chicagoans' stockyard employment, see Herbst, *Negro in the Slaughtering and Meat-Packing Industry*, 182; James R. Barrett, *Work and Community in the Jungle: Chicago's Packinghouse Workers, 1894–1922* (Urbana: University of Illinois Press, 1987);

Walter A. Fogel, *The Negro in the Meat Industry* (Philadelphia: University of Pennsylvania Press, 1970); and Rick Halpern, *Down on the Killing Floor: Black and White Workers in Chicago's Packinghouses, 1904–54* (Urbana: University of Illinois Press, 1997).

41. Boggs, "Nutrition of Fifty Colored Families," 27.

42. CCRR, *Negro in Chicago*, 172, 264; Houghteling, *Income and Standard of Living*, 171–72, 203, 210, 216.

43. CCRR, *Negro in Chicago*, 172. On the lack of space for gardens in the urban North, see Abbott, *Tenements of Chicago*, 178–79; and Opie, *Hog & Hominy*, 57.

44. Davis and Durham interview, Jan. 20, 2010. See also Wilkerson, *Warmth of Other Suns*, 285. Some migrants had enough room adjacent to their homes to keep small livestock like chickens, but this was more the exception than the rule. Davis and Durham interview, Jan. 20, 2010; Black interview, Jan. 21, 2010; Abbott, *Tenements of Chicago*, 123, 178–79; Wilkerson, *Warmth of Other Suns*, 269. Beyond Chicago, see Andrew Wiese, *Places of Their Own: African American Suburbanization in the Twentieth Century* (Chicago: University of Chicago Press, 2004), 19, 69. More broadly on the complex significance of chicken in the African American community, see Williams-Forson, *Building Houses Out of Chicken Legs*.

45. Black interview, Jan. 21, 2010.

46. Black, *Bridges of Memory*, 244.

47. Houghteling, *Income and Standard of Living*, 171, 178, 186–217.

48. Cohn, *God Shakes Creation*, 44; Dickins, "Nutritional Investigation," 35.

49. Sterner, *Negro's Share*, 115.

50. Boggs, "Nutrition of Fifty Colored Families," 24.

51. Data collected from a series of completed survey forms, each titled "My Report to the White House Conference," 1930, Box 80, Folders 3–5, Ernest Watson Burgess Papers, University of Chicago.

52. Lorraine Hansberry, *A Raisin in the Sun* (New York: Vintage Books, [1959] 1995), 31.

53. Houghteling, *Income and Standard of Living*, 171; Boggs, "Nutrition of Fifty Colored Families," 23.

54. Dickins, "Nutritional Investigation," 34; Johnson, *Shadow of the Plantation*, 100. See also Woodson, *Rural Negro*, 5. Alternatively, for firsthand accounts of Alabamans who raised much of their own food, see Sara Brooks and Thordis Simonsen, *You May Plow Here: The Narrative of Sara Brooks* (New York: Norton, 1986), 28–29, 53–58; and Nate Shaw and Theodore Rosengarten, *All God's Dangers: The Life of Nate Shaw* (New York: Random House, 1974), 122, 149, 190–91.

55. Mark D. Hersey, *My Work Is That of Conservation: An Environmental Biography of George Washington Carver* (Athens: University of Georgia Press, 2011), 131, 147, 202–3.

56. Woodson, *Rural Negro*, 5. On the South's abundance, see also Charles Spurgeon Johnson, *Growing Up in the Black Belt: Negro Youth in the Rural South* (Washington, DC: American Council on Education, 1941), 2–3, 8, 47; and Cohn, *God Shakes Creation*, 188. Even those southerners who raised their own livestock and vegetables would have purchased staples such as meat, coffee, sugar, flour, and rice from grocery stores in the nearest town. Hortense Powdermaker, *After Freedom: A Cultural Study in the Deep South* (New York: Viking Press, 1939), 84. See also Brooks and Simonsen, *You May Plow Here*, 27–29, 53–58; Shaw and Rosengarten, *All God's Dangers*, 149.

57. Opie, *Hog & Hominy*, 57; and Vileisis, *Kitchen Literacy*, 142–43.

58. City and country, Chicago and its hinterlands, were still as inextricably linked as they were in the nineteenth century, the former depending on the latter for commodities, the latter depending on the former to distribute and sell its wares. See Cronon, *Nature's Metropolis*.

59. On these markets, see Perry Duis, *Challenging Chicago: Coping with Everyday Life, 1837–1920* (Urbana: University of Illinois Press, 1998), 116, 119, 351.

60. Black, *Bridges of Memory*, 24, 282, 286; See also Black interview, Jan. 21, 2010; Ellis interview; Davis and Durham interview, July 28, 2010; Grossman, *Land of Hope*, 154–55; Michelle Obama, *American Grown: The Story of the White House Kitchen Garden and Gardens across America* (New York: Crown Publishers, 2012), 12.

61. Drake and Cayton, *Black Metropolis*, 579. Peddling food purchased from these markets benefited both buyer and seller, particularly among the working classes, regardless of race. See Chicago Municipal Markets Commission et al., *Report to the Mayor and Aldermen* (Chicago: H. G. Adair, 1914), 51.

62. See Turner, *How the Other Half Ate*, 54–58, 84; and Earl Lewis, *In Their Own Interests: Race, Class, and Power in Twentieth-Century Norfolk, Virginia* (Berkeley: University of California Press, 1991), 92.

63. CCRR, *Negro in Chicago*, 140–41. The extent to which low wages limited the affordability of dining out was reflected in there being nearly twice as many black-owned groceries (many by migrants) than restaurants in 1938. Drake and Cayton, *Black Metropolis*, 438, 454. See also Black, *Bridges of Memory*, 151–52.

64. Grossman, *Land of Hope*, 95. See also Edwards, *Southern Urban Negro*, 124–25, 130. On urban groceries and the increasingly difficult time black grocers had competing with white ethnic grocers and chain stores, see Drake and Cayton, *Black Metropolis*, 443–44, 579; Lizabeth Cohen, *Making a New Deal: Industrial Workers in Chicago, 1919–1939* (New York: Cambridge University Press, 2008), 109, 116–20, 152–54; Turner, *How the Other Half Ate*, 52–54; Vileisis, *Kitchen Literacy*, 160–63.

65. CCRR, *Negro in Chicago*, 169.

66. Richard Wright, "Introduction," in Drake and Cayton, *Black Metropolis*, xxii.

67. Drake and Cayton, *Black Metropolis*, 523.

68. Sandburg, *Chicago Poems*, 4.

69. See esp. Cronon, *Nature's Metropolis*, 211–12, 255–57.

CHAPTER 10: "NO CHEERFUL PATCHES OF GREEN"

1. Anthony Romo, "Interview by Jesse J. Escalante," in *Jesse Escalante Oral Histories*, Global Communities Collection (Chicago: Chicago History Museum, Feb. 7, 1980); U.S. Bureau of the Census, *Fifteenth Census of the United States, 1930* (Washington, DC: National Archives and Records Administration, 1930), Chicago, Cook, Illinois, 48B, E.D. 0956; U.S. Bureau of the Census, *Sixteenth Census of the United States, 1940* (Washington, DC: National Archives and Records Administration, 1940), Chicago, Cook, Illinois, 15A, E.D. 103–663; Edith Abbott and Sophonisba P. Breckinridge, *The Tenements of Chicago, 1908–1935*, Social Service Monographs (Chicago: University of Chicago Press, 1936), 150.

2. Abbott and Breckinridge, *Tenements of Chicago*, 150–51.

3. Mary Faith Adams, "Present Housing Conditions in South Chicago, South Deering and Pullman" (MA thesis, University of Chicago, 1926), 20.

4. Breckinridge and Abbott, "Chicago Housing Conditions, V: South Chicago at the Gates of the Steel Mills," *American Journal of Sociology* 17, no. 2 (1911): 174.

5. Nicolas Kenny, *The Feel of the City: Experiences of Urban Transformation* (Toronto: University of Toronto Press, 2014), 5; Anita Edgar Jones, "Conditions Surrounding Mexicans in Chicago" (MA thesis, University of Chicago, 1928), 55. I use the term *Mexicano* to represent Mexican immigrants and Mexican Americans who self-identified as part of the Mexican community. For more on ethnic community and cooperation within a polluted urban space, see Isabel García, Fernando Guiuliani, and Esther Wiesenfeld, "Community and Sense of Community: The Case of an Urban Barrio in Caracas," *Journal of Community Psychology* 27, no. 6 (1999) 727–40.

6. Max Guzman, "Interview by Jesse J. Escalante," in *Jesse Escalante Oral Histories*; Romo, "Interview"; Sidney Levin, "Interview by Jesse J. Escalante," in *Jesse Escalante Oral Histories*; Jesse John Escalante, "History of the Mexican Community in South Chicago" (MA thesis, Northeastern Illinois University, 1982), 23; Edgar Jones, "Conditions Surrounding Mexicans," 72–73; Kenny, *Feel of the City*, 1.

7. Adams, "Present Housing Conditions," 15; Jorge Hernandez-Fujigaki, "Mexican Steelworkers and the United Steelworkers of America in the Midwest: The Inland Steel Experience (1936–1976)" (PhD diss., University of Chicago, 1991), 27. The specific ethnic groups hired at Illinois steel switched from English, German, Welsh, Belgian, and Dutch to Polish, Bohemian, Slovak, Serbian Bulgarian, Lithuanian, Magyar, Romanian, and Macedonian.

8. Ernest W. Burgess and Charles Newcomb, *Census Data of the City of Chicago, 1920* (Chicago: University of Chicago Press, 1920); Burgess and Newcomb,

Census Data of the City of Chicago, 1930 (Chicago: University of Chicago Press, 1933); U.S. Bureau of the Census, *Abstract of the 15th Census of the United States* (Washington, DC: Government Printing Office, 1933); Lilia Fernandez, *Brown in the Windy City: Mexicans and Puerto Ricans in Postwar Chicago* (Chicago: University of Chicago Press, 2012), 69.

    9. Thomas L. Philpott, *The Slum and the Ghetto: Neighborhood Deterioration and Middle-Class Reform* (New York: Oxford University Press, 1978), 27.

    10. James R. Grossman, *Land of Hope: Chicago, Black Southerners, and the Great Migration* (Chicago: University of Chicago Press, 1989), 113; Adams, "Present Housing Conditions," 70.

    11. H. R. Crohurst and M. V. Veldee, "Report of an Investigation of the Pollution of Lake Michigan in the Vicinity of South Chicago and the Calumet and Indiana Harbors, 1924–1925," in *Public Health Bulletin* (Washington, DC: U.S. Public Health Service, Treasury Department, 1927), 24–31.

    12. Crohurst and Veldee, "Report of an Investigation of the Pollution of Lake Michigan," 35.

    13. Crohurst and Veldee, "Report of an Investigation of the Pollution of Lake Michigan," 4–5, 42.

    14. Melanie Kiechle, "Navigating by Nose: Fresh Air, Stench Nuisance, and the Urban Environment, 1840–1880," *Journal of Urban History* 42, no. 4 (2016): 761. For a more detailed look at stench in areas in and close to the Loop in the late nineteenth century, see Melanie Kiechle, *Smell Detectives: An Olfactory History of Nineteenth-Century Urban America* (Seattle: University of Washington Press, 2017).

    15. Elizabeth Ann Hughes, *Living Conditions for Small Wage Earners in Chicago* (Chicago: City of Chicago Department of Public Welfare, 1925), 9.

    16. Edgar Jones, "Conditions Surrounding Mexicans," 87, 89; Robert Redfield, Robert Redfield Journal, box 59, folder 2, Robert Redfield Papers, Special Collections Research Center, University of Chicago Library, Chicago; Dorothea Kahn, "Mexicans Bring Romance to Drab Part of Chicago in Their Box-Car Villages: 30,000 NOW MAKE CITY," *Christian Science Monitor*, May 23, 1931.

    17. Kahn, "Mexicans Bring Romance."

    18. For more on Chicago Mexican food and memories during this period, see Michael Innis-Jiménez, "Mexican Chicago's Colonia Hull House: Food, Tourism, and Belonging in the 1930s," *Global Food History* 4, no. 1 (2018): 22–39.

    19. Paul Schuster Taylor, *Mexican Labor in the United States: Chicago and the Calumet Region*, University of California Publications in Economics, vol. 7, no. 2 (Berkeley: University of California Press, 1932), 32–33.

    20. Helen Elizabeth Gregory MacGill, "Land Values as an Ecological Factor in the Community of South Chicago" (MA thesis, University of Chicago, 1927), 9, 15–16.

21. Sanborn Map Company, *Chicago, 1905–1951*, vol. F, *1913–May 1946* (Teaneck, NJ: Sanborn), 64.

22. 1930 U.S. Census, Cook County, Illinois, roll 432, 11b–17a; Enumeration District: 2458, FHL Microfilm: 2340167, digital image, Ancestry.com, http://ancestry.com// (accessed Sept. 12, 2017).

23. 1930 U.S. Census, Cook County, Illinois.

24. Alfredo de Avila, "Interview by Jesse J. Escalante," in *Jesse Escalante Oral Histories*; Alfredo de Avila Entry Card, Nonstatistical Manifests and Statistical Index Cards of Aliens Arriving at El Paso, Texas, 1905–1927, Record Group 85, Microfilm Serial A3406, Microfilm Roll 9, Records of the Immigration and Naturalization Service; Hughes, "Living Conditions," 12–13.

25. Hughes, "Living Conditions," 12–13.

26. Jones, "Conditions Surrounding Mexicans," 170–71.

27. Max Guzman, "Interview"; Romo, "Interview," Oct. 22, 1979; Jones, "Conditions Surrounding Mexicans," 72–73; Escalante, "History of the Mexican Community," 23; Levin, "Interview"; Marian Lorena Osborn, "The Development of Recreation in the South Park System of Chicago" (MA thesis, University of Chicago, 1928), 1–2.

28. Levin, "Interview"; Romo, "Interview"; Edgar Jones, "Conditions Surrounding Mexicans," 72–73; Escalante, "History of the Mexican Community," 23; Gilbert Martinez, "Interview by Jesse J. Escalante," in *Jesse Escalante Oral Histories*. Chicago-area Mexican youth frequently referred to working-class European immigrants (other than the Irish and Italians) as Polish regardless of their national origin.

CHAPTER 11: WORK RELIEF LABOR IN THE COOK COUNTY FOREST PRESERVES, 1931–1942

1. Carl Smith, *The Plan of Chicago: Daniel Burnham and the Remaking of the American City* (Chicago: University of Chicago Press, 2006), 87.

2. Smith, *Plan of Chicago*, 87.

3. Brian McCammack, *Landscapes of Hope: Nature and the Great Migration in Chicago* (Cambridge, MA: Harvard University Press, 2017), 218–45.

4. Annual Message of the President of the Forest Preserve District of Cook County Board of Commissioners, Jan. 3, 1921, F37X .C7-R1, Chicago History Museum Research Center (hereafter CHMRC).

5. Stephen Fox, *The American Conservation Movement: John Muir and His Legacy* (Madison: University of Wisconsin Press, 1986), 116; Roderick Nash, *Wilderness and the American Mind* (New Haven, CT: Yale University Press, 2001), 147.

6. Annual Message of the President of the Forest Preserve District of Cook County Board of Commissioners, Jan. 5, 1920, F37X .C7-R1, CHMRC.

7. Annual Message of the President of the Forest Preserve District, 1920.

8. Forest Preserve District Advisory Committee (FPDAC), *Recommended Plans for Forest Preserves of Cook County, Illinois,* Jan. 1929, file III-57-1433, Forest Preserve District of Cook County Collection, Special Collections, University of Illinois at Chicago Library (hereafter FPDCC Collection).

9. FPDAC, *Recommended Plans.*

10. James Feldman, *A Storied Wilderness* (Seattle: University of Washington Press, 2013), 212.

11. FPDAC, *Recommended Plans.*

12. Natalie Bump Vena, "The Nature of Bureaucracy in the Cook County Forest Preserves" (PhD diss., Northwestern University, 2016), chap. 3.

13. FPDAC, *Recommended Plans.*

14. "Select Indiana Man as Superintendent of Forest Preserves," *Chicago Daily Tribune,* Apr. 19, 1929.

15. Charles Sauers, résumé, n.d., file I-27-274, FPDCC Collection.

16. Lizabeth Cohen, *Making a New Deal: Industrial Workers in Chicago, 1919-1939* (Cambridge: Cambridge University Press, 1990), 217.

17. Resolution by Commissioner Walter H. La Buy, Sept. 6, 1932, file I-17-164, FPDCC Collection.

18. Charles Sauers to Commissioner Fleming, Nov. 17, 1932, file I-17-164, FPDCC Collection.

19. Charles Sauers to Miss Cravens, Sept. 11, 1934, file I-27-273, FPDCC Collection.

20. Roberts Mann, "Accomplishments by Forest Preserve District with Relief Labor—1931-1934," n.d., file III-21-292, FPDCC Collection.

21. Obituary of Roberts Mann, n.d., III-22-303, FPDCC Collection.

22. Roberts Mann, "Talk on Park Maintenance," ca. 1935, file III-I-17, FPDCC Collection.

23. Roberts Mann, *Annual Report for Maintenance Department,* 1932, file III-21A-297, FPDCC Collection.

24. Annual Message of the President of the Forest Preserve District of Cook County Board of Commissioners, 1933, F37X .C7-R1, CHMRC; Roberts Mann, "Talk on Park Maintenance."

25. *Camp Skokie Valley Review,* Nov. 16, 1934, file III-8-119, FPDCC Collection.

26. Robert Kingery to the Advisory Board, Jan. 20, 1937, file V-47-647, FPDCC Collection; Charles Sauers, "Brief History of the Origin and Development of the Forest Preserve District of Cook County, Illinois, Part III," *Forest Way,* 1945, file I-47-523, FPDCC Collection.

27. Roberts Mann, "A Paper on the Danger of Over-development under Federal Relief Programs," Sept. 21, 1938, file III-21-285, FPDCC Collection.

28. Robert Kingery to the Advisory Board.

29. Mann, "Danger of Over-development under Federal Relief Programs."

30. Mann, *Annual Report for Maintenance Department.*

31. Mann, "Accomplishments by Forest Preserve District."

32. Clayton Smith to Senator James Hamilton Lewis, Nov. 22, 1937, file I-18-186, FPDCC Collection.

33. Proceedings of staff meeting, Sept. 8, 1941, file III-21-296, FPDCC Collection.

34. Meeting notes of the Conservation Council at the Morrison Hotel, Dec. 20, 1934, file I-27-273, FPDCC Collection.

35. Roberts Mann, "The Silver Lining," ca. 1933, 1, file III-21-292, FPDCC Collection.

36. Cohen, *Making a New Deal,* 278, 281.

37. Mann, *Annual Report for Maintenance Department.*

38. Mickie Fisher, "The Civilian Conservation Corps Camp Chicago-Lemont Located in Willow Springs, Illinois," n.d., 1, file III-59-1469, FPDCC Collection.

39. Neil Maher, *Nature's New Deal: The Civilian Conservation Corps and the Roots of the American Environmental Movement* (New York: Oxford University Press, 2007), 79.

40. Mann, *Annual Report for Maintenance Department.*

41. "Justification of CCC Quarry Company," n.d., file III-36-562, FPDCC Collection.

42. Mann, "Silver Lining."

43. John Tebora to Clayton Smith, Sept. 14, 1937, file I-18-185, FPDCC Collection.

44. McCammack, *Landscapes of Hope,* 203, 245.

45. "Justification of CCC Quarry Company," 2.

46. Memo from Charles Sauers to Division Superintendents, Feb. 6, 1934, file III-9-134, FPDCC Collection.

47. Roberts Mann, "Streamlined Maintenance," ca. World War II, 3, file III-1-18, FPDCC Collection.

48. Neil Maher, *Nature's New Deal,* 70.

49. Mann, "Talk on Park Maintenance."

50. Charles Sauers to Mrs. Coffman, Nov. 8, 1937, file I-27-274, FPDCC Collection.

51. Hal Dardick, "Cook County Forest Preserves to Lose Final Toboggan Slides, Despite Last Battle," *Chicago Tribune,* Apr. 9, 2008.

52. "Justification of CCC Quarry Company."

53. Mann, "Danger of Over-development," 6.

54. Mann, "Danger of Over-development"; Mann, "Accomplishments by Forest Preserve District."

55. Mann, "Accomplishments by Forest Preserve District," 3.

56. Fisher, "The Civilian Conservation Corps Camp."

57. "Justification of CCC Quarry Company," 3.

58. Mann, "Talk on Park Maintenance."

59. "Companies 608 and 618 Make Great Advance in Lagoons and Preserves," *Camp Skokie Valley Review,* Apr. 10, 1936, file III-8-119, FPDCC Collection.

60. Feldman, *Storied Wilderness,* 149.

61. Feldman, *Storied Wilderness,* 150.

62. Mann, "Silver Lining."

63. Minutes of Advisory Committee Meeting, Dec. 8, 1941, file V-47-648, FPDCC Collection.

64. Maher, *Nature's New Deal,* 48.

65. Forest Preserve District of Cook County and Works Progress Administration, "Lesson No. 7: Planting Practices," 1938, file III-32-479, FPDCC Collection.

66. Robert Kingery, John Barstow Morrill, and Charles Sauers, *Revised Report of Advisory Committee to the Cook County Forest Preserve Commissioners,* 1959, file V-53-693, FPDCC Collection.

67. Theodore Sperry, *Report on Proposed Prairie Restoration for Forest Preserve District of Cook County,* 1940, file III-28-422, FPDCC Collection.

68. Roberts Mann, memorandum, July 25, 1935, file III-9-135, FPDCC Collection.

69. Charles Sauers to Al Hornick, Sept. 26, 1938, file I-19-189, FPDCC Collection.

70. "Professional Fire Plan F.P.D.," ca. 1937, file III-17-208, FPDCC Collection.

71. Frank H. Nelson to Mr. N. Y. Alvis, Apr. 27, 1936, file III-17-211, FPDCC Collection.

72. Mann, *Annual Report for Maintenance Department.*

73. Works Progress Administration Information Service, *Better Illinois Communities through W.P.A. Projects,* 1936, pamphlet, 21, file V-21-322, FPDCC Collection.

74. Game Management Conference Minutes, Feb. 14, 1941, file III-21-296, FPDCC Collection.

75. "Heavy Equipment Plays Important Part on Project," *Camp Skokie Valley Review,* Nov. 16, 1934, file III-8-119, FPDCC Collection.

76. Minutes of Cook County Forest Preserve Citizens Advisory Committee meeting, July 24, 1933, file V-47-646, FPDCC Collection.

77. E. J. Lundin to Board President Clayton Smith, Apr. 18, 1940, file I-19-197, FPDCC Collection.

78. "CCC Helps to Make Parks of Swamp Land: Huge Engineering Job about 60 Per Cent Completed," *Camp Skokie Valley Review,* Apr. 2, 1937, file V-19-293, FPDCC Collection.

79. "CCC Helps to Make Parks of Swamp Land."

80. Charles Estes to Mr. Carter Jenkins, July 18, 1940, file III-34-505, FPDCC Collection; Game Management Conference Minutes, Feb. 14, 1941.

81. Roberts Mann, "A Program of Education in Landscape Management," paper presented at the National Conference on State Parks, 1938, 1, file III-21A-297, FPDCC Collection.

82. Mann, "Danger of Over-development."

83. Jens Jensen, "The Beauty of the Skokie," *Chicago Sunday Tribune*, July 31, 1932.

84. Jens Jensen, "Report of the Landscape Architect," in *Report of the Special Park Commission on the Subject of a Metropolitan Park System*, ed. Dwight Perkins (Chicago: City Council of Chicago, 1904), 67.

85. Maher, *Nature's New Deal*, 174, 178.

86. Paul Sutter, *Driven Wild: How the Fight against Automobiles Launched the Modern Wilderness Movement* (Seattle: University of Washington Press, 2009), 4, 7.

87. Maher, *Nature's New Deal*, 9.

88. Aldo Leopold, "Engineering and Conservation," in *The River of the Mother of God*, ed. Susan Flader and J. Baird Callicott (Madison: University of Wisconsin, 1991), 251–53, quoted in Maher, *Nature's New Deal*, 166–67.

89. Leopold, "Engineering and Conservation," 251.

90. Roberts Mann, "Aldo Leopold, Priest and Prophet," 1954, 5, file III-22-300, FPDCC Collection.

91. Aldo Leopold, "Conservation Esthetic," *Bird-Lore* 40, no 2 (Mar.-Apr. 1938): 101–9, quoted in Sutter, *Driven Wild*, 99.

92. Mann, "Aldo Leopold, Priest and Prophet."

93. Roberts Mann, "How the Park Man Should Be Prepared to Meet the Public," 1953, 1, file III-4-82, FPDCC Collection.

94. Charles Sauers to J. Lyell Clarke, Apr. 15, 1940, file III-24-361, FPDCC Collection.

95. Roberts Mann, "State Park Administration: A Paper Given at the Institute in Landscape Management, New York State College of Forestry," n.d., 19, file III-21-284, FPDCC Collection.

96. Chairman and Members of the Advisory Committee to the Cook County Forest Preserve Commissioners, Jan. 8, 1943, file V-47-648, FPDCC Collection.

97. Mann, "State Park Administration," 3.

98. Charles Sauers, "Memorandum on Finances to the Advisory Committee to the Board of Forest Preserve Commissioners from the General Superintendent," 1952, file III-1-5, FPDCC Collection.

CHAPTER 12: MAPS AND CHICAGO'S ENVIRONMENTAL HISTORY

1. Northeastern Illinois Metropolitan Area Local Governmental Services Commission, *Government Problems in the Chicago Metropolitan Area* (Chicago: University of Chicago Press, 1957), 57.

2. Jennifer Koslow, "Public Health," *The Encyclopedia of Chicago*, ed. James R. Grossman, Ann Keating, and Janice L. Reiff (Chicago: University of Chicago Press, 2004), 657–59.

3. William Cronon, *Nature's Metropolis: Chicago and the Great West* (New York: W. W. Norton, 1991).

4. Paul Petreitis, "James H. Rees and the Idea of Chicagoland," unpublished lecture, Chicago Map Society, Nov. 17, 2016.

5. See, for example, J. O. Wright, *Swamp and Overflowed Lands in the United States: Ownership and Reclamation*, U.S. Department of Agriculture, Office of Experiment Stations—Circular 76 (Washington, DC: Government Printing Office, 1907). See also Ann Vilesis, *Discovering the Unknown Landscape: A History of America's Wetlands* (Washington, DC: Island Press, 1999).

6. Steven F. Strausberg, "Indiana and the Swamp Lands Act: A Study in State Administration," *Indiana Magazine of History* 73 (Sept. 1977), 191–203.

7. Margaret Beattie Bogue, "The Swamp Land Act and Wet Land Utilization in Illinois, 1850–1890," *Agricultural History* 25, no. 169 (1951): 169–80.

8. Paul Petraitis, "Lake Michigan Shoreline Data," unpublished manuscript of a lecture, The Newberry Library, May 14, 2014.

9. Mark W. Harrington, *Lake Michigan: Bottle Papers Courses of 1892, 1893, and 1894. Special Drifts and Resultant Currents* (map), in *Surface Currents of the Great Lakes: As Deduced from the Movements of Bottle Papers during the Seasons of 1892, 1893, and 1894* (Washington, DC: U.S. Weather Bureau, 1895), n.p.

10. F. M. Fryxell, *The Physiography of the Region of Chicago* (Chicago: University of Chicago Press, 1927).

11. Louis Joliet, quoted in Robert A. Holland, *Chicago in Maps, 1612 to 2002* (New York: Rizzoli, 2005), 28.

12. H. Roger Grant, "Transportation," in James R. Grossman et al., *Encyclopedia of Chicago*, 827–32.

13. For the early history of Rand McNally, see Cynthia H. Peters, "Rand McNally in the Nineteenth Century: Reaching for a National Market," in *Chicago Mapmakers: Essays on the Rise of the City's Map Trade*, ed. Michael Conzen (Chicago: Chicago Historical Society for the Chicago Map Society, 1984), 64–72.

14. David Woodward, *The All-American Map: Wax-Engraving and Its Influence on Cartography* (Chicago: University of Chicago Press, 1977).

15. Jay P. Pederson, "The History of Wilson Sporting Goods Company," in Pederson, *International Directory of Company Histories*, vol. 24 (London: St. James Press, 1999), 530–32.

16. Interview with former local resident Sandra Nekola (née Senkpiel), Chicago, Feb. 14, 2017.

17. Jane Addams, *Hull-House Maps and Papers: A Presentation of Nationalities and Wages in a Congested District of Chicago, Together with Comments and Essays on Problems Growing Out of the Social Conditions* (New York: Thomas

Y. Crowell, 1895), quote on title page; Samuel Sewell Greeley, *Nationalities Map No. 1[–4], Polk Street to Twelfth . . . Chicago*, detached from Addams, *Hull-House*.

18. Louise C. Wade, "The Heritage from Chicago's Early Settlement Houses," *Journal of the Illinois State Historical Society* 60, no. 4 (Winter 1967): 414.

19. Daniel H. Burnham and Edward H. Bennett, *Plan of Chicago* (Chicago: The Commercial Club, 1909), 32.

20. Burnham and Bennett, *Plan of Chicago*, pl. CVII, 97.

21. Burnham and Bennett, *Plan of Chicago*, pl. CX, 101.

22. A notable contemporary was Arthur Shurcliff's plan for metropolitan Boston, also dating from 1909, developed for the Commonwealth of Massachusetts Metropolitan Improvements Commission.

PART IV: MANAGING (OR NOT) URBAN-INDUSTRIAL COMPLEXITY

1. Rudyard Kipling, quoted in Bessie Louise Pierce, *As Others See Chicago: Impressions of Visitors, 1673–1933* (Chicago: University of Chicago Press, 1933), 251.

2. Sophonisba P. Breckinridge and Edith Abbott, "Housing Conditions in Chicago, IL: Back of the Yards," *American Journal of Sociology* 16, no. 4 (Jan. 1911): 433–35, 437–39; see also Upton Sinclair, *The Jungle* (1906; New York: Norton, 2003), 408.

CHAPTER 13: BLOOD ON THE TRACKS

I wish to thank Theo Anderson, William Barnett, Kathleen Brosnan, Paul Hughes, Ann Durkin Keating, and two anonymous reviewers for their help with this essay.

1. "Crushed by a Train," *Chicago Tribune*, July 18, 1893, 1.

2. "Crushed by a Train."

3. Theodore Steinberg, *Down to Earth: Nature's Role in American History*, 3rd ed. (New York: Oxford University Press, 2013), 155, 169; Suellen Hoy, *Chasing Dirt: The American Pursuit of Cleanliness* (New York: Oxford University Press, 1995); Gretchen A. Condran and Eileen Crimmins-Gardner, "Public Health Measures and Mortality in U.S. Cities in the Late Nineteenth Century," *Human Ecology* 6, no. 1 (1978): 27–54.

4. Thomas Grant Allen, "Chicago Takes Greater Toll of Human Life: Chicago's Loss from Violence," *Chicago Tribune*, Apr. 15, 1906, F2.

5. Another approach to the environmental history of death is to examine the history of bodies. See Ellen Stroud, "Reflections from Six Feet under the Field: Dead Bodies in the Classroom," *Environmental History* 8 (2003): 618–27.

6. Cook County Coroner's Inquest Records, 1872–1911, Illinois Regional Archives Depository, Northeastern Illinois University, Chicago (hereafter CCCIR).

7. Arwen Mohun, *Risk: Negotiating Safety in American Society* (Baltimore, MD: Johns Hopkins University Press, 2013), chap. 7.

8. Joshua Salzmann, *Liquid Capital: Making the Chicago Waterfront* (Philadelphia: University of Pennsylvania Press, 2018), 1–2.

9. William Cronon, *Nature's Metropolis: Chicago and the Great West* (New York: W. W. Norton, 1991), 169–80.

10. Libby Hill, *The Chicago River: A Natural and Unnatural History* (Chicago: Lake Claremont Press, 2000), 121.

11. Daniel Bluestone, *Constructing Chicago* (New Haven, CT: Yale University Press, 1991).

12. William Bross, "Chicago and the Sources of Her Past and Future Growth: A Paper Read before the Chicago Historical Society, Tuesday Evening, January 20th, 1880" (Chicago: Jansen, McClurg, 1880), 3.

13. Louis Sullivan, quoted in Cronon, *Nature's Metropolis*, 14.

14. Matthew Gandy, *Concrete and Clay: Reworking Nature in New York City* (Cambridge, MA: MIT Press, 2002).

15. CCCIR, 1373, Jan. 29, 1874; CCCIR, 155, June 23, 1884.

16. CCCIR, 2589, Mar. 9, 1894; CCCIR, 1125, Jan. 24, 1884.

17. David Von Drehle, *Triangle: The Fire That Changed America* (New York: Grove Press, 2003), 3.

18. CCCIR, 1376, Jan. 31, 1874.

19. CCCIR, 1462, Apr. 15, 1874.

20. CCCIR, 275, Jan. 10, 1894.

21. CCCIR, 2646, Mar. 26, 1894.

22. CCCIR, 1153, Feb. 1, 1884.

23. Jane Addams, *Twenty Years at Hull House* (New York: MacMillan, 1910), 198–99.

24. Thomas Hughes, *American Genesis: A Century of Invention and Technological Enthusiasm, 1870-1970* (New York: Penguin Press, 1989), chap. 5.

25. CCCIR, 1322, Apr. 18, 1884.

26. CCCIR, 1233, Mar. 7, 1884.

27. "To Prevent Death by Asphyxiation," *Chicago Tribune*, May 25, 1892, 6.

28. "Gas Costs Six Lives," *Chicago Tribune*, Oct. 27, 1892, 1.

29. Bessie Louise Pierce, *A History of Chicago*, vol. 3, *The Rise of a Modern City, 1871-1893* (Chicago: University of Chicago Press, 1957), 3–6.

30. Pierce, *A History of Chicago*, 3:3–6.

31. Jonathan J. Keyes, "The Forgotten Fire," *Chicago History 26 (1997): 52-65.*

32. Pierce, *History of Chicago*, 3:11–12.

33. "Fire-Proof Construction," *Chicago Tribune*, Aug. 18, 1875, 8.

34. "Laws Violated at the Iroquois," *Chicago Tribune*, Jan. 2, 1904, 4.

35. Nat Brant, *Chicago Death Trap: The Iroquois Theater Fire of 1903* (Carbondale: Southern Illinois University Press, 2006).

36. "Laws Violated at the Iroquois."

37. "Trace Faults to 'Free List,'" *Chicago Tribune*, Jan. 3, 1904, 2; Christina Cogswell, "Fire Codes in Twentieth Century Chicago: The Iroquois Theater Fire and the Perception of Safety," paper presented at the Conference on Illinois History, Springfield, IL, Sept., 24, 2015.

38. Ordinance: Building Ordinances (Theaters) Amending Clauses IV and V, Chicago City Council Proceedings Files, Document, 3764, Folder 11, Illinois Regional Archives Depository, Northeastern Illinois University, Chicago.

39. Harold M. Mayer and Richard C. Wade, *Chicago: Growth of a Metropolis* (Chicago: University of Chicago Press, 1969), 35.

40. Ted Robert Mitchell, "Connecting a Nation, Dividing a City: How Railroads Shaped Public Spaces and the Social Understanding of Chicago" (PhD diss., Michigan State University, 2009), 3, 120, 135, 179.

41. Upton Sinclair, *The Jungle* (Mineola, NY: Dover, [1906] 2001), 144.

42. Department of Public Health, *Annual Report* (Chicago: City of Chicago, 1870), 66 (hereafter year and page number); 1875, 50; 1881, 81; 1886, 15; 1890, 50; 1901, 64.

43. "The Duty of Railroads to Build Viaducts," *Chicago Tribune*, July 12, 1891, 12.

44. Mitchell, "Connecting a Nation, Dividing a City," 179.

45. CCCIRs, Jan. 1–July 1, 1874.

46. CCCIR, 1359, Jan. 20, 1874.

47. CCCIR, 1521, May 27, 1874.

48. CCCIR, 1438, Apr. 3, 1874.

49. CCCIR, 1479, Apr. 29, 1874.

50. CCCIR, 1381, Feb. 5, 1874.

51. CCCIRs, Jan. 1–July 1, 1874.

52. CCCIRs, Jan. 1–July 1, 1884.

53. CCCIRs, Jan. 1–July 1, 1894.

54. CCCIRs, Jan. 1–July 1, 1894.

55. CCCIR, 1284, Mar. 28, 1884.

56. CCCIR, 2483, Feb. 6, 1894.

57. CCCIR, 2452, Jan. 30, 1894.

58. "Is This Not Murder?," *Chicago Tribune*, Dec. 27, 1889, 1.

59. "Who Should Build Viaducts over Railroad Tracks?," *Chicago Tribune*, July 4, 1891, 10.

60. "Who Should Build Viaducts over Railroad Tracks?"

61. "Who Should Build Viaducts over Railroad Tracks?"

62. "The Supreme Court on Railroad Crossings," *Chicago Tribune*, Jan. 22, 1892, 4.

63. Barbara Welke, *Recasting American Liberty: Gender, Race, Law, and the Railroad Revolution, 1865–1902* (New York: Cambridge University Press, 2001), 36.

64. Allen, "Chicago Takes Greater Toll of Human Life."

65. Allen, "Chicago Takes Greater Toll of Human Life."

66. Linda Nash, *Inescapable Ecologies: A History of Environment, Disease, and Knowledge* (Berkeley: University of California Press, 2006), 83.

67. Hoy, *Chasing Dirt*; and John Duffy, *The Sanitarians: A History of American Public Health* (Chicago: University of Illinois Press, 1992).

68. Allen, "Chicago Takes Greater Toll of Human Life."

69. Allen, "Chicago Takes Greater Toll of Human Life."

70. Mohun, *Risk*, 141–42, 145.

CHAPTER 14: AIR AND WATER POLLUTION IN THE URBAN-INDUSTRIAL NEXUS

1. For an overview of U.S. urbanization, see Lisa Krissoff Boehm and Steven H. Corey, *America's Urban History* (New York: Routledge, 2015), 6–11, 144–47.

2. Robert Lewis, *Chicago Made: Factory Networks in the Industrial Metropolis* (Chicago: University of Chicago Press, 2008), 23–28; Lizabeth Cohen, *Making a New Deal: Industrial Workers in Chicago, 1919–1939* (New York: Cambridge University Press, 1990), 13.

3. Harold L. Platt, *Shock Cities: The Environmental Transformation and Reform of Manchester and Chicago* (Chicago: University of Chicago Press, 2005), 286, 493–98.

4. For overviews of industrialization and social settlement patterns, see Christine Meisner Rosen, *The Limits of Power: Great Fires and the Process of City Growth in America* (New York: Cambridge University Press, 1986), 152–61; Lewis, *Chicago Made*, 29–39; Platt, *Shock Cities*, 117–18, 365; Dominic A. Pacyga, *Chicago: A Biography* (Chicago: University of Chicago Press, 2009), 60–65; Andrew Hurley, "Industrial Pollution," in *The Encyclopedia of Chicago*, ed. James R. Grossman, Ann Durkin Keating, and Janice Reiff (Chicago: University of Chicago Press, 2004), 411–12; David Bensman and Mark R. Wilson, "Iron and Steel," in Grossman et al., *Encyclopedia of Chicago*, 425–27; and Cohen, *Making a New Deal*, 11–52.

5. Dominic A. Pacyga, *Slaughterhouse: Chicago's Union Stockyards and the World It Made* (Chicago: University of Chicago Press, 2015), 90–91.

6. Estelle Latkovich, quoted in "Fifty Year Graduate Shares Fond Memories," *Daily Calumet*, May 23, 1983, as cited by Cohen, *Making a New Deal*, 26.

7. Hurley, "Industrial Pollution," 411–12.

8. David Stradling, *Smokestacks and Progressives: Environmentalists, Engineers, and Air Quality in America, 1881–1951* (Baltimore: Johns Hopkins University Press, 1999), 22; Platt, *Shock Cities*, 117, 251; Chicago Association of Commerce, Committee of Investigation on Smoke Abatement and Electrification of Railway Terminals, *Smoke Abatement and Electrification of Railway Terminals in Chicago* (Chicago: Committee of Investigation, 1915; repr.

Elmsford, NY: Maxwell Reprint Company, 1971), 16–17 (hereafter *Smoke Abatement*).

9. Pacyga, *Chicago*, 39–40; Platt, *Shock Cities*, 247; Christine Meisner Rosen, "Businessmen against Pollution in Late Nineteenth Century Chicago," *Business History Review* 69, no. 3 (1995): 356–57; Paul P. Bird, *Report of the Department of Smoke Inspection City of Chicago* (Chicago: City of Chicago, Department of Smoke Inspection, Feb. 1911), 12–13.

10. "Chicago's Sooty Curse: A North Side District Particularly Unfortunate," *Chicago Daily Tribune*, July 31, 1890, 8.

11. Bird, *Report of the Department of Smoke Inspection*, 12; Rosen, "Businessmen against Pollution," 356–60.

12. Rosen, "Businessmen against Pollution," 360.

13. For differing motivations on compliance, or noncompliance, see Rosen, "Businessmen against Pollution," 361–71.

14. For an overview of waste sinks and waste management systems, see Joel A. Tarr, *The Search for the Ultimate Waste Sink: Urban Pollution in Historical Perspective* (Akron, OH: University of Akron Press, 1996).

15. Platt, *Shock Cities*, 388–89, 404–7; Vinton W. Bacon and Frank E. Dalton, "Professionalism and Water Pollution Control in Greater Chicago," *Water Pollution and Control Federation Journal* 40, no. 9 (Sept. 1968): 1586–88; Metropolitan Water Reclamation District of Greater Chicago, "Mission and Services," https://www.mwrd.org/irj/portal/anonymous?NavigationTarget=navurl:// ac86fd166ae2f8997581bde33ae1034a (accessed Sept. 23, 2017); Carl A. Zimring and Michael A. Byson, "Infamous Past, Invisible Present: Searching for Bubbly Creek in the Twenty-First Century," *IA: Journal of the Society of Industrial Archeology* 39, nos. 1 and 2 (2013): 81; Pacyga, *Slaughterhouse*, 178–79.

16. Bird, *Report of the Department of Smoke Inspection*, 17–18; Platt, *Shock Cities*, 473–79.

17. Stradling, *Smokestacks and Progressives*, 118–20, 124–28.

18. Bird, *Report of the Department of Smoke Inspection*, 26–27. As Platt notes, the May 1905 edition of the *Journal of the American Medical Association* restricted its recommendation on smoke pollution to just one remedy—the taking of children out to the countryside once in a while for fresh air; see Platt, *Shock Cities*, 483.

19. *Smoke Abatement*, 109–10, 178, and for "allies" of smoke, 194–209.

20. *Smoke Abatement*, 282, 1051–52.

21. Platt, *Shock Cities*, 488–89; Stradling, *Smokestacks and Progressives*, 156–60.

22. "Smoke Nuisance Creates Heated Protest Threat," *Chicago Daily Tribune*, Oct. 17, 1948, S2. For an overview of the Donora smog, see Edwin Kiester Jr., "A Darkness in Donora," *Smithsonian Magazine*, Nov. 1999, https://www.smith

sonianmag.com/history/a-darkness-in-donora-174128118/; and Devra Davis, *When Smoke Ran Like Water: Tales of Environmental Deception and the Battle against Pollution* (New York: Basic Books, 2002), 15–20, 31–54.

23. "Air Pollution Creates Slums, Expert Warns," *Chicago Daily Tribune*, Jan. 20, 1949, A9.

24. James Longhurst, *Citizen Environmentalists* (Medford, MA: Tufts University Press; and Hanover, NH: University Press of New England, 2010), 37–39; Gary C. Bryner, *Blue Skies, Green Politics: The Clean Air Act of 1990* (Washington, DC: Congressional Quarterly Press, 1993), 81–85; Scott Hamilton Dewey, *Don't Breathe the Air: Air Pollution Control and U.S. Environmental Politics, 1945–1970* (College Station: Texas A&M Press, 2000), 74–77, 239–44; Indur Goklany, *Cleaning the Air: The Real Story of the War on Air Pollution* (Washington, DC: Cato Institute, 1999), 21–33.

25. David Stradling, "Air Quality," in Grossman et al., *Encyclopedia of Chicago*, 11; Harold Barger, "City Learns S.E. Breezes Are Ill Winds," *Chicago Daily Tribune*, Oct. 11, 1962, S1; William Kling, "Air Pollution Fighters Use Laws, Science," *Chicago Tribune*, Feb. 14, 1966, 9; William Kling, "Air Pollution Fight Is Found to Be Costly to Business," *Chicago Tribune*, Feb. 15, 1966, A9; Joan Pinkerton, "Revised Controls Proposed for Air Pollution Crackdown," *Chicago Tribune*, Aug. 28, 1966, Q1; James Strong, "City Steps Up Pollution Fight," *Chicago Tribune*, Feb. 22, 1967, 7; James Strong, "Investigators Discover Industry Is Not the Only Polluter," *Chicago Tribune*, Feb. 23, 1967, 10; Casey Bukro, "State, City Seek Truce in Pollution Control Civil War," *Chicago Tribune*, Mar. 11, 1979, sec. 3, 16; Casey Bukro, "For Better and Worse, Our Ecology Has Changed," *Chicago Tribune*, Apr. 22, 1980, D8; Robert R. Gioielli, *Environmental Activism and the Urban Crisis: Baltimore, St. Louis, Chicago* (Philadelphia: Temple University Press, 2014), 111, 117.

26. New York City deaths may have been as high as three hundred, see Merril Eisenbud, "Environmental Pollution and Its Control," *Bulletin of the New York Academy of Medicine*, 2nd ser., 4, no. 5 (May 1969): 447. For the lack of deaths from previous inversions in Chicago, see William Kling, "City Ponders Ways to Fight Air Pollution: Experts Say Problem Is Very Costly," *Chicago Tribune*, Feb. 13, 1966, 3.

27. "Attend Clean Air Groups 1st Meeting," *Hyde Park Herald*, July 16, 1969, in Clean Air Coordinating Committee Collection, Series I, "1969–1970 Scrapbook, Part 1," Special Collections (MSCACC75), Richard J. Daley Library, University of Illinois, Chicago.

28. Michael Sneed, "Citizens Committee Establishes Air Pollution Standards," *Chicago Tribune*, July 17, 1969, 83.

29. Gioielli, *Environmental Activism and the Urban Crisis*, 111; Casey Bukro, "Life Line," *Chicago Tribune*, Jan. 10, 1971, NW5.

30. "No Place to Hide" advertisement, *Chicago Tribune*, Jan. 13, 1970, A19; Clarence Petersen, "Light, Dark Sides Figure in 2 Specials," *Chicago Tribune*, Jan. 15, 1970, B27; WBBM-TV News, *Action Guide on Air Pollution: Compiled for "No Place to Hide,"* WBBM-TV, Chicago, 1970 in Hull House Association Records, Box 106, File 1301, "Air Pollution," Special Collections (MSHHA), Richard J. Daley Library, University of Illinois, Chicago.

31. Clarence Petersen, "A Startling Show on Air Pollution," *Chicago Tribune*, Mar. 3, 1970, B15.

32. Gioielli, *Environmental Activism and the Urban Crisis*, 105–6, 115–18.

33. Sheldon Hoffenberg, "Daley Disarms City's Critics," *Lerner Booster Newspapers*, week of Apr. 15, 1970, in Hull House Association Records, Box 106, File 1301, "Air Pollution." For reductions in coal consumption, see Casey Bukro, "Life Line."

34. Dewey, *Don't Breathe the Air*, 242–43; Bryner, *Blue Skies, Green Politics*, 82–83.

35. *Report of the Court Observers Program of the Tuberculosis Institute of Chicago and Cook County*, Feb. 1972, Clean Air Coordinating Committee Collection, "1971–1972," Special Collections (MSCACC75), Richard J. Daley Library, University of Illinois, Chicago.

36. "Edison Reaffirms Clean Air Plan," *Chicago Tribune*, June 27, 1969, 7; James A. Throgmorton, *Planning as Persuasive Storytelling: The Rhetorical Construction of Chicago's Electric Future* (Chicago: University of Chicago Press, 1996), 70–71; John C. Boland, "Light at Last: Commonwealth Edison's Outlook Is Decidedly Brighter," *Baron's National Business and Financial Weekly*, Dec. 22, 1980, 60, 51.

37. Pacyga, *Chicago*, 366–67; Hurley, "Industrial Pollution," 412; Casey Bukro, "Stiff Enforcement Makes Smoke Violations a Rarity," *Chicago Tribune*, Mar. 22, 1979, W1; Bukro, "For Better and Worse"; Casey Bukro, "20 Years Later, Earth Day's Legacy Lingers," *Chicago Tribune*, Apr. 16, 1990, D1; Marc Doussard, Jamie Peck, and Nik Theodore, "After Deindustrialization: Uneven Growth and Economic Inequality in 'Postindustrial' Chicago," *Economic Geography* 85, no. 2 (2009): 183–88.

38. Katherine E. King, "Chicago Residents' Perception of Air Quality: Objective Pollution, the Built Environment, and Neighborhood Stigma Theory," *Population and Environment* 37, no. 1 (2015): 1–21; Scott Blackwood, "These Waters Run Deep," *Chicago Magazine*, Sept. 2017, 95–103, 148–50; Michael Hawthorne, "Chicago River Still Teems with Bacteria Flushed from Sewers after Storm," *Chicago Tribune*, June 23, 2017, http://www.chicagotribune.com/news/local/breaking/ct-chicago-river-still-dirty-met-20170623-story.html.

39. Pacyga, *Slaughterhouse*, 177–200.

CHAPTER 15: CHICAGO'S WASTELANDS

A previous version of this chapter appeared as Craig E. Colten, "Chicago's Waste Lands: Refuse Disposal and Urban Growth," *Journal of Historical Geography* 19 (1994): 124–42 with permission from Elsevier. Although there has been considerable scholarship in the intervening years, this reprint does not update sources. I would like to thank Carolyn Stoga, Doug Moore, Tim Teddy, and David Pseja for their assistance in researching Chicago's historical disposal sites, Julie Snider for preparing the original maps, and Mary Lee Eggart for transforming them to a digital format. STS Consultants, Ltd. in Chicago provided access to soil core records. Louis Cain and William Meyer offered constructive comments on an earlier draft. Anonymous reviewers also contributed insightful criticism that shaped the final version of this paper. Research for this article was funded in part by the Illinois Department of Energy and Natural Resources' Solid Waste Research Fund while I was employed at the Illinois State Museum.

1. General wastes consist of four main components: (1) garbage—biological wastes from homes and restaurants; (2) refuse—mixed wastes that are largely nonbiological; (3) ash—wastes from furnaces and stoves using coal as a fuel; (4) industrial solids—process residue that consists of slag, sludges, and general wastage.

2. See Ian Douglas, *The Urban Environment* (London: Taylor and Francis, 1983); and Andrew Goudie, *The Human Impact on the Natural Environment*, 3rd ed. (Cambridge, MA: Wiley, 1990).

3. A discussion of the highly varied local physical conditions is noted by Martin V. Melosi, *Garbage in the City: Refuse, Reform and the Environment, 1880–1980* (College Station: Texas A&M University Press, 1981), 15.

4. Henry W. Lawrence, "The Greening of the Squares of London: Transformation of Urban Landscapes and Ideals," *Annals of the Association of American Geographers* 83 (1993): 90–118.

5. Michael P. Conzen, "The Progress of American Urbanism, 1860–1930," in *North America: The Historical Geography of a Changing Continent*, ed. Robert D. Mitchell and Paul A. Groves (Totowa, NJ: Rowman and Littlefield, 1987), 347–70; Mona Domosh, "Controlling Urban Form: The Development of Boston's Back Bay," *Journal of Historical Geography* 18 (1992): 288–306; and David Ward, *Cities and Immigrants* (New York: Cambridge University Press, 1972), 116. See also Sam Bass Warner Jr., *Streetcar Suburbs: The Process of Growth in Boston* (Cambridge, MA: Harvard University Press, 1962); and Raymond A. Mohl and Neil Betten, "The Failure of Industrial City Planning: Gary, Indiana, 1906–1910," *Journal of the American Institute of Planners* 38 (1972): 203–14.

6. W. A. Meyer, "Urban Heat Island and Urban Health: Early American Perspectives," *Professional Geographer* 43 (1991): 38–48; and John Sheail,

"Underground Water Extraction: Indirect Effects of Urbanization on the Countryside," *Journal of Historical Geography* 8 (1982): 395–408.

7. K. L. Walwork, *Derelict Land: Origins and Prospects of a Land-Use Problem* (London: David & Charles, 1974); and Craig E. Colten, "Historical Hazards: The Geography of Relict Industrial Wastes," *Professional Geographer* 42 (1990): 143–56.

8. Peirce F. Lewis, "Small Towns in Pennsylvania," *Annals of the Association of American Geographers* 62 (1972): 323–51, discusses the conflict between an unflinching grid and topography.

9. See Eric E. Lampard, "The Nature of Urbanization," in *The Pursuit of Urban History*, ed. D. Fraser and A. Sutcliffe (London: Edward Arnold, 1983), 3–52; and E. K. Muller, "Waterfront to Metropolitan Region," in *American Urbanism: A Historiographical Review*, ed. H. Gillette Jr. and Z. L. Miller (New York: Greenwood Press, 1987), 105–33.

10. Wendy H. Sapan, "Landfilling at the Telco Block, Social, Historical and Archeological Perspectives," *American Archaeology* 5 (1985): 170–74.

11. A useful discussion of locational conflict appears in William B. Meyer and Michael Brown, "Locational Conflict in a Nineteenth-Century City," *Political Geography Quarterly* 8 (1989): 107–22. The expansion of urban sanitary engineering is recounted in Jon A. Peterson, "The Impact of Sanitary Reform upon American Urban Planning, 1840–1890," *Journal of Social History* 13 (1979): 83–103, esp. 94–95; and Stanley K. Schultz and Clay McShane, "To Engineer the Metropolis: Sewers, Sanitation, and City Planning in Late-Nineteenth-Century America," *Journal of American History* 65 (1978): 389–411. Shifting preferences for waste disposal is treated by Joel A. Tarr, "The Search for the Ultimate Sink: Urban Air, Land, and Water; Pollution in Historical Perspective," *Records of the Columbia Historical Society of Washington, DC* 51 (1984): 1–29. Administrative aspects of urban waste management are covered by Melosi, *Garbage in the City.*

12. See Christine M. Rosen, *The Limits of Power: Great Fires and the Process of City Growth in America* (Cambridge: Cambridge University Press, 1986); and William Cronon, *Nature's Metropolis: Chicago and the Great West* (New York: W. W. Norton, 1991).

13. For a brief discussion of suburban extractive activity, see Henry C. Binford, *The First Suburbs: Residential Communities on the Boston Periphery 1815–1860* (Chicago: University of Chicago Press, 1985), 40. Land filling, or dumping, was viewed as a means of making valuable land; see Samuel A. Greeley, "Garbage Disposal and the Economic Recovery of Valuable Constituents of Municipal Wastes," *Municipal and County Engineering* 58 (1920): 22–26.

14. C. E. Colten, "Environmental Development in the East St. Louis Region, 1890–1970," *Environmental History Review* 14 (1990): 92–114.

15. Lampard, "Nature of Urbanization"; Muller, "Waterfront to Metropolitan Region"; Peter O. Muller, *Contemporary Suburban America* (Englewood

Cliffs, NJ: Prentice-Hall, 1981), 37–56; and F. E. I. Hamilton, "Models of Industrial Location," in *Models in Geography*, ed. R. J. Chorley and P. Haggett (London: Metheun, 1967), 361–424.

16. Chicago City Council, Lakeshore Protection Plan, Aug. 10, 1844, Box 48, Document 2180, Chicago City Council Proceedings, Illinois Regional Archives Depository, Northeastern Illinois University Library, Chicago (hereafter Proceedings, IRAD). A more complete discussion of the lakefront problem appears in Robin L. Einhorn, "A Taxing Dilemma: Early Lake Shore Protection," *Chicago History* 18 (Fall 1989): 34–51. See also Lois Wille, *Forever Open, Clear and Free: The Struggle for Chicago's Lake Front* (Chicago: University of Chicago Press, 1972), 26–37.

17. Chicago City Council, Scavengers Ordinance, Apr. 30, 1849, Box 48, Document 5311, Proceedings, IRAD.

18. Chicago City Council, Report of Committee of Police on Bill of William Harrison, Nov. 9, 1852, Box 88, Document 885, Proceedings, IRAD; and Chicago City Council, Petition of Jacob Kuebler, Mar. 9, 1853, Box 92, Document 1278, Proceedings, IRAD.

19. Chicago City Council, Order to Prosecute, May 29, 1854, Box 116, Document 442, Proceedings, IRAD.

20. Chicago City Council, Petition by George P. Hansen, Health Officer, July 23, 1855, Box 144, Document 1037, Proceedings, IRAD.

21. Einhorn, "Taxing Dilemma."

22. Richard G. Berggreen, David L. Grumman, and Paul R. Blindauer, "Polynuclear Aromatic Hydrocarbon Contamination in Downtown Chicago Fill Soils," *Proceedings, Association of Engineering Geologists Annual Meeting*, St. Charles, Illinois, Oct. 1991 (Sudbury, MA: Association of Engineering Geologists, 1991), 519–27, esp. 523. See also "Wants the Lake-Front Cleared," *Chicago Tribune*, Oct. 17, 1890, 3.

23. Berggreen et al., "Polynuclear Aromatic Hydrocarbon Contamination," 523; "South Park Man Plans Reprisals," *Chicago Tribune*, Nov. 28, 1903, 10; U.S. Army, *Annual Reports of the War Department*, vol. 5, *Report of the Chief of Engineers* (Washington, DC: U.S. Army, 1905), 2066.

24. Berggreen et al., "Polynuclear Aromatic Hydrocarbon Contamination," 523.

25. Chicago City Council, Ordinance for Filling Pond, Oct. 2, 1865, Box 255, Document 454, Proceedings, IRAD.

26. 47th General Assembly of Illinois, *Report of the Submerged and Shore Lands Legislative Investigating Committee* (Springfield: Illinois General Assembly, 1911), 143, 177.

27. For a more complete discussion of the Lake Calumet industrial district, see Craig E. Colten, "Industrial Waste in Southeast Chicago: Production and Disposal," *Environmental Review* 10 (1986): 95–105.

28. U.S. Army, "Calumet Harbor," in *Report: Chief of Engineers* (Washington, DC: U.S. Army, 1895), 2591.

29. Melosi, *Garbage in the City*, 42.

30. Most of the construction of parks and other lakefront extensions after 1910 used dredge material; see M. J. Chrzastowski, "The Building, Deterioration and Proposed Rebuilding of the Chicago Lakefront," *Shore and Beach* 59 (1991): 2–10. Dumping of dredged material and industrial waste in the vicinity of water intakes caused controversy and led to the passage of the Mann Act in 1910 (Public Law No. 245). This federal act prohibited the dumping of any material within eight miles of the Illinois and Indiana shorelines. See also John J. Hueberg [?] and W. A. Evans to Jacob M. Dickinson, Secretary of War, June 9, 1909, Pollution and Dumping, 1907–1911 Files, U.S. Army Corps of Engineers, Chicago District.

31. One of the earliest discussions of sanitary landfills or "controlled tips" touted their ability to transform wasteland into useful property; see Donald C. Stone, "Dumping—An Economic Solution of Refuse Disposal," *Public Management* 14 (1932): 350–51. See also E. J. Cleary, "Land Fills for Refuse Disposal," *Engineering News-Record* 121 (Sept. 1, 1948): 270–73; and Douglas, *Urban Environment*, 118.

32. George A. Soper, John D. Watson, and Arthur J. Martin, *A Report to the Chicago Real Estate Board* (Chicago: self-published, 1915), 69–70; and Workers of the Writers' Program of the WPA, *Up from the Mud: An Account of How Chicago's Streets Were Raised* (Chicago: Federal Writers' Program, 1941), 16–17.

33. Chicago Department of Public Works (hereafter CDPW), *Annual Report* (Chicago: Chicago Department of Public Works, 1902), 233.

34. CDPW, *Annual Report*, 1903, 225.

35. CDPW, *Annual Report*, 1876, 130; and Chrzastowski, "Building, Deterioration and Proposed Rebuilding of the Chicago Lakefront," 2–10.

36. CDPW, *Annual Report*, 1876, 130.

37. CDPW, *Annual Report*, 1892, 222.

38. Depth-of-fill was determined by examining the logs of 323 soil probes prepared by STS Consultants, Ltd. of Chicago. Each log indicated the amount of fill that exists over natural soil or bedrock.

39. CDPW, *Annual Report*, 1894, 228.

40. CDPW, *Annual Report*, 1901, 56.

41. CDPW, *Annual Report*, 1902, 233.

42. CDPW, *Annual Report*, 1905, 293.

43. Swill dumps included refuse and scraps of waste food mixed with water for feeding pigs. Joseph M. Patterson, *Report to the Mayor and City Council on the Collection, Removal and Final Disposition of Garbage of the City of Chicago* (Chicago: Chicago City Council, 1906), 1–3. Maps of expanding urbanization appear in Homer Hoyt, *One Hundred Years of Land Values in Chicago* (Chicago: University of Chicago Press, 1933), figs. 29 and 44.

44. Citizens' Association of Chicago, *Garbage Disposal*, Bulletin 28 (Chicago: Citizen's Association of Chicago, 1912), 3; and Chicago City Waste Commission, *Report* (Chicago: Chicago City Waste Commission, 1914), 15.

45. Reduction is a process of heating and compressing garbage to drive out recoverable oils and solvents such as benzene and naphtha. These by-products were sold to underwrite the reduction facilities, as was the solid residue, known as tankage, which served as fertilizer.

46. Chicago City Waste Commission, *Report*, 15; Robert I. Randolph, "The History of Sanitation in Chicago," *Journal of the Western Society of Engineers* 44 (1939): 227–40, see 240; Citizens' Association of Chicago, *Garbage Disposal*; and H. A. Allen, "Chicago's Half-Completed Garbage-Reduction Plant Turns Expense into Income," *Engineering Record*, Feb. 10, 1917, 215–18; CDPW, *Annual Report*, 1907, 348; 1908, 313; 1912, 24. The adoption of a reduction system did not eliminate public disapproval to the siting of offensive municipal activities. There was opposition to the siting of the facility, as with dumps; see Allen, "Chicago's Half-Completed Garbage-Reduction Plant," 215; and CDPW, *Annual Report*, 1905, 293.

47. CDPW, *Annual Report*, 1913–1930.

48. "Chicago's Bubbly Creek Will Bubble No More," *Engineering News*, Oct. 19, 1916, 725–27.

49. Thomas J. Jablonsky, *Pride in the Jungle: Community and Everyday Life in Back of the Yards Chicago* (Baltimore: Johns Hopkins University Press, 1993), 139.

50. U.S. Public Health Service (hereafter USPHS), *The Chicago-Cook County Health Survey* (New York: USPHS, 1949), 204–5.

51. *Olcott's Chicago Land Values* (Chicago: Olcott and Company, 1927, 1944, and 1953); and U.S. Geological Survey, various 7·5 minute topographic maps of Chicago from the 1970s.

52. Murray F. Tuley, comp., *Laws and Ordinances Governing the City of Chicago* (Chicago: Bulletin Printing Co., 1873), 83–84.

53. City of Chicago, *The Chicago Municipal Code of 1922* (Chicago: T. H. Flood, 1922), 654.

54. Greeley, "Garbage Disposal and the Economic Recovery," 23.

55. Soldier Field is the sports arena adjacent to Grant Park. Sim White, *Soldier Field in Chicago* (Chicago: n.p., 1936), 6–9; Luke Cosme, engineer with the Chicago Park District, personal communication, Oct. 25, 1991.

56. CDPW, *Annual Report*, 1930–1935.

57. The almost unbridled enthusiasm for the new technology is exemplified by Cleary, "Land Fills for Refuse Disposal," 270. A more balanced assessment is offered by C. C. Spencer, *Recommended Wartime Refuse Disposal Practice*, U.S. Public Health Service, Supplement no. 173 to the Public Health Reports (Washington, DC: U.S. Public Health Service, 1943), 10. Further discussion is found in Tarr, "Search for the Ultimate Sink"; and Melosi, *Garbage in the City*.

58. Rolf Eliassen, "Why You Should Avoid Housing Construction on Refuse Landfills," *Engineering News Record* 138 (1947): 756–60.

59. USPHS, *Chicago-Cook County Health Survey*, 204–5.

60. Chicago Department of Streets and Electricity, *Annual Report* (Chicago: Chicago Department of Streets and Electricity, 1943), 33, 41.

61. Greeley and Hansen Engineers, *Preliminary Report on Refuse Collection and Disposal Related Matters* (Chicago: Chicago Department of Streets and Electricity, 1948), 6.

62. USPHS, *Chicago-Cook County Health Survey*, 211, 215.

63. Lake Calumet files, Permit No. 4678 and Correspondence, Mar. 7, 1945, U.S. Army Corps of Engineers, Chicago District. See also USPHS, *Chicago-Cook County Health Survey*, 203–9.

64. USPHS, *Chicago-Cook County Health Survey*, 203–9.

65. USPHS, *Chicago-Cook County Health Survey*, 214–15.

66. USPHS, *Chicago-Cook County Health Survey*, 210–11.

67. USPHS, *Chicago-Cook County Health Survey*, 199.

68. Chicago Department of Streets and Electricity, *Annual Report*, 1947, 35.

69. Chicago Department of Streets and Electricity, *Annual Report*, 1950, 34.

70. Chicago Department of Streets and Electricity, *Annual Report*, 1952 and 1953, 34 (each year).

71. Chicago Department of Streets and Electricity, *Annual Report*, 1954, 23.

72. Northeast Illinois Metropolitan Planning Commission (hereafter NIPC), *Refuse Disposal Needs and Practices in Northeast Illinois,* Technical Report 3 (Chicago: NIPC, June 1963), 12.

73. NIPC, *Refuse Disposal Needs and Practices*, 7.

74. NIPC, *Refuse Disposal Needs and Practices*, 70.

75. Clarence W. Klassenitar, "Locating, Designing and Operating Sanitary Landfills," *Public Works* 81 (Nov. 1950): 42–43; and Klassenitar, "Sanitary Fill Standards," *American City* 66 (Feb. 1951): 104–5. In addition, the Illinois State Geological Survey reported that it was common to provide information about local hydrologic conditions to landfill developers in the early 1950s. Frank C. Foley, memo to Illinois State Geological Survey Files, June 10, 1952, Illinois State Geological Survey, Champaign, Illinois.

76. NIPC, *Refuse Disposal Needs and Practices*, 5, 19.

77. See Illinois Department of Public Health, Division of Sanitary Engineering, *Rules and Regulations for Refuse Disposal Sites and Facilities* (Springfield: Illinois Department of Public Health, 1966), 3.

78. CDPW, *Chicago Public Works: A History* (Chicago: CDPW, 1973), 146–51.

79. G. M. Hughes, R. A. Landon, and R. N. Farvolden, *Summary of Findings on Solid Waste Disposal Sites in Northeastern Illinois*, Illinois State Geological Survey, Environmental Geology Notes, no. 45 (Urbana: Illinois State Geological Survey, Apr. 1971); and Colin J. Booth and Peter J. Vagt, "Hydrology and

Historical Assessment of a Classic Sequential-Land Use Landfill Site, Illinois, U.S.A.," *Environmental Geology and Water Science* 15 (1990): 165–78.

80. George M. Hughes, *Selection of Refuse Disposal Sites in Northeastern Illinois*, Illinois State Geological Survey, Environmental Geology Notes, no. 17 (Urbana: Illinois State Geological Survey, 1967), 2, 22–23. This report echoed the sanitary landfill guidelines that called for siting in areas with low permeability. Public water supplies in DuPage, McHenry, and Will Counties derived exclusively from groundwater sources in 1984, while Kane County pumped 84 percent of its public water from wells. James R. Kirk, Kenneth J. Hlinka, Robert Sasman, and Ellis W. Sanderson, *Water Withdrawals in Illinois, 1984*, Illinois State Water Survey, circular 163 (Champaign: Illinois State Water Survey, 1985), 14.

81. Hughes, *Selection of Refuse Disposal Sites in Northeastern Illinois*, 24.

82. NIPC, *Historical Inventory of Solid Waste Disposal Sites in Northeastern Illinois* (Chicago: NIPC, July 1988).

CHAPTER 16: MRS. BLOCK BEAUTIFUL

A previous version of this article originally appeared as Sylvia Hood Washington, "Mrs. Block Beautiful: African American Women and the Birth of the Urban Conservation Movement, Chicago, Illinois, 1917–1954," *Environmental Justice* 1 (2008): 13–23. Permission to reprint granted by Dr. Sylvia Hood Washington, editor-in-chief, *Environmental Justice*.

1. Allan H. Spear, *Black Chicago: The Making of a Negro Ghetto, 1890–1920* (Chicago: University of Chicago Press, 1967), 26.

2. Arnold R. Hirsch, *Making the Second Ghetto: Race and Housing in Chicago, 1940–1960* (Chicago: University of Chicago, [1983] 1998), 3.

3. Hirsch, *Making the Second Ghetto*, 5.

4. Sylvia Hood Washington, *Packing Them In: An Archaeology of Environmental Racism in Chicago, 1865–1954* (Lanham, MD: Rowman and Littlefield/ Lexington Books, 2005), 148.

5. Darlene Clark Hine, *Hine Sight: Black Women and the Reconstruction of American History* (Bloomington: Indiana University Press, 1994), 110.

6. Richard Wright, *Twelve Million Black Voices* (New York: Thunder Mouth Press, 2002), 103.

7. John H. Mims, "How One Neighborhood Solved Its Problems," n.d., Vivian Harsh Collection, Chicago Urban League Papers, Carter G. Woodson Regional Library (hereafter Harsh Collection).

8. Interoffice memorandum to Alva B. Maxey from Frayser Lane, July 26, 1954, Harsh Collection.

9. Hine, *Hine Sight*, 127.

10. These clubs also focused "on raising the cultural, intellectual, and educational status of black women"; Hine, *Hine Sight*, 121.

11. Interoffice memorandum to Alva B. Maxey from Frayser Lane.

12. *A Plan for Community Organization by DuSable District*, n.d., Harsh Collection.

13. *A Plan for Community Organization by DuSable District*.

14. Wright, *Twelve Million Voices*, 106-7.

15. Chicago Urban League, *First Annual Report 1916*, 9, Chicago Urban League Papers, University of Illinois, Special Collections (hereafter CUL Papers). For the detailed study, see Washington, *Packing Them In*.

16. "Block Clubs," n.d., Harsh Collection.

17. "Block Clubs."

18. Chicago Urban League, *20th Annual Report, 1936*, CUL Papers.

19. "Brief Summary of Work of Chicago Urban League, March 1-August 1, 1917," CUL Papers.

20. Chicago Urban League, *10th Annual Report, 1926*, CUL Papers.

21. "Block Clubs."

22. The League's Community Organization Department had several names during its earlier existence and prior to the 1940s was called the Civic Improvement Department.

23. *Civic Improvement Department Report, January 1938*, Harsh Collection.

24. *Civic Improvement Department Report, January 1938*.

25. *Civic Improvement Department Report, January 1938*.

26. *Civic Improvement Department Report, January 1938*.

27. *Civic Improvement Department Report, January 1938*.

28. *Civic Improvement Department Report, January 1938*.

29. Junior Woman's Auxiliary's Constitution, 1, Harsh Collection.

30. "What the Urban League Means to Chicago," 1938, radio interview by Irene McCoy Gaines and Earl B. Dickerson, Irene McCoy Gaines Papers, Chicago Historical Society (hereafter CHS).

31. Cheryl Johnson-Odim, "Irene McCoy Gaines," in *Women Building Chicago, 1790-1990: A Biographical Dictionary*, ed. Rima Lunin Schultz and Adele Hast (Bloomington: Indiana University Press, 2001), 294.

32. Cheryl Johnson-Odim, "Irene McCoy Gaines," in Schultz and Hast, *Women Building Chicago*.

33. The Chicago and Northern District Association of Colored Women, "Story of the Year, 1937-1938," 1, CHS.

34. Chicago and Northern District Association of Colored Women, "Story of the Year," 1.

35. "What the Urban League Means to Chicago."

36. "Old Settlers Who's Who," in *1927 Intercollegian Wonder Book: The Negro in Chicago, 1779-1927* (Chicago: Washington Intercollegiate Club of Chicago, 1927), 127.

37. "Joanna Cecilia Snowden," in *Who's Who in Colored America, 1941–44* (Yonkers-on-Hudson, NY: C. E. Burckel, 1941), 476; "Joanna Cecilia Snowden," in *Who's Who in Colored America, 1933–1937* (Yonkers-on-Hudson, NY: C. E. Burckel, 1937), 485; "Joanna Cecilia Snowden," in *Who's Who in Colored America 1930–1932* (Yonkers-on-Hudson, NY: C. E. Burckel, 1932), 396; *1927 Intercollegian Wonder Book*, 120.

38. "Rachel R. Ridley Obituary," *Chicago Tribune*, May 5, 1986; and "Rachel R. Ridley," undated audio tape, Bethel New Life Papers, Chicago Public Library. Ridley served as the director of the Human Relations Commission from 1980 to 1985.

39. "Rachel R. Ridley."

40. Chicago Urban League Farewell Program Booklet for Rachel Ridley, Aug. 30, 1952, Harsh Collection.

41. Chicago Urban League, *34th Annual Report*, 1950, Feb. 7, 1951.

42. "Block Clubs."

43. "West Side Slums Fire Claims Four," *Chicago Bee*, Jan. 26, 1947.

44. "So. Side Wars on Rats," *Chicago Bee*, June 22, 1947.

45. "Wendell Elbert Green," in *Who's Who in Colored America 1950* (Yonkers-on-Hudson, NY: C. E. Burckel, 1950), 224.

46. "Loraine R. Green," *1927 Intercollegian Wonder Book*, 117.

47. "Loraine R. Green"; and research notes from senior historian, Anne Knupfer, in Knupfer's possession.

48. *1950 Annual Block Beautiful Contest Awards Program Booklet*, n.p., Harsh Collection.

49. Chicago Urban League, *A Sight Worth Seeing*, 1951, Harsh Collection.

50. "Introduction to the Description of the Five Negro Neighborhoods," n.d, Harsh Collection.

51. *A Plan for Community Organization by DuSable District*.

52. *The Langley Avenue Neighborhood Club announces Judgment Day Has Come!*, flier, Harsh Collection.

53. Urban League letter, May 21, 1953, Harsh Collection.

54. Urban League letter to community leaders from Frayser T. Lane, Apr. 5, 1949, Harsh Collection.

55. *Organizing Neighborhood and Community Councils, Summary and Conclusions*, n.d., 1, Harsh Collection.

56. *What Are Basic Needs of a Good Community?*, flier, Harsh Collection.

57. *What Are Basic Needs of a Good Community?*

58. Chicago Urban League, *A Sight Worth Seeing*.

59. *1952 COD [Community Organization Department] Report*, 2, Harsh Collection.

60. *1952 COD Report*, 2.

61. *We Fight Blight*, 1954, Harsh Collection.

PART V: REENVISIONING THE LAKE AND PRAIRIE

1. Jens Jensen, "Siftings," in *Of Prairie, Woods, and Water: Two Centuries of Chicago Nature Writing*, ed. Joel Greenberg (Chicago: University of Chicago Press, 2008), 398.

CHAPTER 17: MAY THEILGAARD WATTS AND THE ORIGINS OF THE ILLINOIS PRAIRIE PATH

1. May Watts, "Future Footpath?," *Chicago Tribune*, Sept. 30, 1963; Rachel Carson, *Silent Spring* (New York: Houghton Mifflin, 1962). On rails-to-trails, see Record Group 1, Series 2, Box 1, Folders 11–12, and Record Group 5, Series 1, Box 4, Illinois Prairie Path Archives, Oesterle Library, North Central College, Naperville, IL (hereafter IPP Archives). Additional spurs known as the Geneva Spur and the Batavia Spur account for the current sixty-one miles of trail. On the expansion of the Illinois Prairie Path, see "History," Illinois Prairie Path, http://www.ipp.org/history/ (accessed Apr. 11, 2020).

2. For path maps, see Record Group 1, Box 1, Folder 7, IPP Archives; and "Trail Map," 2009, Illinois Prairie Path, http://www.ipp.org/trail-map/.

3. Stewart L. Udall, *The Quiet Crisis* (New York: Holt, Rinehart and Winston, 1963); Paul Ehrlich, *The Population Bomb* (San Francisco: Sierra Club, 1968); William H. Whyte, *The Last Landscape* (New York: Doubleday, 1968); Aldo Leopold, *A Sand County Almanac* (New York: Oxford University Press, 1949); Edwin Way Teale, *The American Seasons* (New York: Dodd, Mead, 1951–1965); and May Theilgaard Watts, *Reading the Landscape: An Adventure in Ecology* (New York: Macmillan, 1957).

4. On the University of Chicago, see Box 1, Folders A–C; and Box 2, Folder A, May Watts Papers, Sterling Morton Library, Morton Arboretum, Lisle, IL (hereafter Watts Papers).

5. Watts, "Future Footpath?"

6. On the Appalachian Trail as a model, see Record Group 2, Box 1, Folder 2, IPP Archives. On MacKaye and the Appalachian Trail, see Paul S. Sutter, *Driven Wild: How the Fight against Automobiles Launched the Modern Wilderness Movement* (Seattle: University of Washington Press, 2002).

7. David Stradling, ed., *Conservation in the Progressive Era: Classic Texts* (Seattle: University of Washington Press, 2004), is a solid introduction to conservationists.

8. Watts, "Future Footpath?"

9. Opposition to Chicago's sprawl links Watts to Robert Gottlieb's broadening of the definition of environmentalism from wilderness advocates to activists responding to urban-industrial problems. See Gottlieb, *Forcing the Spring: The Transformation of the American Environmental Movement* (Washington, DC: Island Press, 1993).

10. Watts, "Future Footpath?" Kevin Armitage examines lesser-known conservationists with similarities to Watts in *The Nature Study Movement: The Forgotten Popularizer of America's Conservation Ethic* (Lawrence: University of Kansas Press, 2009). He analyzes nature writers like Ernest Thompson Seton and Anne Botsford Comstock who combined a love of nature with scientific understanding.

11. See "Illinois Bus Ride" and "The Sand Counties," in Leopold, *Sand County Almanac*; "River of Fireflies" in Edwin Way Teale, *Journey into Summer* (New York: Dodd, Mead, 1960); and "Dusty Autumn" in Teale, *Autumn across America* (New York: Dodd, Mead, 1961). For a broader discussion of bulldozers as symbols of development, see Adam Rome, *The Bulldozer in the Countryside: Suburban Sprawl and the Rise of Modern Environmentalism* (New York: Cambridge University Press, 2001).

12. May Watts, "The Herbs of the Prairie," *Cornell Plantations* (Aug. 1975), in Box 7, Watts Papers.

13. Watts, "Future Footpath?"

14. May Watts to Dr. H. R. Selm, 1960, Box 33, Folder B, Watts Papers.

15. Laura Theilgaard McVey to Carol Doty, 1989, Box 26, Folder C, Watts Papers.

16. Family and Personal Mementos, [n.d.], Box 27, Folder A, Watts Papers.

17. Box 1, Folders A–C, and Box 2, Folder A, Watts Papers.

18. Watts to Dr. Selm, 1960, Box 33, Folder B; and W. S. Cooper, "Henry Chandler Cowles," in *Ecology* (July 1935), Box 2, Folder B, Watts Papers.

19. Watts, *Reading the Landscape*, viii.

20. "Raymond Watts: Man of Ideas and Deeds," *Naperville Sun*, July 21, 1966, 6A.

21. Watts used the pseudonym "Face-the-Wind" for her poems. Face-the-Wind, "Vision," *Chicago Evening Post*, Record Group 4, Box 4, Folder 1, IPP Archives.

22. May Watts, "On Improving the Property," Record Group 4, Box 4, Folder 1, IPP Archives.

23. May Watts, *Ravinia: Her Charms and Destiny* (1936) in Box 3, Folder E, Watts Papers.

24. Jens Jensen, *Siftings: The Major Portion of the Clearing and Collected Writings* (Chicago: R. F. Seymour, 1956), 45. See Robert E. Grese, *Jens Jensen: Maker of Natural Parks and Gardens* (Baltimore: Johns Hopkins University Press, 1992), 136–50 on Ravinia, 52 and 121–25 on Cowles.

25. See Carol Doty, "May Watts, Founder of the Illinois Prairie Path, on Its 50th Anniversary," May 18, 2013, Box 26, Folder A; Evelyn Lauter, "May Theilgaard Watts: She Walks in Beauty," *Lake Forest News*, Oct. 31, 1963, Box 35, Folder B; and Ruth Wenner, "Mrs. Watts," 1967, Box 25, Folder C, all in Watts Papers.

26. On "Friends of Our Native Landscape," see Grese, *Jens Jensen*.

27. Nature School, Box 13, Folder A, Watts Papers.

28. Sophie Thielgaard, "The Hem of His Garment: A Testimony," [n.d.], 162, 164–65, Box 26, Folder D, Watts Papers.

29. Carol Doty, "A Tribute to May Theilgaard Watts," Oikos Conference, Morton Arboretum, Apr. 3, 1976, in Box 25, Folder G, Watts Papers. For teaching materials, see Box 4, Folders F–K; Box 15, Folders A–D; Box 16, Folders A–E, Watts Papers.

30. May Watts, "Adventures in Nature Education," *Morton Arboretum Bulletin of Popular Information* (Jan. 1948), Box 6, Folder C, Watts Papers.

31. Wenner, "Mrs. Watts."

32. On the varied classes, see May Watts, "Adventures in Nature Education," *Morton Arboretum Bulletin of Popular Information* (Jan. 1948), Box 6, Folder C, Watts Papers. The book group read Leopold, Udall, and Teale, and also Anna Botsford Comstock, Sarah Orne Jewett, and Mary Austin. See Box 15, Folders A–B, and Box 16, Folder E, Watts Papers.

33. Louisa King to May Watts, Dec. 7, 1946, Box 28, Folder I; May Watts, "Adventures in Nature Education at the Morton Arboretum," *University of Washington Arboretum Bulletin* (Winter 1958), Box 7, both in Watts Papers.

34. May Watts to Mrs. Joseph Cudahy, Jan. 16, 1948, Box 28, Folder E; and Suzette Zurcher address, 1961, Box 27, Folder B, both in Watts Papers.

35. See Linda J. Lear, *Rachel Carson: Witness for Nature* (New York: Henry Holt, 1997).

36. Carson, *Silent Spring*, 1.

37. Alfred Etter, "Reading May Watts," *Morton Arboretum Quarterly* (Autumn 1975), Box 25, Folder F, Watts Papers.

38. May Watts, "Lament for Our Central Park," *Naperville Sun*, Sept. 12, 1974. See also Record Group 4, Box 4, Folder 1, IPP Archives.

39. On the bulldozer as symbol, see Rome, *The Bulldozer in the Countryside*. See also Christopher C. Sellers, *Crabgrass Crucible: Suburban Nature & the Rise of Environmentalism in Twentieth-Century America* (Chapel Hill: University of North Carolina Press, 2012). Sellers disagrees with Rome on whether outside experts or suburbanites themselves pushed for environmental reform; see 4–5. Watts complicates this debate, as both expert naturalist and suburban resident.

40. May Watts, "Our Changing Landscape," [n.d.], Box 16, Folder F, Watts Papers.

41. "Harold Moser, 87, Builder Known as Mr. Naperville," *Chicago Tribune*, Dec. 19, 2001. See "Moser, Naperville Grow Up Together," *Chicago Tribune*, Apr. 15, 1989; and Michael Ebner, "Harold Moser's Naperville," *Illinois Periodicals Online*, Northern Illinois University Libraries, 1999, https://www.lib.niu.edu/1999/iht719939.html.

42. U.S. Department of Agriculture, County Tables, "Farms and Farm Characteristics: 1945," Census of Agriculture Historical Archive, Illinois, 1945,

http://usda.mannlib.cornell.edu/usda/AgCensusImages/1945/01/12/1171/Ta ble-01.pdf; and "Number, Land and Value of Farms: 1974," Census of Agriculture Historical Archive, Illinois, 1974, http://usda.mannlib.cornell.edu/usda/ AgCensusImages/1974/01/13/306/Table-01.pdf; Carson, *Silent Spring*, 3.

43. Watts, *Reading the Landscape*, viii

44. Watts, *Reading the Landscape*, 127–28, viii.

45. Watts, *Reading the Landscape*, 142, 117.

46. Watts, *Reading the Landscape*, 191, 193, 191.

47. Watts, *Reading the Landscape*, 29, 21, 28; Leopold, "Good Oak," in *Sand County Almanac*, 6–18.

48. On prairies along the path, see Record Group 5, Box, 3, Folders 2–9, IPP Archives.

49. Jensen, *Siftings*, 75, 57.

50. Teale, *Autumn across America*, 79–80.

51. Audrie Chase, "May Watts Walk," *Chicago Tribune*, Oct. 7, 1963.

52. Etter, "Reading May Watts."

53. On Openlands, see Record Group 2, Box 2, Folders 12–16 and Box 3, Folders 1–4, IPP Archives.

54. On Turner, see Record Group 4, Box 3, Folders 1–5, IPP Archives. See also Box 20, Folders 20B and 20C, Watts Papers.

55. On the presentations, see Record Group 2, Box 1, Folders 12–13, IPP Archives.

56. On bridges, see Record Group 5, Series 7, Boxes 16–17, and Series 8, Boxes 18–20, IPP Archives.

57. On DuPage County assuming responsibility in 1986, see "History," Illinois Prairie Path, http://www.ipp.org/history/ (accessed Apr. 11, 2020).

58. On Openlands, see Gerald W. Adelmann, "Openlands Project," in *Encyclopedia of Chicago*, 2004, http://www.encyclopedia.chicagohistory.org/pages/ 932.html. Indiana Dunes National Lakeshore became Indiana Dunes National Park in February 2019. "Indiana Dunes National Park: A Long-Sought Summit Achieved," editorial, *Chicago Tribune*, Feb. 22, 2019.

59. "Excitement Grows over Birth of Rare Wild Bison in Illinois," *Chicago Tribune*, Apr. 8, 2015; "Baby Bison Born at Midewin," *Chicago Tribune*, Apr. 28, 2016.

CHAPTER 18: "HARD-NOSED PROFESSIONALS"

1. "The Big Cleanup," *Newsweek*, June 12, 1972, 37.

2. Samuel P. Hays and Barbara D. Hays, *Beauty, Health, and Permanence: Environmental Politics in the United States, 1955–1985* (New York: Cambridge University Press, 1987), remains the standard, national history of American environmentalism, but it is in many ways a synthetic analysis of hundreds of local movements. See also James Lewis Longhurst, *Citizen Environmentalists*,

Civil Society: Historical and Contemporary Perspectives (Medford, MA: Tufts University Press, 2010); Adam Rome, *The Genius of Earth Day: How a 1970 Teach-In Unexpectedly Made the First Green Generation* (New York: Hill and Wang, 2013).

3. For an example of this top-down critique, see Mark Dowie, *Losing Ground: American Environmentalism at the Close of the Twentieth Century* (Cambridge, MA: MIT University Press, 1996).

4. See esp. Southwest Organizing Project to National Wildlife Federation et al., Mar. 16, 1990, http://www.ejnet.org/ej/swop.pdf. On institutionalization within environmental organizations, see Jennifer Thomson, "Surviving the 1970s: The Case of Friends of the Earth," *Environmental History* 22, no. 3 (Apr. 2017): 235–56; Frank Zelko, *Make It a Green Peace! The Rise of Countercultural Environmentalism* (New York: Oxford University Press, 2013).

5. Midas Muffler. "The Quietest Cars on the Road," advertisement, *Life*, June 6, 1960, 63.

6. Robert Crean, "Hi-Yo, Midas! Away!," *Chicago Tribune Magazine*, Nov. 22, 1970, 28–33; "Ambush at Generation Gap," *Time*, May 3, 1971; Kenan Heise, "Gordon Sherman, Exec and Activist," *Chicago Tribune*, May 15, 1987; "History of Midas Inc.," in *International Directory of Company Histories*, ed. Tina Grant, vol. 56. (London: St. James Press, 2004).

7. Crean, "Hi-Yo, Midas! Away!," 29.

8. Crean, "Hi-Yo, Midas! Away!," 30; Dan Jedlicka, "Business Executives Form Group to Investigate Public Issues Here," *Chicago Sun-Times*, Oct. 5, 1969.

9. Margery Frisbie, *An Alley in Chicago: The Ministry of a City Priest* (Franklin, WI: Sheed & Ward, 2002); Heise, "Gordon Sherman, Exec and Activist."

10. Benjamin Soskis, "The Problem of Charity in Industrial America, 1873–1915" (PhD diss., Columbia University, 2010); Olivier Zunz, *Philanthropy in America: A History* (Princeton, NJ: Princeton University Press, 2012).

11. Sean Dobson, "Freedom Funders: Philanthropy and the Civil Rights Movement, 1955–1965," Report Commissioned by the National Committee for Responsive Philanthropy (Washington, DC: National Committee for Responsive Philanthropy, June 2014); Zunz, *Philanthropy in America*, 201–31; Patricia Rosenfield and Rachel Wimpee, "The Ford Foundation: Constant Themes, Historical Variations," *Report of the Ford Foundation History Project* (Tarrytown, NY: The Rockefeller Archive Center, 2015).

12. Ford Foundation Oral History Project, "Gordon Harrison Transcript," 1972, Ford Foundation Records, Rockefeller Archive Center, Sleepy Hollow, NY (hereafter Ford Foundation Records).

13. Gordon Harrison, "A League for Environmental Law," June 1969, Ford Foundation Unpublished Reports no. 005087, Ford Foundation Records.

14. Gabriel Kolko, *Railroads and Regulation, 1877–1916* (Princeton, NJ: Princeton University Press, 1965); Charles A. Reich, *The Greening of America:*

How the Youth Revolution Is Trying to Make America Livable (New York: Random House, 1970); Paul Sabin, "Environmental Law and the End of the New Deal Order," *Law and History Review* 33, no. 4 (Nov. 2015): 965–1003.

15. Gordon Sherman, "Parallel Organizations," ca. late 1960s, Box 18, Folder 17, Businessmen for the Public Interest Papers, Special Collections, University of Illinois, Chicago (hereafter BPI Papers).

16. Alexander Polikoff, quoting Gordon Sherman, BPI, "Third Report: 1971–1972," 1971, Box 37, Folder 20, BPI Papers.

17. Civia Tamarkin, "The White Nights of BPI," *Chicago Tribune Magazine*, Nov. 25, 1979.

18. *Gautreaux v. Chicago Housing Authority*, 296 F. Supp. 907 (N.D. IL, 1969); Dan Jedlicka, "Business Executives Form Group to Investigate Public Issues Here," *Chicago Sun-Times*, Oct. 5, 1969; Tamarkin, "The White Nights of BPI"; BPI, "Second Report: 1970–1971," 1971, Box 37, Folder 20, BPI Papers; BPI, "Third Report: 1971–1972," 1971, Box 37, Folder 20, BPI Papers; Alexander Polikoff, *Waiting for Gautreaux: A Story of Segregation, Housing and the Black Ghetto* (Chicago: Northwestern University Press, 2006).

19. Saul Alinsky, *Rules for Radicals: A Practical Primer for Realistic Radicals* (New York: Random House, 1971); Mark Santow, "Saul Alinsky and the Dilemmas of Race in the Post-War City" (PhD diss., University of Pennsylvania, 2000). For the origins of CAP, see Robert Gioielli, "Not Quite Suburban: Progressive Politics in Postwar Chicago," in *Social Justice in Diverse Suburbs: History, Politics, and Prospects*, ed. Christopher Niedt (Philadelphia: Temple University Press, 2013), 91–104.

20. Ed Chambers and Richard Harmon to Saul Alinsky, Dec. 18, 1969, Box 39, Folder 611, Industrial Areas Foundation Papers, Special Collections, University of Illinois, Chicago.

21. Harza Engineering Company, "A Lake Michigan Site for Chicago's Third Major Airport," Oct. 1967, Box 21, BPI Papers; Jerry Lockwood, "Lake Airport Chronology," n.d., Box 21, BPI Papers; "Lake Airport near Chicago under Protest," *Eau Claire Daily Telegram*, Apr. 10, 1970; "Opposes Lake Airport," *Chicago Tribune*, Jan. 28, 1970; Open Lands Project, "Will a Lake Airport Best Serve the Chicago Area?," n.d., Box 21, BPI Papers.

22. BPI, "Don't Do It in the Lake!," ca. 1969, Box 32, Folder 12, BPI Papers; Lockwood, "Lake Airport Chronology."

23. "Abundant Power from Atom Seen," *New York Times*, Sept. 17, 1954, 5; U.S. Nuclear Regulatory Commission, *Nuclear Regulatory Commission Issuances: Opinions and Decisions of the Nuclear Regulatory Commission with Selected Orders* (Washington, DC: U.S. Nuclear Regulatory Commission, 1978), 568–74; Alfred Eipper, "Nuclear Power on Cayuga Lake," *Boston College Environmental Affairs Law Review* 1, no. 1 (Apr. 1971): 165–90; George Goodman, "David D. Comey Dies at Age 44; Active in Environmental Affairs," *New York Times*, Jan. 7, 1979.

24. Bruce Ingersoll, "The Power Industry Acquires a 'Nader' of Its Own," *Chicago Sun-Times*, Aug. 2, 1970, 34; J. Samuel Walker and Thomas Raymond Wellock, *A Short History of Nuclear Regulation, 1946–2009*, NUREG/BR, 0175, rev. 2 (Washington, DC: U.S. Nuclear Regulatory Commission, 2010), 37–41.

25. Kate Brown, *Plutopia: Nuclear Families, Atomic Cities, and the Great Soviet and American Plutonium Disasters* (Oxford: Oxford University Press, 2013); Walker and Wellock, *A Short History of Nuclear Regulation*, 35.

26. BPI, "Third Report: 1971–1972"; Tom Hall, "Nuclear Energy: The Deadly Gamble," *BPI Magazine*, 1975, Box 37, Folder 23, BPI Papers.

27. Ron Cockrell, *A Signature of Time and Eternity: The Administrative History of Indiana Dunes National Lakeshore, Indiana* (Omaha, NB: U.S. Department of the Interior, National Park Service, Midwest Regional Office, Office of Planning and Resource Preservation, Division of Cultural Resources Management, 1988).

28. BPI, *Annual Report 1976*, 1971, Box 37, Folder 20, BPI Papers; Cockrell, *A Signature of Time and Eternity*; Nancy J. Obermeyer, *Bureaucrats, Clients, and Geography: The Bailly Nuclear Power Plant Battle in Northern Indiana*, Geography Research Paper, no. 216 (Chicago: University of Chicago Press, 1989); "Utility Cancels Nuclear Plant near Indiana Dunes," *New York Times*, Aug. 30, 1981.

29. BPI, *Annual Report 1976*; Goodman, "David D. Comey Dies at Age 44"; Alexander Polikoff, interview with the author, Mar. 3, 2017.

30. George Gunset, "Midas Is Sick: G. B. Sherman," *Chicago Tribune*, Apr. 20, 1971, 7; Hugh McCann, "A Dealer Defection at Midas Muffler," *New York Times*, June 6, 1971; Polikoff interview; Jeremy Sherman, email exchange with author, Feb. 27, 2017.

31. Robert Lifset, *Power on the Hudson: Storm King Mountain and the Emergence of Modern American Environmentalism* (Pittsburgh: University of Pittsburgh Press, 2014), chap 5; BPI, *Annual Report 1973*, Box 37, Folder 20, BPI Papers.

CHAPTER 19: THE CALUMET REGION

1. NHA Feasibility Studies make the case for the "national significance" of the region under study. NHAs are not parks as such, but a "grassroots, community-driven approach to heritage conservation and economic development." See National Park Service, "What Is a National Heritage Area?," https://www.nps.gov/articles/what-is-a-national-heritage-area.htm. (accessed Apr. 6, 2020). Given this approach, the general public of the Calumet region is as much a part of the audience for the study as is the National Park Service. The study's themes and specific language were vetted through a public comment process. Mark Bouman served as primary author, with significant editorial assistance from CHP colleagues Sherry Meyer, Mike Longan, and Bill Peterman; Field Museum colleagues Madeleine Tudor and Mario Longoni; consultants Augie Carlino and Nancy Morgan;

and the entire CHP board. The process was supported by grants from ArcelorMittal, Nicor, and Boeing. Much of the following reflects language used in the Feasibility Study. In drawing attention to the contrast between the modern industrial landscape and the region's natural endowments, this essay omits Feasibility Study sections on Native American heritage, contact with French explorers and development of the fur trade, the traumatic removal of the Pottawatomi and opening of the landscape to European settlement, the assembly of an industrial labor force and the communities and neighborhoods that resulted, and the cultures of conservation and place making that pervade the region today, continuing to kick dust across the lines separating city from nature. The entire study and a description of the process may be accessed at Calumet Heritage Partnership, *Calumet National Heritage Area Initiative Feasibility Study*, http://www.calumetheritage.org/cnha/Calumet_FeasibilityStudy_June2018web.pdf (accessed Apr. 6, 2020).

2. David Nye, *American Technological Sublime* (Cambridge, MA: MIT Press, 1994). See Mark J. Bouman, "A Mirror Cracked: Ten Keys to the Landscape of the Calumet Region," *Journal of Geography* 100 (2001): 104–10, a brief work that considers the region's "juxtaposition" of traits that forms something of the conceptual basis for the Feasibility Study. Scholars of the Calumet as a region should be aware of the work of Alfred H. Meyer, "Circulation and Settlement Patterns of the Calumet Region of Northwest Indiana and Northeast Illinois (The Second State of Occupance—Pioneer Settlement and Subsistence Economies)," *Annals, Association of American Geographers* 46 (1956): 312–56.

3. The film *Everglades of the North: The Story of the Grand Kankakee Marsh* (2013) vividly captures the story of the marsh; see http://www.kankakeemarsh.com/. The production team followed the film with *Shifting Sands: On the Path to Sustainability* (2006), a film about the Calumet region more directly concerning the themes of this essay.

4. Meyer wrote of the term *Calumet* that "its regional use well expresses a chorographic reality coinciding roughly with the Calumet drainage basin and the essential homogeneity of its historic-geographic cultural development"; Alfred H. Meyer, "Toponymy in Sequent Occupance Geography, Calumet Region, Indiana-Illinois," *Proceedings of the Indiana Academy of Science* 54 (1945): 142–59, 144. The meandering history of the Grand Calumet, Little Calumet, and main stem Calumet is best explained in Kenneth J. Schoon, *Calumet Beginnings: Ancient Shorelines and Settlements at the South End of Lake Michigan* (Bloomington: Indiana University Press, 2003), 39–42, and is built on the unpublished work of the historian Paul Petraitis. Also see David M. Solzman, *The Chicago River: An Illustrated History and Guide to the River and Its Waterways*, 2nd ed. (Chicago: University of Chicago Press, 2007), esp. the map on p. 20. The closing of the Indiana mouth of the Grand Calumet is discussed in Powell A. Moore, *The Calumet Region: Indiana's Last Frontier*, Indiana Historical Collections, vol. 39 (Indianapolis: Indiana Historical Bureau, 1959), 10–11.

5. Meyer, "Circulation and Settlement Patterns of the Calumet Region," 322.

6. Cowles said the words at a hearing conducted in Chicago on the advisability of a national park on October 30, 1916; see Stephen T. Mather, *Report on the Proposed Sand Dunes National Park, Indiana* (Washington, DC: U.S. Department of the Interior, 1917), 44.

7. A sketch of *Thismia americana* forms the frontispiece to the magisterial Floyd Swink and Gerould Wilhelm, *Plants of the Chicago Region*, 4th ed. (Bloomington: Indiana University Press, 1994).

8. Charles B. Cory, *The Birds of Illinois and Wisconsin*, Zoological Series 9 (Chicago: Field Museum of Natural History, 1909), 555.

9. Robert Lewis, "Networks and the Industrial Metropolis: Chicago's Calumet District, 1870–1940," in *Industrial Cities: History and Future*, ed. Clemens Zimmerman (Frankfurt: Campus Verlag, 2013), 102. Also see Robert Lewis, *Chicago Made: Factory Networks in the Industrial Metropolis* (Chicago: University of Chicago, 2008), 141–42. For a background to the post–Civil War era rise of industrial urbanization, see, for example, John R. Borchert, "American Metropolitan Evolution," *Geographical Review* 57 (1967): 301–31; Alfred D. Chandler, *The Visible Hand: The Managerial Revolution in American Business* (Cambridge, MA: Belknap Press of Harvard University Press, 1977); Michael P. Conzen, "The Maturing Urban System in the United States, 1840–1910, " *Annals of the Association of American Geographers* 67, no. 1 (1977): 88–108; Donald L. Miller, *City of the Century: The Epic of Chicago and the Making of America* (New York: Simon and Schuster, 1996); Sam Bass Warner Jr., *The Urban Wilderness: A History of the American City* (Berkeley: University of California Press, 1972).

10. Borchert writes that as a boy he lived

on the edge of one of the steepest geographical gradients in the world at that time. On one side of the gradient stood my home town, . . . a typical Corn Belt county seat of 2500. . . . Yet just ten miles north of my home town was the south edge of the new, 100-thousand city of Gary, laid out less than a decade earlier by the U.S. Steel Corporation on the marshes and sand dunes at the south end of Lake Michigan. Just five miles farther north were the gates of the largest steel mills in the world, the economic base of Gary. . . . The train ride from Crown Point to the heart of Chicago took 59 minutes. Through the dirty day-coach windows I watched, on trip after trip, the quick, bewildering transition from my rural home countryside, through a heavy industrial complex that matched the Ruhr and the Pittsburgh-Cleveland axis for world leadership.

John R. Borchert, "A Journey of Discovery," John R. Borchert Memorial Website, http://www.borchert.com/john/journey_of_discovery.htm (accessed Nov. 26, 2017).

11. Charles C. Colby, "Centrifugal and Centripetal Forces in Urban Geography," *Annals, Association of American Geographers* 22 (1933), repr. in Harold M. Mayer and Clyde F. Kohn, eds., *Readings in Urban Geography* (Chicago: University of Chicago Press, 1933), 287-98. Also see John B. Appleton, *The Iron and Steel Industry of the Calumet District*, University of Illinois Studies in the Social Sciences, vol. 13, no. 2 (Urbana: University of Illinois Press, 1927).

12. "Chicago Plant Will Produce New Ford SUV," *Chicago Tribune*, Apr. 27, 2018.

13. Solzman, *Chicago River*, 161.

14. Solzman, *Chicago River*, 32, 181. Also see Richard Lanyon, *Building the Canal to Save Chicago* (Bloomington, IN: Xlibris, 2012).

15. A good contemporary portrait of the industrial river corridor is found in Solzman, *Chicago River*, 170, 174-75.

16. Schoon, *Calumet Beginnings*, 107.

17. Schoon, *Calumet Beginnings*, 81.

18. Schoon, *Calumet Beginnings*, 80, 101-2. For the "astronomical" amounts of sand, see Schoon, *Calumet Beginnings*, 86, 176, where Schoon relates that the Santa Fe Railroad placed an order for 150,000 carloads of sand in 1899. See Fritiof M. Fryxell, *The Physiography of the Region of Chicago* (Chicago: University of Chicago Press, 1927), 48 (quote).

19. Schoon, *Calumet Beginnings*, 98-99; 103; Rollin D. Salisbury and William C. Alden, *The Geography of Chicago and Its Environs*, Geographic Society of Chicago Bulletin, no. 1. (Chicago: Rand McNally, 1900), 61; Fryxell, *Physiography of the Region of Chicago*, 48; Elizabeth A. Patterson, "Michigan City," in *The Encyclopedia of Chicago*, ed. James D. Grossman, Ann Durkin Keating, and Janice L. Reiff (developed by the Newberry Library with the cooperation of the Chicago Historical Society) (Chicago: University of Chicago Press, 2004).

20. Opinions diverged on the U.S. Steel alterations at Gary, Powell Moore calling it "an achievement of epic proportions," while Bradley J. Beckham said that "what took nature thousands of years to mold, man in the guise of progress subverted in a few month"; Moore, *Calumet Region*, 275; Beckham is quoted in James B. Lane, *City of the Century: A History of Gary, Indiana* (Bloomington: Indiana University Press, 1978), 28; and Schoon, *Calumet Beginnings*, 97. On the pumping of sand at Gary, see Schoon, *Calumet Beginnings*, 100.

21. This discussion follows Schoon's excellent map *Surface Geology of the Calumet Area* that forms the frontispiece of *Calumet Beginnings*, as well as the insert map in H. B. Willman and Jerry A. Lineback, *Surficial Geology of the Chicago Region*, Circular no. 160 (Springfield: Illinois State Geological Survey, 1970).

22. For a map of slag deposits, see Robert T. Kay, Theodore K. Greeman, Richard F. Duwelius, Robin B. King, John E. Nazimek, and David M. Petrovski, *Characterization of Fill Deposits in the Calumet Region of Northwestern Indiana*

and Northeastern Illinois, U.S. Geological Survey, Water Resources Investigations Report 96–4126 (De Kalb, IL: U.S. Department of the Interior, 1997), 12–13.

23. Early farmers also leveled the sand ridges and deposited the sand in nearby marshes; Schoon, *Calumet Beginnings,* 97.

24. Some municipalities are reluctant to vacate streets because their allocation of county highway funding is based on length of the overall system. On the draining of the ridges, see Lane, *City of the Century,* 20–21.

25. Greenberg quotes a hunter writing in 1884: "Calumet! What a host of recollections this well-known name will recall to the memory of Chicago sportsmen, old as well as young!"; Joel Greenberg, *A Natural History of the Chicago Region* (Chicago: University of Chicago Press, 2004), 233.

26. Institutions such as the Chicago Academy of Sciences, the Field Museum, and local colleges and universities took an especially strong interest in regional natural history. The Field Museum holds collections from the Calumet region that date back to the 1890s and contain more than twenty thousand objects and specimens from the region, including rare or extirpated species.

27. Geoffrey J. Martin, *American Geography and Geographers: Toward a Geographical Science* (New York: Oxford University Press, 2015), 766.

28. Kenneth J. Schoon, *Shifting Sands: The Restoration of the Calumet Area* (Bloomington: Indiana University Press, 2016), 94.

29. Harold M. Mayer and Richard C. Wade, *Chicago: Growth of a Metropolis* (Chicago: University of Chicago Press, 1969), 352.

30. David Bensman and Roberta Lynch, *Rusted Dreams: Hard Times in a Steel Community* (New York: Norton, 1987); Alan Berger, *Drosscape: Wasting Land in Urban America* (New York: Princeton Architectural Press, 2006); Christine Walley, *Exit Zero: Family and Class in Postindustrial Chicago* (Chicago: University of Chicago Press, 2013).

31. Manufacturing employment dropped 22 percent between 1990 and 2000 in the Calumet region, as mapped by Mark Bouman and reported by Ryan Chew, "Discovering the Calumet," *Chicago Wilderness Magazine,* Spring 2009, 6–11. Also see Andrew Hurley, *Environmental Inequalities: Race, Class, and Industrial Pollution in Gary, Indiana* (Chapel Hill: University of North Carolina Press, 1995).

32. Data, reports, and maps are from the Toxic Release Inventory, Environmental Protection Agency, https://www.epa.gov/toxics-release-inventory-tri -program (accessed Apr. 7, 2020).

33. The CERCLIS list was established by the Comprehensive Environmental Response and Liability Act of 1980.

34. Leah Verjacques, "The Compounded Pain of Contamination and Dislocation," *Atlantic,* June 26, 2017.

35. For more on the Area of Concern, see "Grand Calumet River Area of Concern," Environmental Protection Agency, https://www.epa.gov/great-lakes -aocs/grand-calumet-river-aoc (accessed Apr. 7, 2020).

36. Craig E. Colten, "Chicago Waste Lands: Refuse Disposal and Urban Growth, 1840–1990," *Journal of Historical Geography* 20, no. 2 (1994): 124–42. On Paxton II, see "Cleanup Programs: Paxton 2," Environmental Protection Agency, http://www.epa.state.il.us/land/cleanup-programs/paxton-2.html (accessed Apr. 7, 2020).

37. The consent decree may be found at Environmental Protection Agency, https://www.epa.gov/enforcement/bp-whiting-settlement-flaring (accessed Apr. 7, 2020).

38. Berger, *Drosscape*, 65.

39. Nina Sandlin et al., *Beyond Recycling: Materials Reprocessing in Chicago's Economy* (Chicago: Center for Neighborhood Technology, 1993); Lynne M. Westphal and J. G. Isebrands, "Phytoremediation of Chicago's Brownfields: Consideration of Ecological Approaches and Social Issues," in *Brownfields 2001 Proceedings* (Chicago, Il). For a brief description of Chicago's brownfields initiatives, see City of Chicago, *Brownfields Forum: Recycling Land for Chicago's Future*, Final Report and Action Program, Nov. 1995; Christopher A. De Sousa, *Brownfields Redevelopment and the Quest for Sustainability* (Oxford: Elsevier, 2008), 240–45.

40. Berger, *Drosscape*, 239.

41. The Southeast Environmental Task Force grew in 1989 from local efforts to stop the siting of a garbage incinerator at the former location of Wisconsin Steel. The retired schoolteacher Marian Byrnes, its extraordinary leader, also had leadership roles in Citizens United to Reclaim the Environment, Calumet Stewardship Initiative, and Calumet Heritage Partnership. Parts of Van Vlissingen Prairie were renamed in her honor in recognition of her years of effort to save open space and reduce pollution in the Calumet region. A reflection on her legacy may be found in the archives of Calumet Stewardship, Fall 2010, http://calumetstewardship.org/news/archives/2010-fall#.Wmz8g3lG1LN. The Task Force's struggles against petcoke are reported with photographs by Terry Evans in Elly Fishman, "Mountain of Trouble," *Chicago Magazine*, Aug. 2015, https://www.chicagomag.com/Chicago-Magazine/August-2015/KCBX-pet-coke/. On the growth of the environmental justice movement in Chicago, see David Naguib Pellow, *Garbage Wars: The Struggle for Environmental Justice in Chicago* (Cambridge, MA: MIT Press, 2002). Obama's complicated relationship with the residents of Altgeld Gardens is reviewed in William Wan, "At the Housing Project Where Obama Began His Career, Residents Are Filled with Pride—and Frustration," *Washington Post*, Jan. 8, 2017.

42. For a brief discussion of the Lake Calumet Airport situation in the context of airport planning in general, see Mark J. Bouman, "Cities of the Plane: Airports in the Networked City," in *Building for Airports: Airport Architecture and Design*, ed. John Zukowsky (Munich: Prestel Verlag; and Chicago: The Art Institute of Chicago, 1996).

43. Michigan City native, geographer, and birder James Landing's "Conceptual Plan for a Lake Calumet Ecological Park: Chicago, Illinois" may be accessed at Illinois Department of Natural Resources, https://www.dnr.illinois.gov/grants/documents/wpfgrantreports/1986119w.pdf (accessed Apr. 7, 2020).

44. National Park Service, Midwest Region, *Calumet Ecological Park Feasibility Study: A Special Resource Study Conducted in the Calumet Region of Northeast Illinois and Northwest Indiana* (N.p.: U.S. Department of the Interior, 1998), https://www.csu.edu/cerc/documents/calumetecologicalparkstudy.pdf.

45. The Chicago Park District has recently created the Big Marsh Bike Park on top of a slag field.

46. Mark Bouman and Sherry Meyer, then the president and vice president of the Calumet Heritage Partnership, respectively, were interviewed about the Heritage Area project by Keith Kirkpatrick on Lake Shore Public Media's *Lakeshore Focus*, episode 1003. The interview may be accessed at https://www.youtube.com/watch?v=v6fwKRXYyPo (accessed Apr. 7, 2020).

47. William Cronon, *Nature's Metropolis: Chicago and the Great West* (New York: W. W. Norton, 1991), 385.

Contributors

James R. Akerman is the director of the Hermon Dunlap Smith Center for the History of Cartography and Curator of Maps at the Newberry Library, Chicago.

William C. Barnett is an associate professor and chair of the Department of History at North Central College, Naperville, Illinois.

Mark Bouman is the director of the Chicago Region Program at the Field Museum, Chicago.

Kathleen A. Brosnan is the Travis Chair of Modern American History, University of Oklahoma.

Craig E. Colten is the Carl O. Sauer Professor of Geography and Anthropology at Louisiana State University.

Steven H. Corey is the dean of the School of Liberal Arts and Sciences and professor of history at Columbia College Chicago.

Matthew Corpolongo holds a PhD in history from the University of Oklahoma and is an architectural historian with the Oklahoma Department of Transportation.

Colin Fisher is a professor and the chair of the Department of History at University of San Diego.

Robert Gioielli is an associate professor of history at the University of Cincinnati, Blue Ash College.

Lynne Heasley is an associate professor at the Institute of the Environment and Sustainability at Western Michigan University.

Michael Innis-Jiménez is an associate professor and the director of graduate studies in American studies at the University of Alabama, Tuscaloosa.

Theodore J. Karamanski is a professor of history and the director of the Public History Program at Loyola University Chicago.

Ann Durkin Keating is the C. Toenniges Professor of History at North Central College, Naperville, Illinois.

Daniel Macfarlane is an associate professor of environment and sustainability at Western Michigan University.

Katherine Macica is a doctoral candidate in Loyola University, Chicago's joint doctoral program in American history and public history.

Brian McCammack is an assistant professor of environmental studies and the chair of American studies at Lake Forest College.

Robert Morrissey is an associate professor of history at University of Illinois, Urbana-Champaign.

Peter Nekola is a visiting professor of philosophy at Luther College in Decorah, Iowa.

Harold L. Platt is professor emeritus of history at Loyola University, Chicago.

Joshua Salzmann is an associate professor of history at Northeastern Illinois University, Chicago.

Natalie Bump Vena is an assistant professor in the Department of Urban Studies, Queens College, City University of New York.

Sylvia Hood Washington is a member of the governor of Illinois's Environmental Justice Commission. She is the cochair of the Illinois EPA's EJ Advisory Board and editor in chief of the journal *Environmental Justice*.

Index

95, 100, 102, 180, 278, 280, 282,
294; engineers, 54; Environmental
Protection Agency, 270; federal
intervention and the lakefront, 228;
Forest Preserve District and, 154–55,
158, 161–63; garbage survey, 233–35;
Illinois and Michigan Canal (I&M
Canal) and, 180; Indiana Dunes and,
282, 294; land surveyors, 27; landfill
guidelines, 237; lawsuits associated
with civil rights and housing, 277;
Nader and, 275–76; Native American
land cessions and, 27; nuclear power
and, 280–82; presidents, 28, 56–58,
95, 102, 122, 297; Progressive Era
conservationists and, 258; public
works projects and, 86; Ridley and,
249; river mandates, 219; settlement
of the Great Lakes and, 52–54, 56–58;
survey of Chicago skies, 212; swampy
areas around the southern part of
Lake Michigan characterized by, 178;
The Wilderness Society and, 162;
Tunnel and Reservoir Plan (TARP)
and, 89–90; U.S. Supreme Court,
61, 95, 242, 253, 282. *See also* Air
Pollution Control Act of 1955; Air
Quality Act of 1967; Atomic Energy
Commission (AEC); *Brown v. Board
of Education* (1954); Civilian Conservation Corps (CCC); Civil Rights
Act of 1964; Civil Works Administration (CWA); *Gautreaux v. Chicago
Housing Authority* (1996); Nuclear
Regulatory Commission (NRC);
Pipeline and Hazardous Materials
Safety Administration (PHMSA);
Plessy v. Ferguson (1896); Works
Progress Administration (WPA)
government, intergovernmental partnerships: air pollution control programs,
213; creating Board of Health, 36;
emergency relief and conservation
agencies, 155, 163; flood control and,
88–90; Illinois and Michigan Canal
(I&M Canal) construction and modifications, 66, 71–72; related to airport
debates and developments, 279, 298–
99; removing drosscape components
from the Calumet region, 296–98;

Sanitary and Ship Canal construc-
tion, 65, 72, 75–77; tending to disease,
32–33. *See also* Calumet Land Use
Plan; forest preserves; Tunnel and
Reservoir Plan (TARP)
government, international agreements.
See Boundary Waters Treaty; Chicago Diversion; Great Lakes Water
Quality Agreement; Great Lakes
Water Treaty of 1932; Great Lakes-St.
Lawrence River Water Resources
Basin Compact; Great Lakes-St. Lawrence Waterway; St. Lawrence Seaway
and Power Project; St. Lawrence
Seaway; Treaty of Chicago
government, local: airports and, 264, 278–
80, 298; aldermen, 61, 85, 155, 197,
216–17, 250; City of Chicago, 37, 64,
71, 85, 155, 197, 199, 299; corruption,
9, 47, 83, 85–89, 196–97; mayors, 4, 5,
12, 51, 62, 85, 87, 89, 113, 215–16, 247,
250–51, 252, 278, 298; ordinances related to animals, 29, 31, 39–40, 43–48;
ordinances related to fire, 196; ordinances related to railroads, 197, 201;
ordinances related to rapid industrial
urbanization, 189; ordinances related
to smoke, 207, 212; ordinances related
to waste and disease, 29, 31, 46–47,
207, 224, 226, 232, 234–35; privatism, 31, 33–34; racism, 240–43, 253;
response to cholera, 37. *See also* 1919
race riot; Boone, Levi; Busse, Fred;
Calumet Land Use Plan; Chicago Air
Pollution Appeals Board; Chicago
Board of Health; Chicago Board of
Sewerage Commissioners; Chicago
Bureau of Social Surveys; Chicago
City Council; Chicago City Hydraulic
Company; Chicago Commission
on Race Relations (CCRR); Chicago
Council of Social Agencies; Chicago
Department of Air Pollution Control;
Chicago Department of Energy and
Environmental Protection; Chicago
Department of Environmental Con-
trol; Chicago Department of Public
Health; Chicago Department of
Public Works; Chicago Department
of Smoke Inspection (DSI); Chicago